Cellular Automata

Mathematics and Its Applications

Volume 460

Cellular Automata

A Parallel Model

edited by

M. Delorme

and

J. Mazoyer

Laboratoire de l'Informatique du Parallélisme,
ENS Lyon,
Lyon, France

KLUWER ACADEMIC PUBLISHERS
DORDRECHT / BOSTON / LONDON

A C.I.P. Catalogue record for this book is available from the Library of Congress.

ISBN 978-90-481-5143-1

Published by Kluwer Academic Publishers,
P.O. Box 17, 3300 AA Dordrecht, The Netherlands.

Sold and distributed in North, Central and South America
by Kluwer Academic Publishers,
101 Philip Drive, Norwell, MA 02061, U.S.A.

In all other countries, sold and distributed
by Kluwer Academic Publishers,
P.O. Box 322, 3300 AH Dordrecht, The Netherlands.

Printed on acid-free paper

CONTENTS

Preface

We all feel slightly uneasy when we think about our brain, how it works, how it can store such a tremendous number of information and retrieve them almost instantly, how it can reach conclusions or decisions from a number of facts and observations, very often in almost no time, how in an even more surprising way it can create new ideas which are not direct consequences of the known facts. We also feel a bit uneasy and powerless when we see children learn, learn how to behave, learn to recognize objects, learn their languages, and adapt themselves to unknown, unpredictable and surprising circumstances. We feel uneasy for, clearly, we do not understand how all the mental processes involved in retrieving, recognizing, reasonning, decision-making, learning, adapting, creating do work.

Of course, a lot of work has been dedicated to mimicking, or simulating, or representing in various formal logic systems leading to computational models all the functions performed by the brain. This leads to a number of interesting and useful pieces of software which allow many computer-based devices to be called "intelligent": they are indeed intelligent, since they can "understand" restricted fragments of natural languages, written or spoken, they can analyze images and recognize some simple shapes, they can help a robot move on a given scene, avoid obstacles, and find its ways to reach a given point. But I dare to say that all these interesting and sometimes extremely cleaver achievements constitute as many proofs of the fact that brain does not work like computers, performing one operation at a time, in a sequential manner. All the recognition algorithms, based on some scanning of the image to be recognized, are much to slow to explain how human beings, and animals too, recognize a danger on an extremely small fraction of second in a situation which is, at most, analogous to a situation which they have met already.

The mystery lives in a few hundred-thousands neurons packed in a few cubic centimeters of space delimited by the skull. A lot of work has been done too, to understand how these very small cells behave: each cell is quite limited in space and computation power. It receives signals or messages from neighboring cells. And, according to laws which are not completely known, it emits signals to its neighbors. The chemical nature of these signals has been extensively studied and proved. Experiments have been made with real, actual, neurons of rats, connected to form a planar net, proving at least that some information can be carried from one border point of the net to another one, or that some information injected at one point can be spread to the whole net.

I believe, personally, that one of the most influential piece of text of this century has been article of McCulloch and Pitts of 1942 in which they describe the brain as a set of formal "neurons", where each neuron is an extremely simple device which may be in a very small number of states, which receives messages from neighboring "neurons" and according to which state it is in, which messages it has received, goes into another state and emits a message. Both John von Neumann who used in order to describe electronic circuits a language directly inspired from McCulloch and Pitts, and Stephen Coole Kleene when he gave the definition of finite automata which is still used today acknowledge they debt to neurobiologists who were McCulloch and Pitts.

The model of cellular automata was created at that time: it is a very natural one. Take a graph, any graph, which may be the discrete line, $\mathbb{Z}$, or the infinite square grid, $\mathbb{Z}^2$, and install at each node a very small automaton: this automaton receives information from its neighbors, and reacts, moves from one state to another, and sends information to its neighbors, according to some transition law. The problem is: how do they behave, will these small automata reach a same state, some sort of global stable situation, or oscillate between a finite number of "limit" states? The first people to make use of this model were physicists, for obvious reasons. Many "physical" situations are, straightforwardly, represented as cellular automata: water drops on a windshield, electrons in a magnetic field, all kind of particles in various situations. Physicists built cellular automata representing the phenomena they were interested in, and ran them observe their behaviors, attempted to classify them.

For a long time, there was absolutely no theory of these objects. At least, one "positive" theory. A variant of cellular automata, invented by the mathematician, John Conway, a cellular automata on $\mathbb{Z}^2$, was there to prove that more or less any reasonnable question which can be raised about the behavior of 2D cellular automaton cannot be answered, the answer is "undecidable". All the results about cellular automata were "negative" ones, undecidability results, or results of decidability with such a high complexity that they were no hope of seeing the solution in any human future.

In France, a few people in Grenoble, lead by François Robert, a specialist of discrete iterations and discrete dynamical systems, started working on one dimensional cellular automata. This is certainly the simplest case, each cell has two neighbors and two limit cells have only one. M. Cosnard, E. Goles and M. Tchuente among others were able to prove results in that case, to say when some such 1D cellular automata have a stable limit global state and in how many steps it may be reached. Jacques Mazoyer went a little farther, in his dissertation and further writings. Working on the 1D cellular automata known as solutions of the "Firing Squad problem", which had been for years a puzzle to many researchers, he formalized a notion of "signal" allowing to understand what is happening and how all the "members of the Firing Squad" can simultaneously reach the same state "Fire". In the Laboratory of Parallelism, LIP, of the Ecole Normale Supérieure de Lyon, a whole team of researchers has spent a lot of time and energy, around J. Mazoyer, to clarify these ideas and enlarge their field applications, especially to 2D cellular automata or cellular automata on graphs, especially Cayley graphs. Their work resulted in new undecidability results, which may be not surprising, but are hard to prove using tilings of the plane and also methods to build 2D cellular automata with a given behavior. I am very happy to write this introduction to a book which stems out from a Spring School of Theoretical Computer Science, held in 1996 in Saissac and beautifully organized by M. Delorme and J. Mazoyer.

A series of Spring Schools initiated in 1973, held any year since 1973, on a different subject, in a different location, by different people, have deeply influenced research in Theoretical Computer Science in France for 25 years.

But not all these Spring Schools give rise to a published volume. Jacques Mazoyer and his collaborators from LIP, his fellow colleagues from various european laboratories have done an exceptional effort of presentation of the latest state of the art in the field of cellular automata.

It is my pleasure to thank them and to hope that this volume will be used by a number of people from all origins, mathematicians, physicists, biologists, computer scientists and others: one can find in it both the most recent methods to build and use in practical circumstances cellular automata, and questions to think about for years to come, about very nature of computing, the weakness of our current models and machines as compared to the major problem of what is really parallel computation and/or how does our brain really works.

Maurice Nivat,
Professor, University Paris 7, Denis Diderot,
July 14th, 1998.

[illegible] to understand what is happening and how all the [illegible] of [illegible] "Squid" can simultaneously reach the same [illegible] the [illegible] of Pandikhan, [illegible] of the Rod [illegible] de Superficies de [illegible] [illegible] of researchers has spent [illegible] time and energy [illegible] J. Mazoyer [illegible] clarify these ideas [illegible] through [illegible] applications [illegible] in 2D cellular automata or [illegible] graphs [illegible]. Their work resulted in [illegible] understanding [illegible] results [illegible] may be [illegible] surprising, but are hard to prove [illegible] methods [illegible] 2D cellular automata [illegible] [illegible] from [illegible] of Theoretical Computer Science [illegible] beautiful [illegible] J. Delorme and J. Mazoyer.

[illegible]

[illegible] Delorme [illegible] colleagues [illegible] state of the art [illegible] cellular automata.

[illegible] to thank [illegible] could be used [illegible]

Maurice Nivat

Professeur, Université Paris 7 Denis Diderot,

July [illegible] 1998

Part 1
A general survey

First, basic definitions and results on cellular automata are introduced. Some of them, well-known, are not treated in any following chapter, but still have to be pointed out. Let us cite some properties of the global functions, questions connected to classifications, and the formal mastery of the phase space in the one-dimensional additive case, for example. A notion of signal, essential to build cellular automata, is analyzed. Then some simulations are presented which play fundamental roles when cellular automata are to be compared with other models or between them, and thus are particularly implied in concepts such as universality or self-reproduction.

Second, the "Game of Life" is presented and serves to illustrate some items previously pointed out. The game of Life is, maybe, the most popular cellular automaton (and probably the only one with a fanclub!) . Who did not gaze at it as a screen-saver on a computer? Its success is due to the fact that it is the first "visible automaton", which produces - from a very simple local rule - very complex global behaviors, giving rise to powerful and general questions and results, on universalities for example. Here, in fact, Turing universality and intrinsic universality are discussed.

Part I
A general survey

AN INTRODUCTION TO CELLULAR AUTOMATA

Some basic definitions and concepts

M. DELORME
LIP ENS Lyon
46, Allée d'Italie, 69364 Lyon Cedex 07, France

1. Introduction

At the beginning of this story is John von Neumann. As far back as 1948 he introduced the idea of a *theory of automata* in a conference at the Hixon Symposium, September 1948 (von Neumann, 1951). From that time on, he worked to what he described himself not as a theory, but as "an imperfectly articulated and hardly formalized "body of experience"" (introduction to "The Computer and the Brain", written around 1955-56 and published after his death (von Neumann, 1958)). He worked up to conceive the first cellular automaton (he is also said to have introduced the *cellular* epithet (Burks, 1972)). He also left interesting views about implied mathematics, including logics, probabilities, leading from the discrete to the continuous (von Neumann, 1951; von Neumann, 1956; von Neumann, 1966).

It is not the place to write the fascinating history nor to point out the role, in the development of the occidental civilization, of the dream, idea, concept, realizations of (automata) machines or games. Let us only recall the cornerstones that are the works of Lulle, Leibnitz, Pascal, Descartes, Vaucanson, Babbage ... But, in 1948, the minds were under the influence of the Turing (universal) machines (Turing, 1936), the first neural networks (McCullough and Pitts, 1943) and also the natural and artificial automata of cybernetics (Wiener, 1961) (an enthusiastic review of which von Neumann published in "Physics Today" (Mosconi, 1989)).

Through some of the different texts he left as (von Neumann, 1951; von Neumann, 1958; von Neumann, 1956; von Neumann, 1966), it is clear that von Neumann deeply gave his attention to the comparisons between *natural automata*, as the brain or other adaptable or evolutive systems, and *artificial automata*, as, especially, the computers (at that time just constructed). He was very interested in their respective complexities, which both, but in different ways, come under analog and digital procedures. The main points at issue were, on the one hand, the reliability of the computers, in danger to fail under increasing complexity, and, on the other hand, the power of the complexity of living organisms which seems to be a condition for their ability to reproduction. Thus von Neumann looked for an artificial system which would be as powerful as a universal Turing machine (that meant be *universal for computation*), would be able of reproduction and, even more, to construct, in a sense which has to be and will be made clear later, some significant inner components (that meant be *universal for the construction*).

J. von Neumann did not succeed to build the first model in three dimensions he tried, called the *cinematic model* by Burks and described in (Burks, 1972). Then, on a suggestion of S. Ulam, he made up his mind to conceive a "2-dimensional paper automaton", with a *cellular* or *crystalline* structure, and he managed to get a cellular automaton, universal for computation and construction, and also self-reproducing. Actually the device was completed by Burks (Burks, 1972), afterwards improved by a lot of people (Thatcher, 1964), (Codd, 1968) for example, and it has had a rich posterity, as the following of the volume will partly show it. Actually, two large branches will expand. Along the first one, cellular automata are studied as *parallel models of computation* and *dynamic systems*. That leads to algorithms setting-up, language or pattern recognition, complexity theory, classifications ... Along the second one, cellular automata are conceived as *models for natural processes* in physics, chemistry, biology, economy ... Two purposes are pursued in this modeling process: to simulate phenomena on cellular automata, but also to try to predict phenomena in studying properties of relevant cellular models.

Now we will give basic definitions necessary to understand what are cellular automata, as well as to work with. But we also want to point out some efficient, and sometimes problematic, concepts as *signal*, *simulation* and *universality*. We will, especially, put to light different notions of universality which emerge more or less explicitly with time, but also recall some problems and results - nowadays classic in the field - that does not form the subject of chapters in this volume.

2. Cellular automata: main definitions

Informally, a cellular automaton or cellular space is an abstract object, with two intrinsically tied components. First a *regular*, *discrete*, *infinite* network, which represents the *architecture*, the *universe* or the *underlying space structure* of the cellular automaton, depending the use it will be made of it. Second, a finite automaton, a copy of which will take place at each node of the net. Each so decorated node will be called a *cell* and will communicate with a finite number of other cells, which determine its *neighborhood*, geometrically uniform. This communication, which is *local*, *deterministic*, *uniform* and *synchronous* determines a *global evolution* of the system, along *discrete time steps.*

2.1. CLASSIC CELLULAR AUTOMATA

We will first give the standard formal definition, then indicate some more or less usual variants.

Definition 1

A d-dimensional cellular automaton (or d-CA), $\mathcal{A}$, is a 4-uplet $(\mathbb{Z}^d, S, N, \delta)$, *where :*

- *S is a finite set, the elements of which are the* states *of $\mathcal{A}$,*
- *N is a finite ordered subset of $\mathbb{Z}^d$, $N = \{\vec{n_j}/\vec{n_j} = (x_{1j}, \ldots, x_{dj}), j \in \{1, \ldots, n\}\}$, called the neighborhood of $\mathcal{A}$,*
- *$\delta : S^{n+1} \mapsto S$ is the local transition function or local rule of $\mathcal{A}$*[1].

Among the states are, sometimes, distinguished states s, called *quiescent* states, such that $\delta(s, \ldots, s) = s$.

2.2. NEIGHBORHOODS

Let $\mathcal{A}$ be a cellular automaton $(\mathbb{Z}^d, S, N, \delta)$. The neighborhood of a cell c (including the cell itself or not, in accordance with convention) is the set of all the cells of the network which will locally determine the evolution of c. It is finite and geometrically uniform.

In principle, a neighborhood can be any ordered finite set, but, actually, some special ones are mainly considered. In case of $\mathbb{Z}^d$, classic neighborhoods are the von Neumann's and the Moore's ones. They are known as the *nearest neighbors neighborhoods*, and defined according to the usual norms and the associated distances. More precisely, if $\vec{z} = (z_1, \ldots, z_d)$, $\|\vec{z}\|_1$ will denote $\sum_{i=1}^{d} | z_i |$, $\|\vec{z}\|_\infty$ will denote $Max\{| z_i | \ /i \in \{1, \ldots, d\}\}$, d_1, d_∞ the associated distances, and we get:

- Von Neumann neighborhood: $N_{VN}(\vec{z}) = \{\vec{x}/\vec{x} \in \mathbb{Z}^d \text{ and } d_1(\vec{z}, \vec{x}) \leq 1\}$ with a given order,
- Moore neighborhood : $N_M(\vec{z}) = \{\vec{x}/\vec{x} \in \mathbb{Z}^d \text{ and } d_\infty(\vec{z}, \vec{x}) \leq 1\}$, with a given order.

Some other interesting neighborhoods appear in the literature, the Cole's ones (Cole, 1969), or some others in (Smith, 1971) for example which are

[1]This definition takes into account the cell and its neighborhood. It is a quite common choice, but not a universal one: Cole (Cole, 1969), for example, does not consider the cell in its nighborhood.

represented with the Von Neumann's and the Moore's ones on Figure 1, when the cells are considered on the vertices or, more usually, by duality, as the unit squares of the grid.

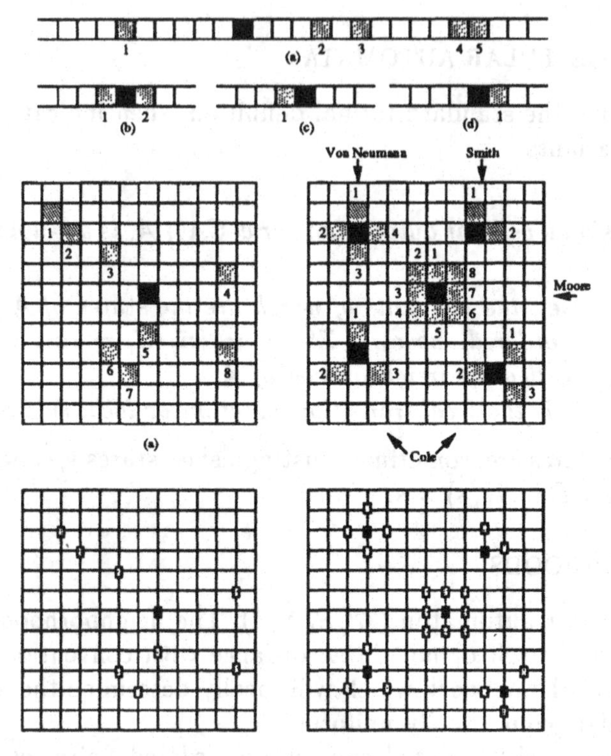

Figure 1. Examples of classic neighborhoods (the numbers mark the cells ranks). In dimension 1, (b) represents, in the two-way case, von Neumann's or Moore's neighborhood, which then coincide, while (c) and (d) represent the first neighbors neighborhood in case of one-way cellular automata. In both dimensions, (a) represents any neighborhood.

Moreover the following result holds, which means that, *from a computation point of view*, all neighborhoods can be understood as *equivalent*, which is generally not the case when complexity or architecture are to be taken into account (see for example (Codd, 1968), (Smith, 1971), (Lindgren and Nordahl, 1990)).

Fact 1

Every d-cellular automaton can be simulated by a d-cellular automaton with the nearest neighbors neighborhood.

Another neighborhood on 1-cellular automaton is used which consists of the closest right (left) cell of a given cell. It characterizes cellular automata called *one way cellular automata* (OCA for short), which form a special interesting class (see (Choffrut and Čulik, 1984), and *Part 3: Computational Powe*).

The *radius* r of a neighborhood is defined as the distance between the cell of which it is the neighborhood and the farthest one in the neighborhood. So, for one-dimensional cellular automata, the radius of the nearest neighbors neighborhood is $r = 1$ and the number of cells is $2r + 1 = 3$. This terminology and the corresponding notation are often used in case of symmetric neighborhood in dimension 1.

The neighborhoods for which the ratio between the radius and the number of states is exponential are not considered in the complexity domain, because, in this framework, NP-complete problems may be solved in polynomial time. Nevertheless *an interesting problematics is the optimization of the constraint "size of the neighborhood/number of states"*.

2.3. REPRESENTATIONS

Outside the usual transition table or matrix, the local transition function can be displayed in different and more efficient ways. The first one, due to Wolfram, is well known and refers to the rules of its, now historical, classification.

Representation 1 *Wolfram numbers*

We can describe δ by a word obtained in the following way: first of all a linear order on the elements of the δ-domain S^{n+1}, interpreted as words of length $n + 1$ on S, is chosen. Then δ is given by a word the letters of which are the successive δ-values of the ordered words of the δ-domain. When working with 1-dimensional cellular automata, $S = \{0, 1\}$ and the lexicographic order on $\{0, 1\}^{n+1}$, we get a word which can be understood as the binary development of a positive integer. This integer is the Wolfram number of the rule δ.

So, for 1-dimensional cellular automata with $\{-1, 0, 1\}$ neighborhood (cell included in the middle), $\{0, 1\}$ set of states, the lexicographic order $\{w_0, \ldots, w_7,\}$ on $\{0, 1\}^3$, if $\delta(w_i) = s_i$, $s_0 \ldots s_7$ represents δ. And conversely, any positive integer less than $256 = 2^8$ defines a 1-CA of the last type. For example, as $54 = 2^1 + 2^2 + 2^4 + 2^5$, the "54 rule" is given by the table below:

000	001	010	011	100	101	110	111
0	1	1	0	1	1	0	0

Another representation is very fruitful for 1-CA with nearest neighborhood N, $|N| \geq 3$. It is connected to the idea to link a 1-CA $(\mathbb{Z}, S, N, \delta)$ to a finite automaton with S^n as set of states, letters in S, and the transition function δ' defined by $\delta'(su, \delta(sut)) = ut$ for $s, t \in S$, $u \in S^{n-1}$, the underlying graph of which is a De Bruijn graph. The properties of De Bruijn graphs (as, for example, being connected and hamiltonian) can be successfully exploited to get decidability results on CA as in (Head, 1989), (Sutner, 1991). See also *Linear Cellular Automata and De Bruijn Automata* in this volume.

Representation 2 *De Bruijn graphs*

In the case of dimension 1, *we also get description of* δ *aid of a graph defined as follows: its vertices correspond to, and are labeled by, the words of length* n *on* S, *and there is an arrow from* su *to* ut, *labeled by* $\delta(sut)$.

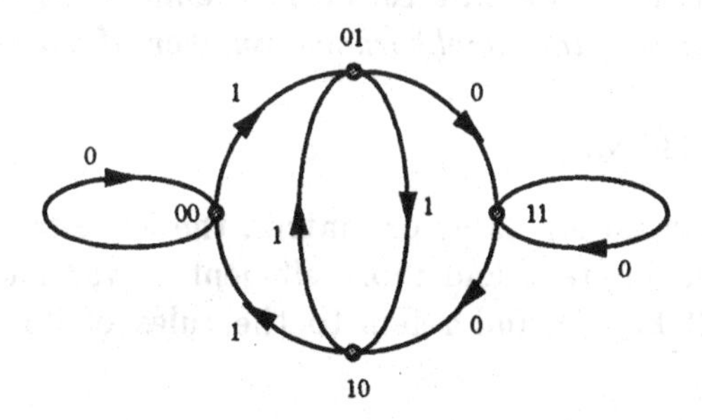

Figure 2. The De Bruijn representation of the "54 rule", $r = 1$

The above representations do not answer any concern for optimization. Some results can be found about minimization of the transition tables in (Durand, 1994).

2.4. CONFIGURATIONS AND GLOBAL FUNCTION

The space component already comes to light with the neighborhood, but it takes its full significance with the notions of configurations and global function. We will first precise definitions and then recall some results.

2.4.1. *Configurations, behavior, global function, dynamic system*

If we conceive a cellular automaton $\mathcal{A} = (\mathbb{Z}^d, S, N, \delta)$ as a machine which evolves with time, knowing it at time t is knowing the state of each cell at this time. This is formalized through an application $c_t^{\mathcal{A}} : \mathbb{Z}^d \mapsto S$, called a *configuration*, or *instantaneous description* or *global state* of $\mathcal{A}$ at time t. The *behavior* or *action* or *evolution* of $\mathcal{A}$ is a sequence $(c_t^{\mathcal{A}})_{t\geq 0}$ of configurations determined by the given initial configuration $c_0^{\mathcal{A}}$ and the derivation rule described as follows:

$c_t^{\mathcal{A}} \vdash c_{t+1}^{\mathcal{A}}$ if and only if $c_{t+1}^{\mathcal{A}}(\vec{z}) = c_{t+1}^{\mathcal{A}}(z_1, \ldots, z_d)$ is given by $\delta(c_t^{\mathcal{A}}(z_1, \ldots, z_d), c_t^{\mathcal{A}}(z_1 + n_{11}, \ldots, z_d + n_{d1}), \ldots, c_t^{\mathcal{A}}(z_1 + n_{1n}, \ldots, z_d + n_{dn}))$, where, $\vec{z}$ represents the standard cell and $(\vec{n}_1, \ldots, \vec{n}_n)$ the neighborhood. This relation actually gives rise to an application $G_{\mathcal{A}}$ from the set $C_{\mathcal{A}}$ of $\mathcal{A}$-configurations into itself, which associates to each configuration $c^{\mathcal{A}}$ the configuration $G_{\mathcal{A}}(c^{\mathcal{A}})$, defined by :

$$G_{\mathcal{A}}(c^{\mathcal{A}})(\vec{z}) = \delta(c^{\mathcal{A}}(\vec{z}), c^{\mathcal{A}}(\vec{z}+\vec{n}_1), \ldots, c^{\mathcal{A}}(\vec{z}+\vec{n}_n)).$$

The function $G_{\mathcal{A}}$ so obtained is the *global function* of $\mathcal{A}$. It plays an essential role in practicing cellular automata, especially when they are considered as dynamic system. Indeed $(\mathbb{Z}^d, G_{\mathcal{A}})$ is a dynamic system[2], but a very special one due to its *local* definition via the local transition function and to its homogeneity.

A sequence $(c_t^{\mathcal{A}})_{t \geq 0}$ of configurations of a cellular automaton $\mathcal{A}$ is designated, according to the context, as the *orbit* of $c_0^{\mathcal{A}}$ when cellular automata are considered as dynamic systems, or as a *computation* on $c_0^{\mathcal{A}}$ when cellular automata are seen as computation models or even as a *propagation* or a *positive motion* in a more physical point of view.
In the computation case, each derivation $c_t^{\mathcal{A}} \vdash c_{t+1}^{\mathcal{A}}$ is called a *computation step*. Their evaluation determines the computation time of a computation (see *Computations on Cellular Automata*) or the time complexity of a problem or a language (see *Cellular Automata as Languages Recognizers*).

Incidentally, we have to notice that the notion of computation does not distinguish *computation time* and *communication time* between neighbor cells. This does not pose problem here because our aim is more to exhibit parallel algorithms than to conceive physical or "real" machines.

2.4.2. *Special configurations*

Cellular automata are not "real" machines, in the sense that they don't exist physically but have to be simulated. So, when experiment has to be done or some representations of their evolutions are to be given, some types of *feasible or significant* configurations are of special interest, which also are amenable for mathematical treatment. Among them, the following ones play essential roles.

- *Finite configurations*
  Let q_e be a quiescent state for some cellular automaton $\mathcal{A}$, and c a configuration. The *support* of c is the set $Supp(c) = \{\vec{z}/c(\vec{z}) \neq q_e\}$. Then a *finite configuration* is a configuration the support of which is finite.
  As data are there finite, these configurations are essential in the computation or language recognition areas, and so play a basic role in computation, computation universality and complexity.

[2] $G_{\mathcal{A}}$ is then often called the evolution or next state transformation, configurations being the states of the system.

Let us remark that when several quiescent states exist, which is often encountered, one of them is explicitly reserved for the definition of finite configurations.

– *Periodic configurations*

A *periodic configuration* is a configuration c for which there exists $\vec{p} \in \mathbb{Z}^d$ such that, $c(\vec{z}+\vec{p}) = c(\vec{z})$ for each $\vec{z} \in \mathbb{Z}^d$.
Note that, except the *quiescent configuration* c_{q_e} (such that $c_{q_e}(\vec{z}) = q_e$ for each $\vec{z} \in Z^d$) periodic configurations are not finite. They are worthy to study because they naturally come for automata on rings (which are those we are able to "observe" on computers) and more generally n-dimensional tori, but also because they constitute "good" subsets of $S^{\mathbb{Z}^d}$, in fact countable dense ones (as well as the finite ones) for the canonical topology recalled in the next section.
Let us here pass some remark on the terminology. Actually, we should name *space-periodic* the above periodic configurations, in order to distinguish them from *time-periodic* configurations which are these configurations c the orbits of which are periodic, which means that there exists a time t_0 and a positive integer p such that, for all positive integers k, $G_{\mathcal{A}}^{t_0+kp}(c) = G_{\mathcal{A}}^{t_0}(c)$, where $G_{\mathcal{A}}^{t}(c)$ denotes the result of the t-th iteration of $G_{\mathcal{A}}$ on c. This last notion is very natural when cellular automata are considered as dynamic systems. Clearly space periodic configurations are also time-periodic, but the converse is, generally, not true.

– *Garden of Eden*

The so called configurations are those which have no predecessor according to the global function. So they can only be taken as initial configurations (which explains the terminology), and, consequently, only "given" by external input. They appeared in the context of construction universality and self-reproduction, and, in fact, originated the systematic study of global functions (Moore, 1962), (Myhill, 1963) and (Čulik *et al.*, 1989).

2.4.3. *Some results and open problems*

Let us consider a finite set S, a positive integer d, $d \geq 1$. If S is endowed with the discrete topology (for which all subsets are open), the set $S^{\mathbb{Z}^d}$ of applications from $\mathbb{Z}^d$ into S, canonically endowed with the product topology, is a compact metric space. A *shift of vector* $\vec{v}$, $\vec{v} \in \mathbb{Z}^d$, is an application $\sigma_{\vec{v}}$ from $S^{\mathbb{Z}^d}$ into itself, such that, for each c, $c \in S^{\mathbb{Z}^d}$, each $\vec{z}$, $\vec{z} \in \mathbb{Z}$, $\sigma_{\vec{v}}(c)(\vec{z}+\vec{v}) = c(\vec{z})$. Then one of the source of interest in cellular automata can be found in the basic following result, proved independently of the same one (but in case $d = 1$) by Hedlund (Hedlund , 1969).

Theorem 1 *Richardson's theorem, (*Richardson, 1972*)*
An application $f : S^{\mathbb{Z}^d} \longrightarrow S^{\mathbb{Z}^d}$ which commutes with shifts is a continuous function if and only if it is the global function of some cellular automaton.

Some classical properties of the global functions, as surjectivity, injectivity and bijectivity, have been intensively studied. A first result is the Moore-Myhill theorem, often known as the Garden-of-Eden theorem.

Theorem 2 *Garden-of-Eden theorem, (*Moore, 1962*), (*Myhill, 1963*)*
A cellular automaton with a quiescent state is surjective if and only if it is injective when restricted to finite configurations.

Another property of cellular automata has been intensively studied, from different areas but especially when looking cellular automata as models of physical phenomena (and particularly in microscopic physics where processes can be reversible), the *reversibility*. A cellular automaton is said to be *reversible* when its global function G is bijective and G^{-1} is also realized by a cellular automaton.
A consequence of Richardson's and Moore-Myhill's theorems is that *bijectivity, reversibility and injectivity* are equivalent for cellular automata.

We can now sum up the main other results in Figure 3, taken from (Durand, 1994) where they are reviewed and/or proved.

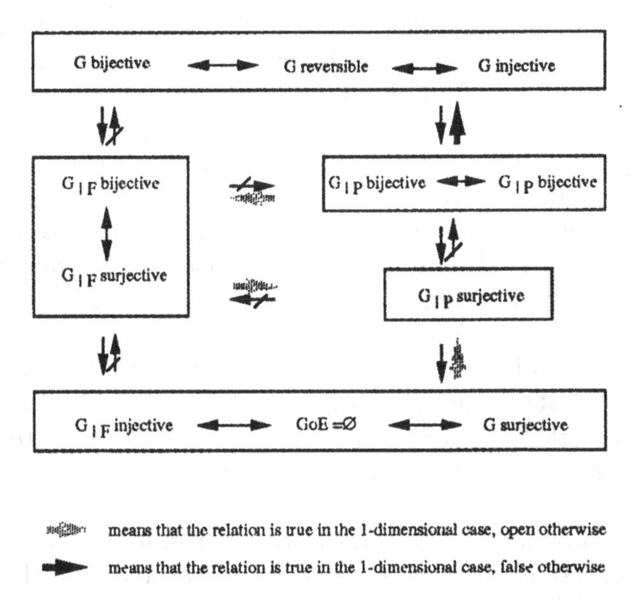

Figure 3. $G|_F$ denotes the functions restricted to finite configurations, $G|_P$ those restricted to the periodic ones and GoE the Garden-of-Eden configurations.

Notice that in the above cited paper are also mentioned decidability and undecidability results, which are to be stressed. First they reveal the existing gap between 1- and 2-dimensional cellular automata: actually many problems are decidable in dimension 1 while they are undecidable in dimension 2. Secondly, the undecidability proofs, involved enough, put to light tight links between cellular automata and tilings, and so exhibit another powerful proof method, to be opposed to the diagonalization-like methods. Let us simply set out the basic result by Kari, which is exemplary.

Theorem 3 *Kari's theorem (*Kari, 1994*), (*Durand, 1994*)*
The surjectivity and reversibility problems are undecidable for 2*-dimensional cellular automata.*

We will now pay more attention to the configurations evolutions.

2.5. DYNAMICS

2.5.1. *Space-time diagrams*

A cellular automaton deals with information. It works on given information, converts, creates, cancels, stores and delivers some. It can be very useful to represent all these transformations which occur along discrete times. For cellular automata of low dimension, especially 1 and, to lesser extent, 2, we get very expressive graphical representations, their *space-time diagrams* in $\mathbf{N} \times \mathbb{Z}^d$, obtained in attributing a special pattern or color to each state, and consequently to each cell of the automaton. Thus, the evolution of the automaton on a given starting configuration (also designated as initial configuration) can be seen as a stacking-up of decorated lines in $\mathbb{Z}^2$, or of decorated planes in $\mathbb{Z}^3$ depending on whether the automaton dimension is 1 (see Figure 4) or 2.

These examples, although very simple, put already to light two possible kinds of computations : bounded and unbounded ones. We will come back on what that imply, especially in the definitions and properties of complexity classes, in chapter *Cellular Automata as Languages Recognizers.* A lot of tasks on cellular automata naturally impose bounded computation areas, as it can be seen through this volume.

It is quite easy to imagine that graphical representations are suffering from technology limits (in space - most simulations are done on rings of cells - and time), nevertheless they are powerful and rich enough to make understand or imagine phenomena and algorithms open to cellular automata treatments.

Outside the production of attractive or strange images, and the question of the connections between cellular automata and fractals that were naturally set up ((Willson, 1984), (Čulik and Dube, 1989), more recently (Peitgen *et al.*, 1998) and references there), the observation and analysis

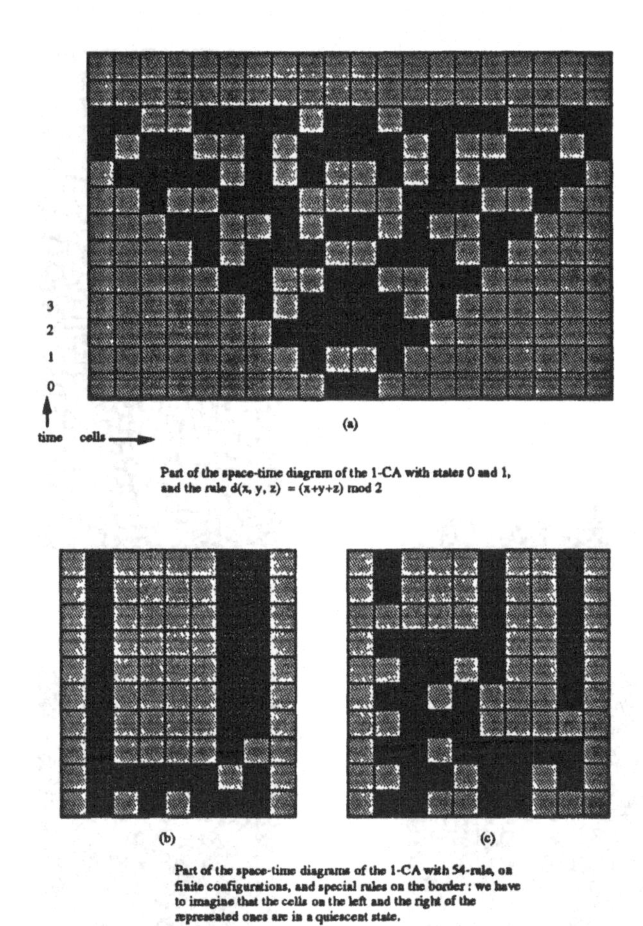

3
2
1
0
time
cells
(a)
Part of the space-time diagram of the 1-CA with states 0 and 1,
and the rule d(x, y, z) = (x+y+z) mod 2
(b)
(c)
Part of the space-time diagrams of the 1-CA with 54-rule, on
finite configurations, and special rules on the border : we have
to imagine that the cells on the left and the right of the
represented ones are in a quiescent state.

Figure 4. Example of space-time diagrams on finite configurations. On (a) the computation is unbounded while it is bounded on (b) and (c).

of the complexity, or degree of intricacy, of such space-time diagrams for 1-dimensional cellular automata is at the foundation of some classifications for these automata. The first and widespread one by Wolfram (Wolfram, 1986) (recalled in chapter *Topological Definitions of Deterministic Chaos*), focuses on the behavior of 1-dimensional cellular automata with two or three states on periodic configurations, and rests on the "observation" of cellular automata evolutions on random initial configurations. Unfortunately, these space-time diagrams observations don't allow any formal definition of the evolution complexity, so the corresponding heuristic classification remains coarse even if it is suggestive enough and, especially, gives evidence of some "edge", sometimes evoked as "the edge of chaos", or of a possible computation-universality for some automata. See Figure 5.

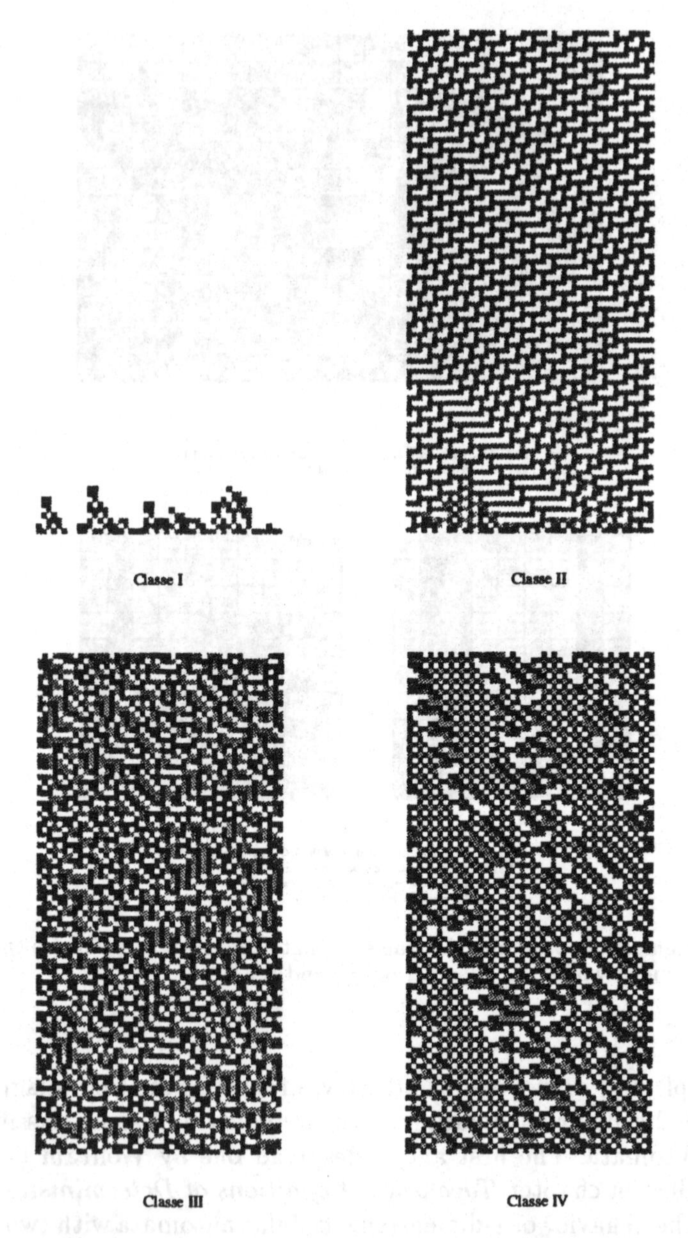

Figure 5. Other examples of space-time diagrams, characteristic of the four Wolfram classes.

The study of these diagrams has also inspired concepts and methods efficiently used in cellular automata investigations, as, for example, the ones of signal and grouping. We will later come back to these notions, but let us already note that an attempt of algebraic classification, founded on a grouping notion, could give a new look on the complex world of cellular automata (Mazoyer and Rapaport, 1998). Let us also remark that practicing these diagrams has also certainly lead to the idea of a better use of some "inactive" computational areas as, for example, in computations developed in (Mazoyer, 1996) and in chapter *Computations on grids.*

2.5.2. *Phase space*

Another way to analyze the global function of a cellular automaton $\mathcal{A}$ is to structure the configurations set into a graph according to the above derivation relation. More precisely the following graph, called the *phase space* of $\mathcal{A}$, is considered: its vertices are the configurations, and there is an arc from c to c' if and only if $c' = G_{\mathcal{A}}(c)$.
Actually in this space we get two sorts of orbits. Those corresponding to time-periodic configurations c which split into two parts : a first one such that $G^t_{\mathcal{A}}(c) \neq G^{t'}_{\mathcal{A}}(c)$ for all $t' < t$, called the *transient* part of c, and a second one which is a cycle, sometimes called the *period* of c. And those corresponding to orbits without cycle. So the phase space appears as a set of graphs (a forest), which obviously are the orbits of the Garden-of-Eden configurations of the automaton!

It is generally not possible to completely handle the phase space of a cellular automaton. Nevertheless, for *additive* or *linear* cellular automata, the phase space restricted to the periodic configurations can be completely described via some polynomials canonically associated to the global functions (Martin *etal.*, 1984), as it is the case for the examples of Figure 6. Incidentally, let us note here that additive cellular automata have been much studied : their "arithmetic" nature allows to think them easier to understand although they may have arbitrary complex behaviors (along with (Martin *etal.*, 1984), see (Aso and Honda, 1985), (Jen, 1988), (Sutner, 1991), (Allouche *et al.*, 1996), (Allouche *et al.*, 1998) and relevant references in these papers). Some attempts to generalize linear cellular automata are set about, with bilinear ones for example (Bartlett and Garzon, 1993), (Bartlett and Garzon, 1995).

The above mentioned difficulty leads to pay attention to the canonical topological structure (evoked in 2.4.3.) of the configurations set (then also named phase space) and especially to some of its subsets considered as giving some good approximation of the global asymptotic cellular automaton behaviour on some configurations. Among them appears *the limit set*, introduced by Wolfram (Wolfram, 1986), which represents the set of configurations such that their predecessors set (for the global function) is a tree of infinite depth.

Definition 2 *Limit set*

Let $\mathcal{A} = (S, \delta)$ be a d-cellular automaton, and $G_{\mathcal{A}}$ its global function.

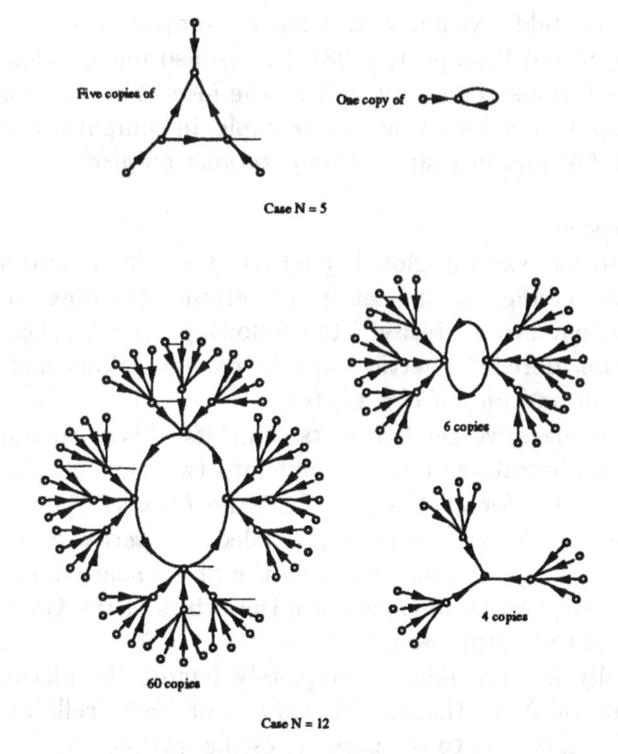

Figure 6. Phase space of the cellular automata defined on N-cells rings, the configuration of which are given, at time t, by $c^{(t)}(x) = \sum_{j=0}^{N-1} a_j^{(t)} x^j$ with $a_j^{(t)} = a_{j-1}^{(t-1)} + a_{j+1}^{(t-1)}$ and $a_j^{(t)} \in Z/2Z$.

The limit set of $\mathcal{A}$ *is* $\Omega(\mathcal{A}) = \cap_{t\geq 0}\Omega^t$ *with* $\Omega^0 = S^{\mathbb{Z}^d}$ *and* $\Omega^t = G_{\mathcal{A}}(\Omega^{t-1})$ *for* $t \geq 0$.

This notion (which makes sense because $\Omega(\mathcal{A})$ is non empty, due to the fact that the space is a compact one) allows to separate cellular automata into two classes: class 1 contains all cellular automata for which there exists i such that $\Omega^i = \Omega$ and class 2 the others, which is made relevant by the following result proved in (Čulik *et al.*, 1989)[3]: *$\mathcal{A}$ belongs to class 2 if and only if there exists a countable intersection D of dense open sets such that* $\Omega \cap (\cup_{i\in\mathcal{N}} G^i_{\mathcal{A}}(D))= \emptyset$. Most non trivial properties of the limit sets are recursively undecidable, even when the sets are restricted to finite configurations (see (Čulik *et al.*, 1989)).

[3]Another classification, may be the first formal one, was established in (Čulik and Yu, 1988) for special finite configurations. Later, S. Ishii in (Ishii, 1992), proposed a new classification based on a dynamic property of orbits of "almost every" configurations.

When we regard a configuration as a bi-infinite word on the finite alphabet S, we can consider its finite factors. The finite factors of the elements of $\Omega(\mathcal{A})$, for 1-CA's $\mathcal{A}$, make up a language, *the limit language* of $\mathcal{A}$, that has been intensively studied. It was proved that there exist 1-cellular automata with regular, context-free, context sensitive, recursive-enumerable and non-recursive-enumerable limit languages ((Hurd, 1987), (Čulik *et al.*, 1989), (Goles *et al.*, 1993)), all results that contribute to answer a question set up by Wolfram (Problem 13 in (Wolfram, 1985)). Let us remark that reversible cellular automata have rational limit sets: the whole set of words. As there exist universal reversible cellular automata ((Morita and Imai, 1996), (Dubacq, 1995)), there exist universal cellular automata with rational limit sets. But these sets are trivial. It is not always the case: in (Goles *et al.*, 1993) a cellular automaton is built, which is universal and has an *explicit* non trivial rational limit language.

2.5.3. *The work in progress*

Problems studied in the dynamics field are mainly of two sorts. On the one hand, general, mathematical properties of the global function are investigated, as surjectivity, injectivity, bijectivity and shift invariance, as it was already mentioned. On the other hand, topologies or metrics are defined on the phase space and their properties checked, especially those which seem to be relevant to found significant classifications. See the chapter *Topological Definitions of Deterministic Chaos*, and (Blanchard *et al.*, 1997). All these authors study the phase-space, but their approaches are quite different. People around Cattaneo start from cellular automata (so from the dialectic local/global) and look, among other investigations, for a notion of chaos which better corresponds to some "cellular" intuition, while people around Kůrka apply results on dynamic systems to the particular ones that are cellular automata.

To finish, let us emphasize that, in this area, contrary to what happens in the computation one, all configurations are taken into account, especially infinite ones.

2.6. VARIOUS TYPES OF CELLULAR AUTOMATA

Some types of cellular automata are of special interest, on various argument.

- *Totalistic cellular automata* play a particular role. They are 1-dimensional cellular automata (S, δ) such that there exists some function $f : \mathbf{N} \longrightarrow \mathbf{N}$ defining δ, that is more precisely for automata with the first neighbors neighborhood : $\delta(s_1, s_2, s_3) = f(s_1 + s_2 + s_3)$.
- We have already mentioned *linear cellular automata.* Some aspects of them are studied in chapters *Linear Cellular Automata and De Bruijn*

Automata and *Cellular Automata, Finite Automata and Number Theory*. One of their main property is to fulfil the ***superposition principle***, which means that their global function is an endomorphism. In this class we find, in particular, cellular automata the space-time diagrams of which are superposition of Pascal's triangles modulo integers (see (Reimen, 1992)).

- Other types of cellular automata are determined by restrictions due to their functions, as languages recognizers for example. In this latter case the input mode also lead to different types of automata as can be seen in *Part 3: Computational Power*.

2.7. CELLULAR AUTOMATA ON CAYLEY GRAPHS

A natural generalization of cellular automata has been ***graphs of automata***, obtained from undirected graphs (finite or infinite, but of finite degree), in putting at each vertex of the graph a copy of one finite automaton, the graph edges defining the connections between the automata. But, such a radical generalization gives birth to difficulties, some of which are discussed in chapter *An Introduction to Automata on Graphs*. In particular the regularity of the graph as the uniformity of the neighborhood are lost, and a cell does not know its environment. A way to avoid it is to narrow the definition to a special class of graphs, the Cayley graphs. Cayley graphs were introduced by Cayley (Cayley, 1878) for "drawing groups". His research for representing groups originated a lot of beautiful problems and results in group theory, and in studying cellular automata on Cayley graphs we can take advantage of it. Moreover, we know that each finite graph can be regarded as an induced subgraph of a Cayley graph (*Let Y be a graph with n vertices and let G be a group. Then some Cayley graph of G has Y as an induced subgraph if* $|G| \geq (2+\sqrt{3})n^3 \approx 3.74n^3$ (Godsil and Imrich, 1987)). Finally, as Z^d can be interpreted as a Cayley graph, we also get with cellular automata on Cayley graphs a direct and natural generalization of standard cellular automata.

But the idea of considering cellular automata on Cayley graphs can also be justified "from the outside". Some modelizations, in Physics or in Biology, use different tilings of their spaces, as tilings by triangles or hexagons for example, and people thought it was interesting to study the power of cellular automata on such structures. It turned out to be of low interest as shows the Róka's result just below. But other tilings remain to be studied.

Some work has already been done on these objects (Machí and Mignosi, 1993), (Róka, 1998). In the former one the Moore-Myhill theorem is extended to automata on Cayley graphs of groups of non exponential growth.

Theorem 4 *(Machí and Mignosi, 1993)*
Let $\mathcal{G}$ be the Cayley graph of a group of non exponential growth. Then, for a cellular automaton on $\mathcal{G}$, there exist Garden-of-Eden configurations if and only if there exist two mutually erasable configurations[4].

We also will here emphasize the following result, which could be rephrased in the computer science frame: *every reasonable plane architecture is equivalent to the grid.*

Theorem 5 *(Róka, 1998)*
For some given notion of simulation, the cellular automata on the Cayley graphs that define any archimedian tilings of the plane are equivalent.

2.8. OTHER GENERALIZATIONS

If cellular automata seem to be very efficient as computation models, and if they give interesting results in the field of modeling, some of their features imply that they are not so well adapted to biological processes. If a very big number of identical cells is efficient for computing, the rigidity and even the regularity of the structure is far from the biological reality.
So we can imagine extensions of these automata to more malleable models. See, in this volume, the global cellular automata of Čulik, *Decision problems on global cellular automata.*.

Moreover it is also possible to replace the finite automaton (S, δ) by other types of machines (as pushdown automata, Turing machines, ...).

3. Signals

When working with cellular automata, a very useful and familiar notion is the one of *signal.* It is a fundamental notion or tool, but, paradoxically, it seems difficult to get a satisfactory formalization of it, perhaps because it arises from an intricate mixture of different levels of understanding. Actually, we can imagine a signal as the virtual track of some specific information transmitted from some cells to some others, along a specific way, through the universe, if we understand the universe as the space-time diagram of some initial configuration.

In fact, we have essentially two ways to cope with cellular automata: we have either to build an automaton achieving some task, or to discover or interpret what an automaton is performing. Let us take the case of 1-dimensional cellular automata. In the first case we can try to imagine what would be the automaton space-time diagram, and to find some way

[4] This terminology was also the terminology of Moore and Myhill, which was gradually transformed with time into the one used in 2.

to geometrically structure this space to set up the result, what gives birth to geometric diagrams. Some significant real lines of the plane appear that can then be called *signals*, and, when they are *constructible* by cellular automata, they can be transformed into *cellular automata signals.* In the second case, we can observe some distributions or patterns of specific states which seem to be significant and can be considered as making up *cellular automata signals.*

Let us be a little more precise.

3.1. ONE DIMENSIONAL SIGNALS

We are here interested in conceiving 1-dimensional cellular automata. The reader will find in chapters *Computations on Cellular Automata* and *Computations on grids* sophisticated illustrations of signals in (geometric) diagrams on which are based cellular automata. But let us give a first simple idea.

3.1.1. *Signals in geometric diagrams*

Suppose we intend to design a cellular automaton which marks the integer powers of 2. The geometric diagram (Figure 7) put to light efficient straight lines (A with equation $x = 0$, Δ with equation $y = 3x$) or segments (parts of the straight lines $\vec{D_{2^n}}$ with equation $y = x + 2^n$ and $\vec{D'_{2^n}}$ with equation $y = -x + 2^n$ between A and Δ), which are then interpreted as signals in a dynamical understanding of the behaviour of a cellular automaton.

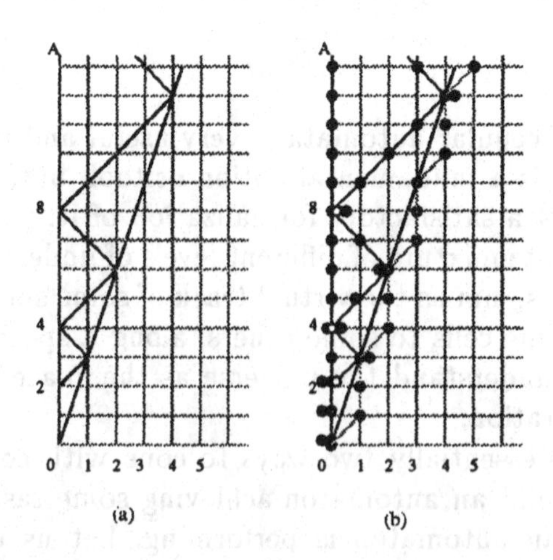

Figure 7. Geometric (a) and cellular (b) signals to obtain the powers of two on the cell 0.

And so it is very natural to say: to get the 2^n positions, $n \in \mathrm{N}$, on the cell number 0, it is sufficient to send, from this very cell, some *signals*: the first one will mark the cell 0, the second one will evolve as Δ, the third one will oscillate between A and Δ, at "maximal speed", and will give a wanted integer each time it arrives on cell 0. Its meeting points with Δ are the points $(2^{n-1}, 3(2^{n-1}))$, $n \geq 1$ (see Figure 7(a)).

If we are going back to the line of cells, we can imagine signals as information handled as an indivisible unit, able to stay on cells or move between cells (as quickly as possible would be the best), and, sometimes, when meeting others, to modify their moves directions or, even, to give birth to new ones. These moves are controlled by means of local rules and finite (what is essential) sets of states.

Our example, given in (Choffrut and Čulik, 1984), is simple, and it is not difficult to design a cellular automaton the behaviour of which allows to get the 2^n on the first cell, as shows the space-time diagram of the Figure 8.

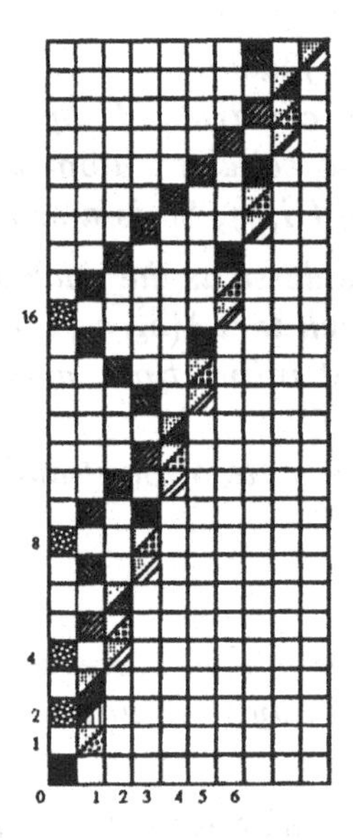

Figure 8. Space-time diagram for signals to obtain the powers of two on the cell 0.

3.1.2. *Simple or basic cellular automata constructible signals*

In the above example, we observe that signals on the geometric diagram are straight lines or segments of straight lines, which can be seen as the best trajectories of information between cells. Moreover, as time is not reversible, these lines are strictly increasing functions of it. Finally, as we are considering cellular automata with 1-neighborhood, the course of an atom of information will to be found inside a cone with the initial cell as vertex, so, modulo a possible rotation of 45°, we can limit our attention to sets of cells in $\mathbf{N} \times \mathbf{N}$, and the atom will stay on a cell or go to one of its consecutive neighbors. This leads to the first following definitions.

Definition 3 *Signals*

1. *A signal is a set* $\{(f(t),t)\}_{t\in\mathbf{N}}$, *where* f *is a function from* $\mathbf{N}$ *into* $\mathbf{N}$ *such that* $(f(t+1),t+1)$ *is* $(f(t)-1,t+1)$, $(f(t),t+1)$ *or* $(f(t)-1,t+1)$. *Such a signal is said to be rightward (resp. leftward) if, for each* t, $t \in \mathbf{N}$, $(f(t+1),t+1) \in \{(f(t),t+1),(f(t)+1,t+1)\}$ *(resp.* $\{(f(t),t+1),(f(t)-1,t+1)\}$*).*
2. *A signal is said to be simple or basic if the sequence* $(f(t+1)-f(t))_{t\in\mathbf{N}}$ *is ultimately periodic.*

We are interested in signals constructible by cellular automata. What does that mean?

Definition 4 *Constructible signals*

A signal Σ *is said to be constructible by cellular automata (or CA-constructible) if there exists a cellular automaton* $(S,\delta,\sharp,G,s_e,S_0)$ *where* $\sharp, G$ *et* s_e *are special states of* (S,δ), S_0 *is a subset of* S, *such that*

- $\sharp$ *is the border state, considered as the unchanging cell-0 left neighbour,*
- s_e *is a quiescent state such that* $\delta(s_e,s_e,s_e)=\delta(\sharp,s_e,s_e)=s_e$,
- G *is a state such that, at initial time, all cells are in state* s_e *except the cell 0 that is in state* G,
- *a cell* k, *at time* t, *is in a state of* S_0 *if and only if* (k,t) *is in* Σ.

Then we have:

Fact 2

Basic signals are CA-constructible. Moreover, if a signal Σ *is generated by* $(S,\delta,\sharp,G,s_e,S_0)$ *in such a way that all sites outside* Σ *are in state* s_e, *then,* Σ *is finite or basic.*

Let Σ be a basic signal, *per* its lowest period starting at t_0 and $\sigma = f(t+per)-f(t)$ for any t, $t > t_0$. The rational number $\frac{per}{\sigma}$ is called the *slope* of Σ. It is more usual to represent a signal Σ of slope v by a straight line in a geometric diagram than by means of states on its space-time diagram. In a dynamical understanding of a signal we can see this slope as its speed, measured from the starting time of its periodic part.

These basic signals play a very important role in algorithmics on cellular automata, as it can be seen in *Part 2: Algorithmics.* But we may wonder whether they allow to build more complex ones. A beautiful and historical example is to be found in the chapter *Computations on Cellular Automata,* where "Fisher's parabola" is displayed. We will here and now develop a simpler example, that already put to light some difficulties.

3.1.3. *Other cellular automata constructible signals*
To properly answer the above question, two attributes of cellular signals have to be introduced: their ratio and speed. The *ratio* of a signal Σ is an increasing function ρ, from **N** into **N**, which gives, for each cell n, the time required by Σ to reach the cell n, for the first time, from the origin. The *speed* of Σ to reach the cell n is then $v(n) = \frac{n}{\rho(n)}$. We immediately observe that the maximal possible speed in case of von Neumann neighborhood is 1, but actually it depends, in a non trivial way, on the neighborhood.

Now let us come back to the production of the 2^n, $n \in N$, for showing how it can be transformed to give rise to a (cellular) signal of exponential ratio. Actually, that is illustrated on the geometric diagram of the Figure 9. The signal Σ' is constructed thanks to two others, D and Δ', in the following way. The three start from the origin, Σ' and D follow the same way up to $(2,4)$, there, they split: the first one goes on up, the second one goes at speed 1 up to meet, at $(3,5)$, Δ', which has followed the indicated track. At $(3,5)$ D comes back to the left at speed 1, up to meet Σ', then they run together up for one time unit, then at speed 1 for one time unit and they part as after their first meeting (the first one going up, the second one on the right at speed 1), while Δ' is going from $(3,5)$ to $(4,5)$ where it takes speed 1/3. This process will be repeated as shown on Figure 9(a).
Actually we start from the first "natural" geometric diagram, Figure 9(a), where the basic points are the (x_n, y_n), with $x_n = n+2^{n-1}-1$, $y_n = 2^n + 2^{n-1} - 1$ for $n \geq 2$. But it is easy to observe that the segments between (x_n, y_n) and (x_{n+1}, y_{n+1}) don't have the same slope, while it is the case for the segments delimited by the points (x'_n, y_n) and (x'_{n+1}, y_{n+1}) where $x'_n = x_n + 1$, the slope of which is 3 (Figure 9(b)).

It remains to transform the last geometric diagram in a cellular geometric diagram, we have to find a way to mark the points (x'_n, y_n). This is done via a new signal D' as it is shown on Figure 9(b). Let us notice that D' knows that it has to die on (x'_n, y_n) one time after meeting Σ', and to be replaced from there by a signal of speed 1/3. And finally a better solution is given on Figure 10.

But does that mean that there exists a 1-cellular automaton generating this signal of ratio $n \longmapsto 2^n$?

Remark 1

*First of all, let us emphasize that the problem to know whether for a given geometric diagram there exists a cellular automaton which realizes it is undecidable. The basic reason is that, generally, it is impossible to determine if the number of different signals which can have to be superposed is finite. Actually, would this problem be decidable, would the problem to know whether a state emerges in the evolution of a cellular automaton be decidable, which is not (see (*Smith, 1972*)).*

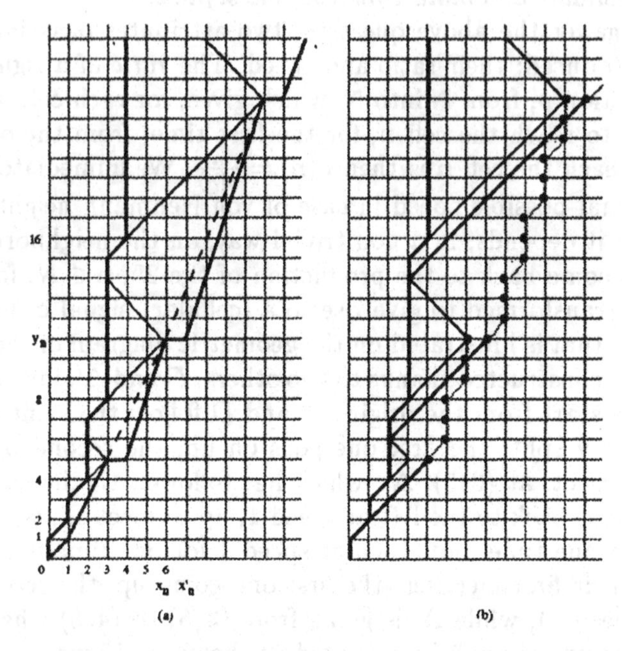

Figure 9. Geometric (a) and cellular (b) signals to obtain a cellular signal of ratio $n \mapsto 2^n$.

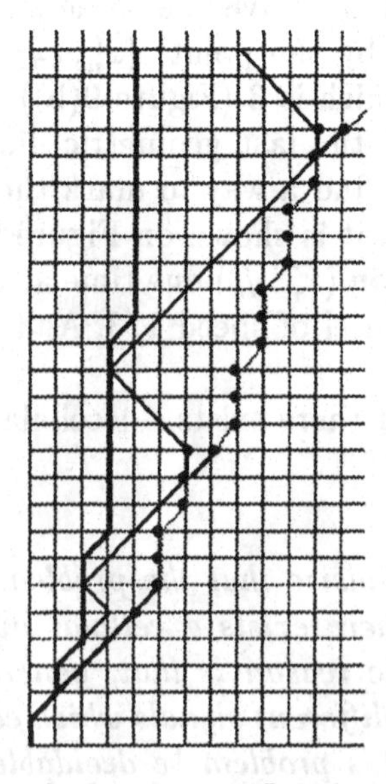

Figure 10. Cellular signal of ratio $n \mapsto 2^n$ and signals needed to build it.

So quite each case is a singular one, and it is not always easy to design a cellular automaton (assuming that such an automaton exists) even governed under an expressive cellular geometric diagram. It is not the case for our example, and a space-time diagram of a suitable automaton is to be found on Figure 11.

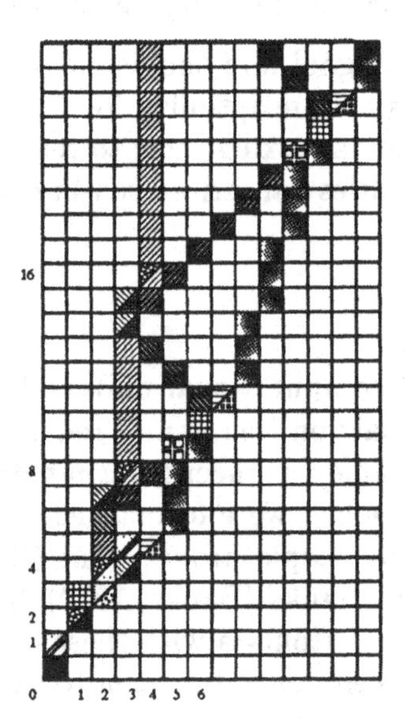

Figure 11. Part of a space-time diagram representing the construction of a signal of ratio $n \longmapsto 2^n$.

Let us mention that it is possible to build cellular signals of quadratic ratios, ratios involving roots and logarithms. The interested reader can refer to (Mazoyer and Terrier, 1998). To finish this section we will still cite an interesting result of this paper, which shows that there exist gaps among the ratios of constructible signals.

Theorem 6 *A gap theorem*

For any constructible signal of ratio $n \longmapsto \rho(n)$, either $\rho(n) - n$ becomes constant or the signal is lower bounded by a signal of ratio $n + log_c(n)$ (c is a constant, $c \in \mathbf{N}$), which means that $n + log_c(n) \leq \rho(n)$ for every n, $n \in \mathbf{N}$.

3.2. SIGNALS IN DIMENSION D, $D > 1$

Although different notions of signals commonly occur in the literature for dimension 2 (significant examples are to be found in (Codd, 1968) and the gliders of the "Game of Life" - see chapter *The Game of Life: Universality revisited* - can be interpreted as signals), nothing formal is known in this area, which is still to open up. It is already difficult to see inside the space-time diagram of a 2-dimensional cellular automaton. Moreover we can imagine all sorts of trajectories (curves) depending on the chosen neighborhood, or waves fronts or volumes ... In fact, when "signals" are touched on in higher dimensions, it is often via projections on the plane, which is a way of bypassing the difficulty, as in (Delorme *et al.*, 1998).

Actually the notion of signal and the use it is done of are closely connected to the problematics of time optimization, especially the problematics of *real time*, that is, roughly speaking, how to realize objectives *as soon as possible*. Then some difficulties can arise even in dimension 1.

4. Universalities

As it was already noticed computational universality and some biological-like robustness were essential von Neumann's requirements for the device he was in search of. So he became a pioneer for *computational universality* and *construction universality*, which were formalized, and deeper investigated, a little later by Burks and people around him (Burks, 1966), as well as by Codd (Codd, 1968) and Banks (Banks, 1970). Their main work was to build 2-dimensional cellular automata with these properties and few states, or to find conditions for a cellular automaton to satisfy them.

Outside construction universality that will be considered later, people are, in our context, essentially or ultimately interested in computational universality. But there are several notions of computational universality, depending on the fact that they are set up by means of simulations or not, and, in the former case for example, depending upon the fact that simulations are between computation models of different nature, essentially Turing machines and cellular automata (extrinsic simulations), or between cellular automata (intrinsic simulations), or depending on the configurations considered, on the encoding or decoding understandings ...

Intrinsic universality, corresponding to some intrinsic simulations, was pointed out later in (Albert and Čulik, 1987). It also expresses the power of some cellular automaton with respect to other (or a class of) cellular automata, and so attests a sort of model (inner or auto-) coherence and potential expressive power.

In the sequel we present the notions of universality which are at work in the basic historical universality results. Most of them are founded on implicit simulation notions, generally "ad hoc". We put to light some related difficulties and refer to chapter *The Game of Life: Universality revisited* for illustration and a deeper discussion.

4.1. COMPUTATIONAL UNIVERSALITY

What is at stake in this area is to study the computational power of the model. Finite configurations are then basic objects. Here appears with a noticeable clarity the power given to cellular automata by their infinite component (their space), but also the difficulties it arises, in particular because a computation step implies the instantaneous transformation of the infinite number of cells, so that some isolated cell can have hidden perverse effects on computation areas generated by some other very distant finite part of the initial configuration.

Roughly speaking, a cellular automaton will be computation-universal if it is able to compute any Turing-computable (or recursive) partial function. But what means "computing a function ψ" for a cellular automaton ? Still roughly, that means that there exists some initial configuration, part of which can be interpreted as an argument for ψ, that possibly evolves to an halting one, part of which can be interpreted as the ψ-image of this very argument. Three problems already arise: *how to encode the argument (in particular in such a way that the result does not appear in this encoding), how to recognize the end of a computation, how to decode the image*? The main difficulty is to suitably encode finite data into an infinite set of sites. This direct way was taken by von Neumann or more precisely by his main continuators Burks, Thatcher, Codd and Banks. Another way to tackle the problem has been to directly compare cellular automata and Turing machines computational power by means of simulations.

4.2. DIRECT COMPUTATIONAL UNIVERSALITY

4.2.1. *Computation universality by Codd*

We will first recall the thought processes in the sixties. They end up with devices - actually special configurations of some cellular automaton -, computing the partial recursive functions. We will follow Codd (Codd, 1968), who set up formal definitions rendering thoughts, and their explicit achievements, at the time. These definitions are not the more general and have to be used with caution because they let subsist some difficulties (which do not affect the effective constructions). Note that Codd focused on 2D-cellular automata, but that what he established is also available for 1-dimensional ones.

Computation-universal cellular automata We now consider cellular automata with a quiescent state ν and global function G. Let us start with some technical definitions and notations.

1. If two configurations c and c' are disjoint ($Supp(c) \cap Supp(c') = \emptyset$), $(c \sqcup c')$ denotes the configuration defined by $(c \sqcup c')(x) = c(x)$ on $Supp(c)$, $(c \sqcup c')(x) = c'(x)$ on $Supp(c')$, and $(c \sqcup c')(x) = \nu$ on the complement (in the space) of $Supp(c) \cup Supp(c')$.
2. A *subconfiguration* c' of a configuration c is a configuration such that $c_{|Supp(c')} = c'_{|Supp(c')}$.
3. A configuration c is called *passive* when it is a fixpoint for G, that is $G(c) = c$, and *completely passive* when all its subconfigurations are passive.
4. A configuration c is said to pass on information to a configuration c', disjoint of c, when there exists a time t such that $G^t(c \sqcup c')_{|Supp(G^t(c'))} \neq G^t(c')_{|Supp(G^t(c'))}$.
5. A configuration c is a translation of a configuration c' if there exists a vector τ such that $c'(x) = c(x - \tau)$ for each x in the space.

To take into account the functions data, special sets of *finite configurations* are distinguished, called *Turing domains.* Let $\mathcal{A}$ be a cellular automaton, a Turing domain for $\mathcal{A}$, denoted $T_{\mathcal{A}}$, is a set of $\mathcal{A}$-configurations which satisfies the following conditions:

1. they are finite configurations and there is an effective (recursive) injective application $\iota_{\mathcal{A}}$ from N into $T_{\mathcal{A}}$,
2. each configuration in $T_{\mathcal{A}}$ is completely passive and, moreover,
3. the configurations in $T_{\mathcal{A}}$ are collectively passive, which means that for each finite subset of $T_{\mathcal{A}}$, each set of translations which yields disjoint translates, the union of these translates is completely passive.

Definition 5 *Computation-universal cellular automaton*

A cellular automaton $\mathcal{A}$ is said to be computation-universal *if there exists a Turing domain for $\mathcal{A}$ and if, for each Turing-computable partial function ψ from* $\mathbf{N}$ *to* $\mathbf{N}$, *there exists an $\mathcal{A}$-configuration c, $c \notin T_{\mathcal{A}}$, which* computes ψ, *that is that there exists a cell α, $\alpha \in Supp(c)$ and a non-quiescent state s_{halt}* [5] *such that for each n in the ψ-domain:*

1. $G^t(c \sqcup \iota_{\mathcal{A}}(n))(\alpha) = s_{halt}$,
2. $G^{t'}(c \sqcup \iota_{\mathcal{A}}(n))(\alpha) \neq s_{halt}$ *for all* $t' \leq t$
3. $G^t(c \sqcup \iota_{\mathcal{A}}(n))_{|\overline{\cup_{f \in T_{\mathcal{A}}} Supp(f)}}$ *does not pass information to* $G^t(c \sqcup \iota_{\mathcal{A}}(n))_{|\cup_{f \in T_{\mathcal{A}}} Supp(f)}$,
4. $G^t(c \sqcup \iota_{\mathcal{A}}(n))_{|\cup_{f \in T_{\mathcal{A}}} Supp(f)} = \iota_{\mathcal{A}}(\psi(n))$.

As $T_{\mathcal{A}}$ is a set of finite configurations and $\iota_{\mathcal{A}}$ is one-to-one, there is no problem of decoding. So $\mathcal{A}$ is such that each Turing-computable partial function is computed by means of one of its configurations.

[5] Another suitable way is to consider fix points as halting configurations.

Universal computers We can wonder whether it could be done by only one of them. This leads to the notion of universal computer, which is a particular configuration of some cellular automaton.

Definition 6 *Universal computer*

Let $\mathcal{A}$ be a cellular automaton with a Turing domain $T_{\mathcal{A}}$. An $\mathcal{A}$-configuration c, $c \notin T_{\mathcal{A}}$, is called a universal computer *if, for any Turing-computable partial function ψ, there exist d in $T_{\mathcal{A}}$ and a translation defined by a vector τ such that the translate d_τ of d by τ is disjoint of c, not in $T_{\mathcal{A}}$, and that $c \sqcup d_\tau$ computes ψ.*

Clearly, the existence of a universal computer for a cellular automaton makes it a computation-universal cellular automaton. But it is not known whether a computation-universal cellular automaton has necessarily a universal computer. Actually,

Fact 3

The historical "universal" 2-dimensional cellular automata (with von Neumann neighborhood), as the von Neumann's one (29 states), the Codd's one (8 states) and the Banks' one (4 states) are cellular automata with universal computers.

Their design is not so simple (and even needs books - for the first ones - to be described and proved right). The interested reader is invited to read (Codd, 1968), (Banks, 1970) and (Burks, 1972).

Another way to cope with the computable partial functions was to find suitable cellular automata for each of them or, at least for the basic ones. That was recently done by J. Mazoyer.

4.2.2. *Computation universality via the Kleene's recursive functions*

We know that the Turing computable functions are exactly the class of partial recursive functions defined as the smallest class of partial functions that contains the constant function **0**, the successor function, the projections and is closed under substitution, recursion and minimization (see for example (Cutland, 1980)). Very special algorithmic techniques on grids, as it can be seen in the chapter *Computations on grids*, are used to establish the following result.

Fact 4 *(*Mazoyer, 1996*)*

Each partial recursive function can be computed on a 1-dimensional cellular automaton.

We have no more a "universal device", but a *property of the cellular automata computation model*, that we can - a little improperly - render by 1-*dimensional cellular automata are computationally universal.*

Let us now come to notions and results founded on simulations, first between different models and, afterwards, inside the model.

4.3. COMPUTATION UNIVERSALITY VIA TURING MACHINES

It seems obvious that if every one-bi-infinite-tape Turing machine (or even better a one-bi-infinite-tape universal Turing machine) can be simulated by one cellular automaton, "cellular automata" have, at least, the Turing machines computational[6] power and so can be considered as a "computation universal model".

The main question is then what do we mean by a simulation? A lot of problems are solved by means of simulations (models comparisons, various optimization achievements, complexity issues, ...). But, although the expected results deeply depend upon them, simulations which realize, under conditions, the connection between two devices (or two classes of devices) are ad hoc or, often, underhanded. In particular, different notions of universality are attached to different sorts of simulations.

4.3.1. *Simulations*

A Turing machine which computes a partial function from **N** to **N** can be considered as a 4-tuple $(\mathcal{C}, \mathcal{T}, \mathcal{D}, \mathcal{E})$, where $\mathcal{C}$ denotes its configurations set, $\mathcal{T}$ its global transition function (mapping each configuration on its successor via the transition function of the machine (supposed deterministic), $\mathcal{D}$ is a function from $\{0,1\}^+$ into $\mathcal{C}$ which sends each word (coding a non negative integer) on a configuration which will be an initial configuration for the machine, and $\mathcal{E}$ a partial function from $\mathcal{C}$ to $\{0,1\}^+$ which will associate to each halting configuration a word on $\{0,1\}$ (and so, a non negative integer).

We can certainly get an analogue description for cellular automata. But a great difference with Turing machines for which configurations are finite[7] is, precisely, that for cellular automata configurations are infinite objects. Thus, finite words have to be set up on infinite spaces. This difficulty leads first to limit the set of configurations to the one of finite configurations, which requires automata with quiescent states. A computing cellular automaton can be represented by $(\mathcal{C}_f, \mathcal{G}, \mathcal{D}, \mathcal{E})$.

Now, to formalize a simulation of a Turing machine by a cellular automaton, we need a coding of the configurations of the first one into the finite configurations of the second one.

[6]Recall that from the computability point of view Turing machines and cellular automata - the latter ones here considered on finite configurations - are equivalent, but from the complexity point of view, cellular automata are much more efficient. See Part 3 in this volume and (Wagner and Wechsung, 1986) for example.

[7]Usually, or at least in the computability area.

Definition 7 *Simulation of a Turing machine by a cellular automaton with a quiescent state*

Let $(\mathcal{C}_{\mathcal{M}}, \mathcal{T}_{\mathcal{M}}, \mathcal{D}_{\mathcal{M}}, \mathcal{E}_{\mathcal{M}})$ *and* $(\mathcal{C}_{f\mathcal{A}}, \mathcal{G}_{\mathcal{A}}, \mathcal{D}_{\mathcal{A}}, \mathcal{E}_{\mathcal{A}})$ *be descriptions for a Turing machine* $\mathcal{M}$ *and a cellular automaton* $\mathcal{A}$ *(with a quiescent state). We say that an injective recursive application* Γ *from* $\mathcal{C}_{\mathcal{M}}$ *into* $\mathcal{C}_{f\mathcal{A}}$ *is a simulation of* $\mathcal{M}$ *by* $\mathcal{A}$*, or that* $\mathcal{A}$ *simulates* $\mathcal{M}$*, if the following conditions are satisfied:*

- *for each* u *in* $\{0,1\}^+$ *there exist* t, t' *such that* $\Gamma(\mathcal{T}_{\mathcal{M}}{}^{t}(\mathcal{D}_{\mathcal{M}}(u)))$ *is* $\mathcal{G}_{\mathcal{A}}{}^{t'}(\Gamma(\mathcal{D}_{\mathcal{M}}(u)))$*,*
- *for each halting configuration* c *in* $\mathcal{C}_{\mathcal{M}}$*, there exists* t *such that* $\mathcal{G}_{\mathcal{A}}{}^{t}(\Gamma(c))$ *is a configuration* c' *in the* $\mathcal{E}_{\mathcal{A}}$*-domain such that* $\mathcal{E}_{\mathcal{A}}(c') = \mathcal{E}_{\mathcal{M}}(c)$*.*

In fact, when we are interested in the simulation complexity, we set up a more uniform definition in assuming the existence of Γ and of two integers k et k' such that for each u in $\{0,1\}^+$, $\Gamma(\mathcal{T}_{\mathcal{M}}{}^{k}(\mathcal{D}_{\mathcal{M}}(u)))$ is $\mathcal{G}_{\mathcal{A}}{}^{k'}(\Gamma(\mathcal{D}_{\mathcal{M}}(u)))$. That makes the simulation configuration-independent.

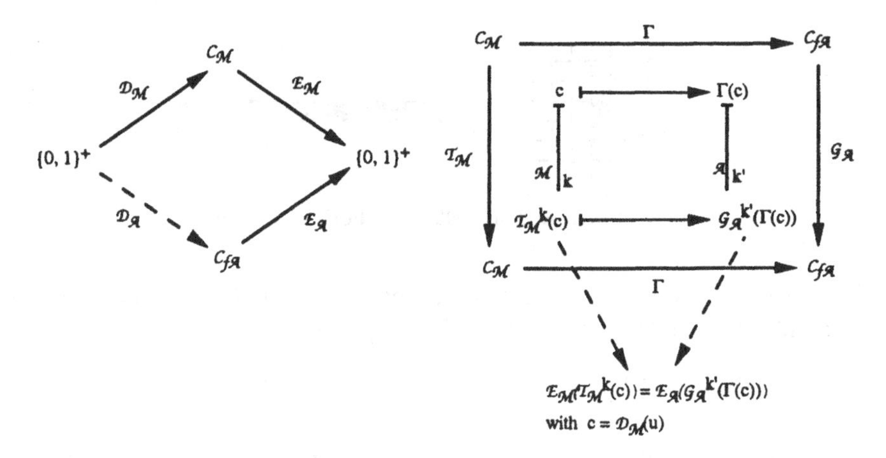

Figure 12. A simulation of Turing machines by cellular automata.

This notion is used to get results as the following ones, which are historical in this trend. In (Smith, 1971), Smith looks for "small" (in number of states and in number of neighbor cells) cellullar automata able to compute Turing computable functions. Let us mention here that Codd (Codd, 1968) got a computation-universal 2-states automaton by simulation of its (8-states, von Neumann neighborhood) automaton, but with a neighborhood of 85 cells!

Smith also proved the noticeable fact that dimension 1 is sufficient for computational universality. Let us now list some of his results which also evaluate the simulations times.

Theorem 7 *(*Smith, 1971*)*

1. *For each (m, n) Turing machine (m symbols, n states), there exists a 2-dimensional cellular automaton with $max(m+1, n+1)$ states and a neighborhood of seven neighbors, which simulates it real-time ($k = 1 = k'$ in the above simulation).*
2. *For each (m, n) Turing machine, there exists a 2-dimensional cellular automaton with $max(2m+1, 2n+2)$ states and the von Neumann neighborhood, which simulates it in 3 times real-time ($k = 1, k' = 3$).*
3. *For each (m, n) Turing machine (m symbols, n states), there exists a 1-dimensional cellular automaton with $max(m+1, n+1)$ states and a neighborhood of six neighbors, which simulates it real-time ($k = 1 = k'$ in the above simulation).*
4. *For each (m, n) Turing machine, there exists a 1-dimensional cellular automaton with $m + 2n$ states and the von Neumann neighborhood, which simulates it in at most 2 times real-time.*

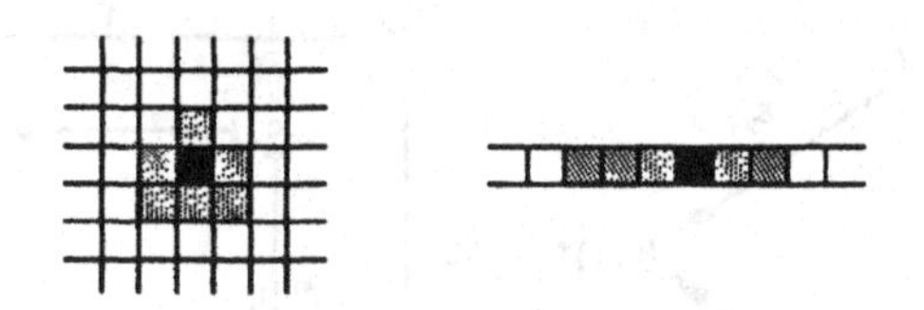

Figure 13. Seven and six cells neighborhoods of Smith.

These results, together with the existence of a $(6, 6)$ universal Turing machine (Minsky, 1967), imply the following corollaries:

Theorem 8 *(*Smith, 1971*)*

1. *There exists a 2-dimensional cellular automaton computation -universal with 7 states and a neighborhood of seven neighbors.*
2. *There exists a 2-dimensional cellular automaton computation -universal with 14 states and the von Neumann neighborhood.*
3. *There exists a 1-dimensional cellular automaton computation -universal with 7 states and a neighborhood of six neighbors.*
4. *There exists a 1-dimensional cellular automaton computation -universal with 18 states and the von Neumann neighborhood.*

A lot of results have been obtained since, in the same vein, aiming to evaluate and optimize the simulations in number of states (the search for small universal cellular automata is still alive), or to restrict the class in which candidates to be universal lay (see 4.4.4.).

Other proofs of the universal computational power of cellular automata can be found. Let us cite, for example, that *every two-pushdown automaton can be simulated by a 1-dimensional cellular automaton* (Goles *et al.*, 1993), which allows the wanted conclusion because a two-pushdown automaton is equivalent to (has the computational power of) a Turing machine.

Note that in dimension 2, the proof of computation-universality of the "Game of Life" by Conway is in a similar vein, as well as the proof via boolean circuits given in chapter *The Game of Life: Universality Revisited.*

Carrying on with work and reflections initiated by Wolfram (Wolfram, 1986), some people try to find out connections between computational universality and complexity (Wagner and Wechsung, 1986).

The above results lean on simulations of sequential models expressing computational universality, which masks or neglects the parallel nature of cellular automata. Moreover it is interesting to compare the expressive power of cellular automata on their own. This leads to *intrinsic universality.*

4.4. INTRINSIC UNIVERSALITY

The basic question is: do there exist cellular automata able to simulate all cellular automata, at least of a given class? It is not a problem to canonically code the finite automaton underlying each cellular automaton, and to get a recursive enumeration of them. The problem arises with the configurations, which are infinite ... So, once more, if we restrict the devices to their finite configurations, we can imagine a solution analogous to the solution for Turing machines: find a cellular automaton $\mathcal{U}$ such that, given the code of any automaton $\mathcal{A}$ of a given class and the code of a finite configuration c of $\mathcal{A}$, we can get, in a configuration of $\mathcal{U}$, the halting configuration of $\mathcal{A}$ on c when it exists (a serious problem will then be to satisfactorily formulate the notion of halting configuration).

4.4.1. *Intrinsic universality in the style of Čulik*

As it should now be understood, universality is not an intrinsic notion! Most of time it is, at least, indissociable from a simulation notion and a class of automata. So we will first give the definition of a suitable simulation.

Definition 8 *I_1-Simulation*

Let CA be a class of cellular automata and CA_f the set of their finite configurations. We will call I_1-simulation (for intrinsic simulation in sense 1) any triple (Φ, χ, π) of recursive functions, Φ from CA to itself, χ from $CA \times C_{\mathcal{A}f}$ into $C_{\mathcal{A}f}$, π from N *to* N*, such that for each automaton A, each finite configuration c for A, if c_t is the configuration obtained in A after t computation steps starting with c, then $\chi(A, c_t)$ is the configuration ob-*

tained in $\Phi(A)$ after $\pi(t)$ computation steps starting with $\chi(A,c)$, that is $G^{\pi(t)}_{\Phi(A)}(\chi(A,c)) = \chi(A, G^t_A(c))$.

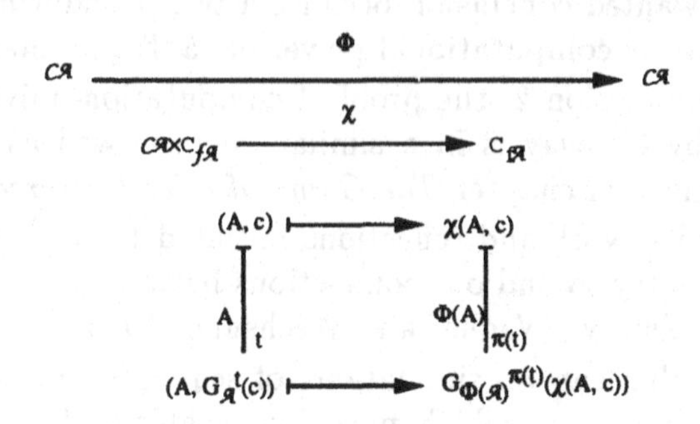

Figure 14. A first simulation: I_1-Simulation between cellular automata.

To this sort of simulation is associated a notion of universality, the I_1-universality.

Definition 9 *I_1-universal cellular automaton*

A cellular automaton $\mathcal{U}$ is said to be I_1-universal for the class $\mathcal{CA}$ if there is an I_1-simulation (Φ, χ, π) such that, for each A in $\mathcal{CA}$, $\Phi(A) = \mathcal{U}$.

A first result in this area was the following one (Albert and Čulik, 1987), which asserts the existence of a I_1-universal cellular automaton for the class of the 1-dimensional-one-way cellular automata.

Theorem 9 *Albert-Čulik theorem*

There is a 1*-dimensional cellular automaton (*14 *states, first neighbors neighborhood), which I_1-simulates any totalistic* 1*-dimensional-one-way cellular automaton.*

This result was extended to any 1-dimensional cellular automaton in (Martin, 1994). And it is worthy to notice that a consequence of this result is that, by composing simulations, computation-universality only needs one dimensional totalistic automata.

But, nevertheless, this last notion of intrinsic universality is not really satisfying because of the restriction to the finite configurations (but if we are interested in computing, they are difficult to by pass) and because the components of the simulation are not homogeneous to the model: they can not be interpreted as global functions of cellular automata. So we are going to set up another more suitable definition of simulation, more like the one given for Turing machines.

4.4.2. *Intrinsic universality in an other style*

This naturally starts with the definition of a new notion of simulation.

Definition 10 *I_2-simulation*
Let $\mathcal{CA}$ a class of cellular automata, $\mathcal{C}$ their configurations set. We say that a cellular automaton $\mathcal{B}$ in $\mathcal{CA}$ I_2-simulates $\mathcal{A}$ of $\mathcal{CA}$ if there exists an injective application Γ from $\mathcal{C}$ into itself such that for each configuration c of $\mathcal{A}$ there exist t, t' such that $\Gamma(G^t_{\mathcal{A}}(c)) = G^{t'}_{\mathcal{B}}(\Gamma(c))$.

In fact we consider uniform simulations in which there exist Γ and integers k and k' such that $\Gamma(G^k_{\mathcal{A}}(c)) = G^{k'}_{\mathcal{B}}(\Gamma(c))$, and, in particular, the case where $k = 1$ ((Róka, 1998)).

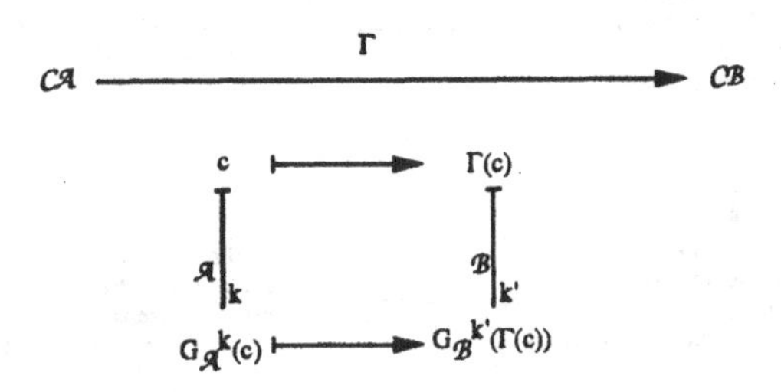

Figure 15. Another simulation: I_2-Simulation between cellular automata.

To this simulation is also attached a notion of universal cellular automaton.

Definition 11 *I_2-universal cellular automaton*
A cellular automaton of the class $\mathcal{CA}$ is I_2-universal for this class if it I_2- simulates each cellular in $\mathcal{CA}$.

Without any further hypothesis on Γ or on the configurations, we take the risk not to find any I_2-universal cellular automaton. So we have to be a little more precise. If we restrict the configurations to finite ones and if we demand Γ to be recursive, then we have examples. But better can be done, and for a more complete discussion we refer the reader to the following chapter dedicated to the Game of Life, which is there proved to be, in fact, intrinsically universal.

We will just put to light that if we ask in the above definition that $k = 1, k' = T$ with some T independent of the configuration and that Γ were the global function G of some cellular automaton of $\mathcal{CA}$, the condition of the definition would be $G(G_{\mathcal{A}}(c)) = G^T_{\mathcal{B}}(G(c))$, then $G \circ G_{\mathcal{A}} = G^T_{\mathcal{B}} \circ G$. Such a commutation property is connected with the notion of grouped cellular automata we will present in the next section.

4.4.3. *Connections between the above notions of universality*

We have brought to the fore some (main) notions of universality, without looking for exhaustiveness, but to recall that universality for cellular automata is multifaceted, designates properties which, most of time, are not equivalent and whose connections, when they exist, are often founded on delicate codings and implicit restrictions on hypotheses. Up to now most of results refer to computational universality in their different senses. But the intrinsic universality issue seems to be noteworthy. We have also wanted to draw attention to the fact that sometimes universality characterizes special configurations, cellular automata, classes of cellular automata or even classes of automata on particular configurations.

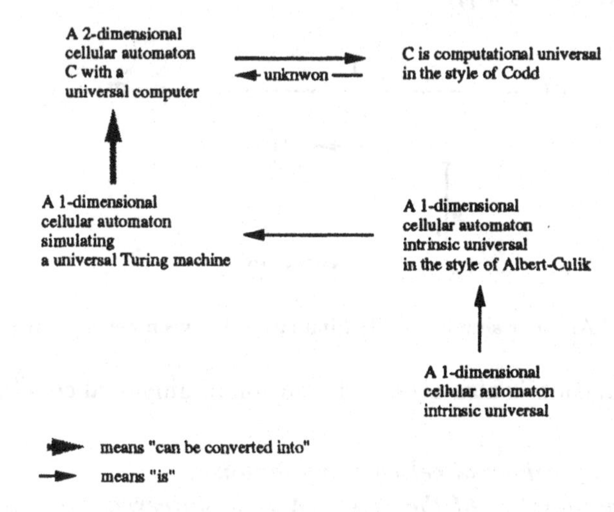

Figure 16. Connections between different notions of universality.

4.4.4. *Minimal computation universal cellular automata*

To finish with universality, let us notice that the search of universal automata with minimal number of states still goes on. But, often, less states seems to imply some compensation for example a bigger neighborhood (dimension 2, 2 states but a neighborhood with 85 cells in (Codd, 1968)), or some background (dimension 2, 2 states, von Neumann neighborhood and a background (Banks, 1970)). See also (Lindgren and Nordahl, 1990) which recall or prove significant results, and (Gajardo , 1998) which is recent in the race. A *background* for a 1-dimensional cellular automaton is a configuration c which is both time and space periodic. Time-periodic means that $(G^t(c))_{t\geq 0}$ is periodic, and, if σ denotes the shift on configurations (defined by $(\sigma(x))_i = x_{i+1}$), c is space periodic when the sequence $(\sigma^n(c))_{n\geq 0}$ is periodic.

Another notion of universality, rather historical than widely or even rigorously developed, but historically interesting, is construction universality, which is often (and often wrongly) mixed up with (self-)reproduction.

4.5. CONSTRUCTION UNIVERSALITY

Once more, let us come back to von Neumann who was interested in understanding the logical organization of natural complex systems, especially systems able of auto-reproduction. He wondered what is necessary for an "automaton" to reproduce itself, without losing any of its properties, especially its self-reproducing ability. This concern, together with the one of computational power he required, leads him to conceive its automaton, and to set up this idea of construction universality.

4.5.1. *Construction universality in the von Neumann's style*

Still following Codd we will say that a configuration c of an automaton $\mathcal{A}$ builds some $\mathcal{A}$-configuration c' if there exists some time t such that c' is a subconfiguration of $G^t_{\mathcal{A}}(c)$, disjoint of c and such that $G^t_{\mathcal{A}}(c) - c'$ does not pass on information to c'. This definition set aside fix points, which would be trivial, and assures its autonomy to the constructed configuration. But these conditions are not sufficient to avoid the risk of triviality (think to the shift or to $(\{0,1\},\delta)$ with $\delta(x,y,z) = x + y + z \pmod 2$ on the initial configuration $0^\omega 10^\omega$). It is why computation abilities were also required.

Let $\mathcal{U}$ be a computation-universal cellular automaton (in the Codd sense) and suppose there exists a set C of configurations such that each of them computes a Turing-computable partial function and such that, conversely, each Turing-computable partial function were computed by a member of C. If every element of C is constructible by some configuration disjoint from C, then $\mathcal{A}$ is said to be *construction-universal.* When one fixed configuration constructs each member of C, it is called a *universal constructor*, and is certainly non trivial since it constructs significant configurations. It is really powerful when it is also a universal computer. But let us notice that a constructor can be a universal constructor without being able to build itself.

Fact 5

The 2-dimensional von Neumann's and Codd's cellular automata are construction universal, with universal computer-constructors, which happen to also be self-reproducing.

Before coming to self-reproduction, let us remark that this notion of construction universality has been interpreted in a different way, in fact including a condition of self-reproduction.

4.5.2. *Construction universality in other styles*

Let us say that a configuration c' is a copy of a configuration c if there exists a translation τ such that $c' = c_\tau$.

Some authors consider that a cellular automaton $\mathcal{A}$ is construction universal when there exists a finite configuration u of $\mathcal{A}$ satisfying the following conditions:

- there is a time t such that $G^t_{\mathcal{A}}(u)$ contains two disjoint copies of u,
- for each finite configuration c of $\mathcal{A}$, disjoint of u, there is a time t such that $G^t_{\mathcal{A}}(u \sqcup c)$ contains a copy of c, disjoint of c.

Under this definition, we can assert that a cellular automaton with Garden-of-Eden configurations can not be construction-universal, and, in that frame, the "Game of Life" is an example of a cellular automaton computation-universal and self-reproducing but not construction-universal.

Actually, the idea which has rather held researchers attention was, and still is, the self-reproduction one, although what is understood under this name often is woolly enough.

4.6. SELF-REPRODUCTION

Von Neumann was mainly interested in the ability of complex natural, living systems, to withstand some damages (in repairing them or not) and, so, to keep their capacities. This certainly motivated his investigations on construction and reproduction, and the conception of his device.

Afterwards many works have been dedicated to "self-reproduction". It is question of configurations producing copies of themselves or replications keeping or not their ability of (re)production. The attraction for the terminology is certainly due to its biological connotation, although it, often, only covers some replication ability.

4.6.1. *Self-reproduction in the style of Moore*

A first paper by Moore (Moore, 1962) in this domain points out some difficulties about the notion (which can be trivial: let us consider the configuration with 1 on cell 0 and 0 everywhere else under the right shift), shows the limitations induced by the finiteness of the representation and by the underlying space itself.

Definition 12 *Self-reproducing configuration*

- *A configuration c' of some cellular automaton $\mathcal{A}$ is said to contain n copies of some configuration c if c' has n, $n \in N$, disjoint subconfigurations, each of which being a copy of c.*
- *An initial configuration c is said to be capable of reproducing n offspring by time T, if there is a time T', $T' < T$, such that $G_{\mathcal{A}}{}^{T'}(c)$ contains at least n copies of c*

– *A configuration c is said to self-reproduce or to be self-reproducing if, for each positive integer n there is a time T such that c is capable of reproducing n offspring by time T.*

We have to keep in mind that this definition does not put aside the risk of triviality. Moreover the following Moore's result shows that the population generated by this former process of self-reproduction does not satisfy the, generally assumed, exponential growth's law of living organisms.

Theorem 10
If a self-reproducing configuration is capable of reproducing $f(T)$ offspring by time T, then there exists a positive real number k such that $f(T) \leq kT^2$.

The paper by Moore ends up in putting to light the Garden-of-Eden configurations and proving a sufficient condition for Garden-of-Eden configurations existence via erasing configurations.

This definition is still used by Smith, who gives in (Smith, 1991) a non constructive, but short, proof of the existence of non-trivial self-reproducing configurations. The proof rests on the existence of universal Turing machines, the classical Kleene's recursion theorem (Rogers, 1967) and the technique of "wiring-in" Turing machines into cellular automata already used in (Smith, 1971).

Theorem 11
There exists a 1-dimensional cellular automaton $\mathcal{A}$ and a $\mathcal{A}$-configuration c such that c self-reproduces and is computation-universal.

Another interest of the theorem proof, noted by the author himself, gives some light on the notion of self-reproduction at work: "*the proof reduces the problem of self-construction to a computation problem, which means that no machinery beyond ordinary computation theory is required by self-reproduction*".

But, if computation-universality is a sufficient condition for self-reproduction, it is not a necessary one, as it is shown by the well-known example of Langton (Langton, 1984). It is a 2-dimensional cellular automaton, with few states, a simple non-trivial configuration c_0 of which (a loop of $15x-cells$, $10y-cells$, for tens of thousands for von Neumann or Codd ...) is capable of constructing itself periodically, or which appears periodically in the space as subconfiguration. The number of copies increases with time, indefinitely, moving further and further away, on the border (and in a given space part of it) of a special regular pattern, which limits the future evolution of the c-copies. In particular, the full configuration context of such a copy prevents it to behave as its very ancestor, who generates a very rigid population ... Can it then be spoken of "self-reproduction"?

Nevertheless, what is interesting here is that two principles are clearly at work in the evolution of c_0: an external one, the local cellular automaton transition function (or "universe law") and an inner one, the structure of $Supp(c_0)$ which allows it to construct itself in this universe.

4.6.2. *Self-reproduction continued*
The development of Artificial Life has reactivated researches about self-reproduction, as can be seen in the proceedings of the Santa Fe Institute Studies in the Sciences of Complexity. Let us also point out some works by and around K. Morita, who, for example, designed in (Morita and Imai, 1996) a cellular automaton with only 12 states, in which different shapes (worms or loops) can self-reproduce.

5. Groupings

The idea of changing the time-scale in considering grouped cells appears in (Fischer, 1965) and the stemming from method, called "grouping method", is used in chapter *Computations on Cellular Automata*, precisely to explain the Fischer's algorithm for prime numbers. Here, we will first present the notion of *grouped cellular automaton*, in case of dimension 2 and Moore's neighborhood which is significant, of easy access and can also easily be transposed in any dimension and any reasonable neighborhood. Afterwards we will show how this notion allows to found an algebraic classification of cellular automata. Finally, we will present a notion of grouping cellular automaton which leads mechanically to grouped automata.

5.1. GROUPED AUTOMATA

Let $\mathcal{A}$ be a 2-dimensional cellular automaton, E a connected subset of a given $\mathcal{A}$-configuration. If the states of E are known, it is possible to know the states of a subset of E, after one or several units of time. For example, if E is a $m \times m$-square, $m \geq 3$, at the next time, the states of a $(m-2) \times (m-2)$-square inside the previous one are known, see figure 17. This leads to functions, that in turn allow to define the grouped automaton stemming from $\mathcal{A}$.

Definition 13
We denote $\tilde{\delta}_m$ the application which associates to m^2 states of $\mathcal{A}$, $(m-2)^2$ states of $\mathcal{A}$, defined as follows:

– *if $m = 3$, $\tilde{\delta}_m = \delta$,*
– *if $m > 3$ and, if we denote the m^2 states*

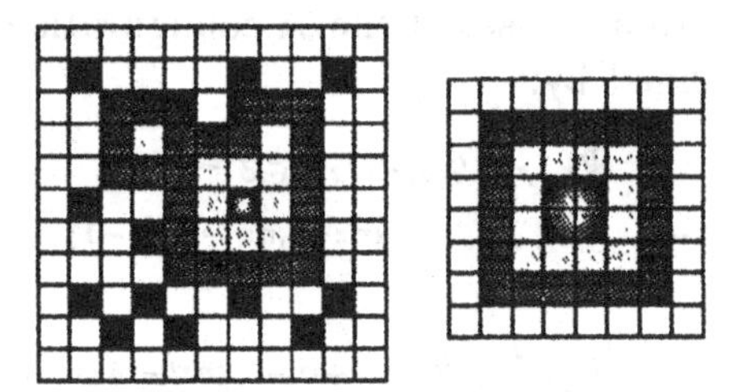

Figure 17. Marked cells at time t, $t+1$ and $t+2$.

$$\begin{pmatrix} q_{1,1} & \dots & q_{1,m} \\ \vdots & & \vdots \\ q_{m,1} & \dots & q_{m,m} \end{pmatrix},$$

then their $\tilde{\delta}_m$-image is

$$\begin{pmatrix} \delta\begin{pmatrix} q_{1,1} & q_{1,2} & q_{1,3} \\ \vdots & & \vdots \\ q_{3,1} & q_{3,2} & q_{3,3} \end{pmatrix} & \dots & \delta\begin{pmatrix} q_{1,m-2} & q_{1,m-1} & q_{1,m} \\ \vdots & & \vdots \\ q_{3,1} & q_{3,2} & q_{3,3} \end{pmatrix} \\ \vdots & & \vdots \\ \delta\begin{pmatrix} q_{m-2,1} & q_{m-1,2} & q_{m,3} \\ \vdots & & \vdots \\ q_{m,1} & q_{m,2} & q_{m,3} \end{pmatrix} & \dots & \delta\begin{pmatrix} q_{m-2,m-2} & q_{m-2,m-1} & q_{m-2,m} \\ \vdots & & \vdots \\ q_{m,m-2} & q_{m,m-1} & q_{m,m} \end{pmatrix} \end{pmatrix}.$$

Let us observe that if we know nine $n \times n$ squares, in a configuration as the one of a cell with its Moore's neighborhood, we know the states of the central square after n successive applications of convenient functions $\tilde{\delta}_m$ with $m \leq 3n$. That determines a new function $\delta^{n,n}$ from $\left(S^{n^2}\right)^9$ into S^{n^2}, which is the composition of n $\tilde{\delta}_m$-type functions. Actually, $\delta^{n,n}$ is $\tilde{\delta}_{n+2} \circ \tilde{\delta}_{n+4} \circ \ldots \circ \tilde{\delta}_{3n}$.

Definition 14 *n-grouped automaton*

Let $\mathcal{A} = (S, \delta)$ be a 2-dimensional cellular automaton. The 2-dimensional cellular automaton $\left(S^{n^2}, \delta^{n,n}\right)$ is called the n-grouped automaton of $\mathcal{A}$, and will be denoted $G_n(\mathcal{A})$.

This definition is purely syntactic. Writing the states of $G_n(\mathcal{A})$ as matrices is a convenient way to describe them, that allows to naturally simulate the evolutions of $\mathcal{A}$ and $G_n(\mathcal{A})$. From a configuration c of $\mathcal{A}$ and an origin point (k, l) cleverly chosen, we get a tiling of c in $n \times n$ squares, each of them representing a $G_n(\mathcal{A})$-cell in a given state, then we get a configuration of $G_n(\mathcal{A})$. Actually, k and l being two given integers, there exists a bijection, we will denote $\Phi_{k,l,n}$, from the set of the $\mathcal{A}$-configurations set onto the set of the $G_n(\mathcal{A})$-ones, defined by:

$$\Phi_{k,l,n}((q_{(z,z')})_{z,z'\in Z} =$$

$$\left(\begin{pmatrix} q_{nz-k,nz'-l} & \dots & q_{nz-k,nz'-l+(n-1)} \\ \vdots & & \vdots \\ q_{nz-k+(n-1),nz'-l} & \dots & q_{nz-k+(n-1),nz'-l+(n-1)} \end{pmatrix}\right)_{z,z'\in Z}$$

It holds:

Proposition 1

Let k, l, n, be given integers, $n > 0$. Then, for each configuration c of $\mathcal{A}$, each integer t, $t \geq 0$,

$$\Phi_{k,l,n}(c^{nt}) = (\Phi_{k,l,n}(c))^t.$$

There are many ways to interpret this last result. Actually, it means that $G_n(\mathcal{A})$ simulates $\mathcal{A}$ with an acceleration of factor n. Moreover space interpretations (homotheties) are also possible as it can be seen in (Delorme *et al.*, 1998), where the method is applied to transform some bundle of discrete parabolas into an other one. We will come back later on an other grouping method, set up in order to improve some simulations results. But, it seems that the notion of grouped automaton is interesting for itself because it could give a good means of comparing cellular automata via their grouped instances. So, we will now indicate how this grouping idea give birth to an algebraic way of classifying one-dimensional cellular automata.

5.2. AN ORDER ON CELLULAR AUTOMATA

Looking at space-time diagrams of 1-dimensional cellular automata from the point of view of complexity brings to try to grasp some "dominant features" and then to "forget some details " (or what can appear some time as such). That leads for example to pay attention to groups of sites rather than sites themselves and so, sometimes, to put to light macroscopic significant behaviors. See Figure 18.

We can easily translate the above subsection definitions to dimension 1. The functions $\tilde{\delta}_m$ are now defined from S^m into S^{m-2}, and if we consider blocks of n cells, the n-grouped automaton $G_n(\mathcal{A})$ of $\mathcal{A} = (S, \delta)$ is still (S^n, δ^n) with $\delta^n = \tilde{\delta}_{n+2} \circ \tilde{\delta}_{n+4} \circ \ldots \circ \tilde{\delta}_{3n}$. As we obtain the n-grouped automaton defined in (Mazoyer and Rapaport, 1998), from now on, $G_n(\mathcal{A})$ will be denoted by $(S, \delta)^n$. In the latter paper a binary relation is set up on the class of 1-dimensional cellular automata by $(S_1, \delta_1) \preceq (S_2, \delta_2)$ if and only if there are non negative integers i and j such that $(S_1, \delta_1)^i \subseteq (S_2, \delta_2)^j$, which means that there exists an injection ϕ from S_1 into S_2 that satisfies $\delta_2(\phi(q_1), \phi(q_2), \phi(q_3)) = \phi(\delta_1(q_1, q_2, q_3))$.

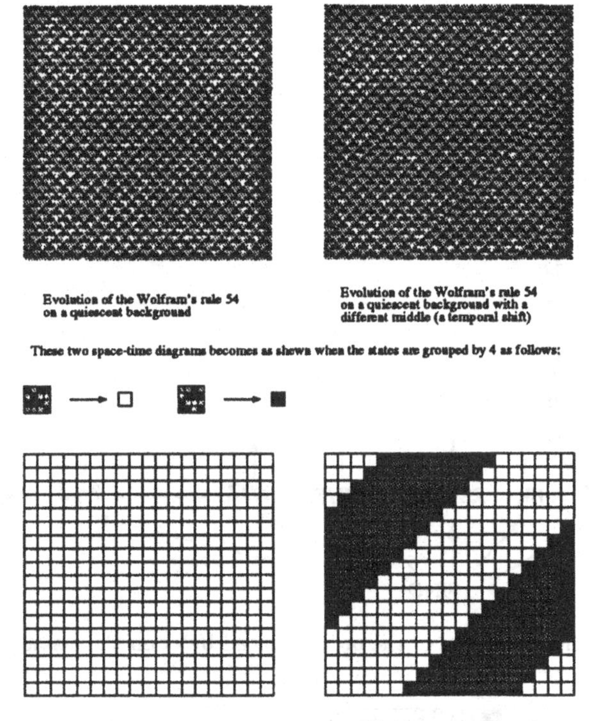

Figure 18. Grouping for the Wolfram's 54 rule.

The relation $\preceq$ is a preorder and, if $\sim$ is the canonically associated equivalence relation, $\preceq$ induces an order on the equivalence classes, which is in the process of being studied, and already proved undecidable. Some results summarized below are represented on Figure 19.

- There is a minimum element, but no maximum.
- On the first level of infinite width, some classes as NIL, PER, R_{shift}, L_{shift} capture some interesting properties.
- The existence of two infinite bounded chains, with an upper bound, which depends on a solution of some firing squad problem.
- The existence of two infinite incomparable chains.

We will now consider an other notion of grouping.

5.3. GROUPING CELLULAR AUTOMATA

When observing the space-time diagrams of some language recognizers for example, the idea arises that the decision could be known earlier in suitably "twisting" them, or at least part of them. The concern for getting the possible transformations mechanically and inside the cellular automata field ends up in the notion of grouping cellular automaton (Heen, 1996).

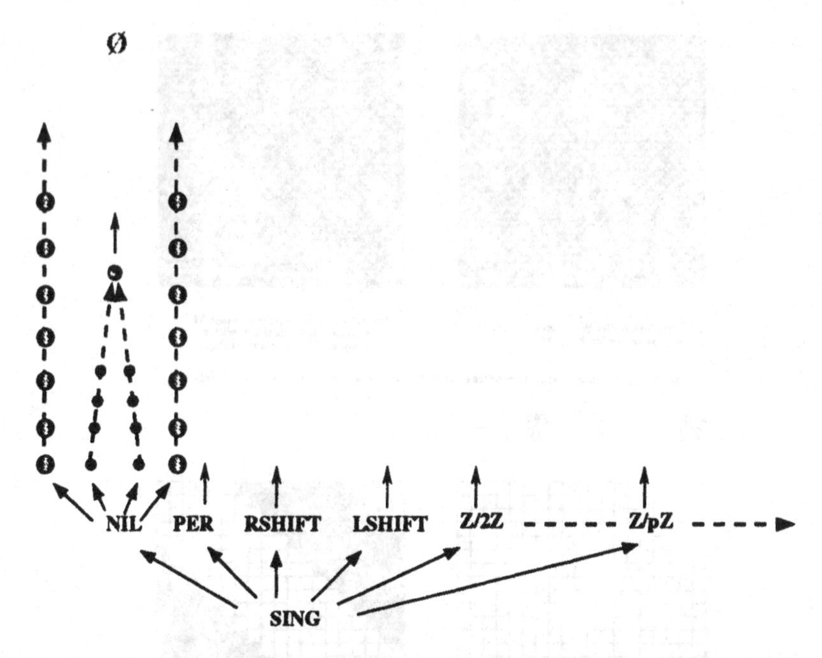

Figure 19. Some properties of the order induced by $\preceq$ on the classes modulo $\sim$.

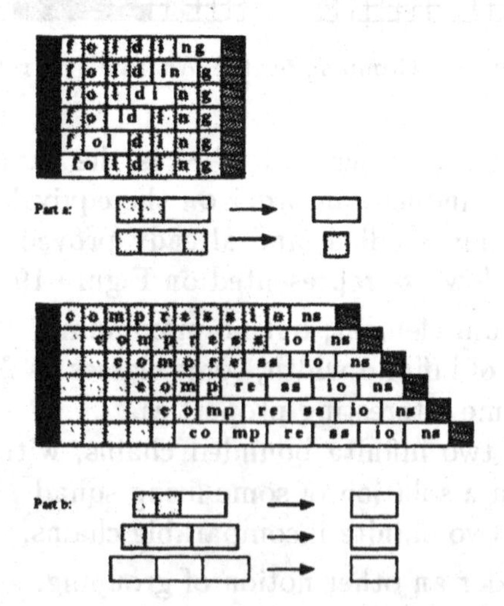

Figure 20. Some examples of folding areas which accelerate the recognition of words.

So, examining the space-time diagrams of some one-dimensional cellular automaton $\mathcal{A}$ put to light some efficient states groupings, the layout of which draws geometric pieces of the cellular space, that are in turn taken as states for the "grouping cellular automaton" $\mathcal{P}_{\mathcal{A}}$, and finally $\mathcal{P}_{\mathcal{A}}$ is made acting on $\mathcal{A}$, its pieces being then "filled" with $\mathcal{A}$-states in such a way that the obtained grouped cellular automaton, $\mathcal{A}\otimes_{\mathcal{A}}\mathcal{P}_{\mathcal{A}}$, gives the expected result.

But, unhappily, that sort of grouping does not allow to define some intrinsic simulation.

6. A last remark on terminology

Cellular automata are given various names through the literature according to the way they are used, mainly as computation models or models of natural phenomena, but also in well defined research areas. Let us cite tessalations structures, iterative circuits (Burks, 1966), pattern-manipulation systems (Smith, 1971), iterative arrays (Cole, 1969) ...

Moreover, another type of mixture can arise. So, as *machine* can be understood a cell, the finite automaton (S, δ), the set of "active" cells or the global machinery $(\mathbb{Z}^d, S, N, \delta)$, as well as some configuration. And the *cellular space* $(\mathbb{Z}^d)$ can be interpreted as a *universe*, a *world*, while the δ function designates the *laws* of this universe and the corresponding configurations *automata living in it.*

References

Albert J. and Čulik II K. A simple universal cellular automaton and its one-way totalistic version. *Complex Systems.* **Vol. no. 1**: 1-16, 1987.

Allouche J.-P., v. Haeseler F., Peitgen H.-O. and Skordev G. Linear cellular automata, finite automata and Pascal's triangle. *Discrete Appl. Math.* **Vol. no. 66**: 1–22, 1996.

Allouche J.-P., v. Haeseler F., Peitgen H.-O., Petersen A. and Skordev G. Automaticity of double sequences generated by one-dimensional cellular automata. To appear in *Theoretical Computer Science.*

Aso H. and Honda N. Dynamical Characteristics of Linear Cellular Automata. *Journal of Computer and System Sciences.* **Vol. no. 30**: 291–317: 1985.

Banks E. R. Universality in Cellular Automata. *I.E.E.E. Ann. Symp. Switching and Automata Theory.* Santa Monica, **Vol. no. 11**: 194–215, 1970.

Bartlett R. and Garzon M. Monomial Cellular Automata. *Complex Systems.* **Vol. no. 7**: 367–388, 1993.

Bartlett R. and Garzon M. Bilinear Cellular Automata. *Complex Systems.* **Vol. no. 9**: 455–476, 1995.

Blanchard F., Kůrka P. and Maass A. Topological and measure-theoretic properties of one-dimensional cellular automata. *Physica D.* **Vol. no. 103**: 86–99, 1997.

Burks E. *Essays on Cellular Automata*, University of Illinois Press, 1972.

Burks E. *Theory of Self-reproduction*, University of Illinois Press, Chicago, 1966.

Byl J. Self-reproduction in small cellular automata. *Physica D.* **Vol. no. 34**: 295–299, 1989.

Cayley A. Theory of groups. *American Journal of Mathematics.* **Vol. no. 1**: 50–52, 1878.

Choffrut C. and Čulik II K. On real-time cellular automata and trellis automata. *Acta Informatica.* **Vol. no. 21**: 393–407, 1984.

Codd E.F. *Cellular Automata*, Academic Press, New York, 1968.

Cole S. Real-time computation by n-dimensional iterative arrays of finite-state machine. *IEEE Trans. Comput.* **Vol. no. C-18**: 349–365, 1969.

Čulik II K. and Yu S. Undecidability of CA classification schemes. *Complex Systems.* **Vol. no. 2**: 177–190, 1988.

Čulik II K., Pachl J. and Yu S. On the limit sets of cellular automata. *SIAM J. Comput.* **Vol. 18 no. 4**: 831–842, 1989.

Čulik II K., Dube S. Fractal and Recurrent Behavior of Cellular Automata. *Complex Systems.* **Vol. no. 3**: 253–267, 1989.

Cutland N. *Computability*, Cambridge University Press, 1980.

Delorme M., Mazoyer J. and Tougne L. Discrete parabolas and circles on 2D-cellular automata. To appear in *Theoretical Computer Science.*

Dubacq J.-C. How to simulate Turing machines by invertible one-dimensional cellular automata. *International Journal of Foundations of Computer Science*, **Vol.6 no. 4**: 395–402, 1995.

Durand B. Global properties of 2D-cellular automata. in *Cellular Automata and Complex Systems*, Golès E. and Martinez S. Eds, Kluwer, 1998.

Fischer P.C. Generation of primes by a one dimensional real time iterative array. *Journal of A.C.M.* **Vol. no. 12**: 388–394, 1965.

Gajardo A. Universality in a 2-dimensional cellular space with a neighborhood of cardinality 3. *Preprint*, 1998.

Garzon M. Cyclic Automata. *Theoretical Computer Science*, **Vol. no. 53**: 307–317, 1987.

Garzon M. Cayley Automata. *Theoretical Computer Science*, **Vol. no. 103**: 83–102, 1993.

Garzon M. *Models of Massive Parallelism*, Springer, 1995.

Godsil C. and Imrich W. Embedding graphs in Cayley graphs. *Graphs and Combinatorics*, **Vol. no. 3**: 39–43, 1987.

Goles E., Maass A. and Martinez S. On the Limit Set of some Universal Cellular Automata. *Theoretical Computer Science*, **Vol. no. 110**: 53–78, 1993.

Head T. One-dimensional cellular automata: injectivity from unambiguity. *Complex systems*, **Vol. no. 3**: 343–348, 1989.

Hedlund G. Transformations commuting with the shift. In: *Topological Dynamics*, eds. Auslander J. and Gottschalk W. G. (Benjamin, New York, 1968), 259; Endomorphisms and automorphisms of the shift dynamical system. *Math. Syst. Theor.* **Vol. 3 no. 320**, 1969.

Heen O. Economie de ressources sur Automates Cellulaires. *Diplôme de Doctorat*, Université Paris 7 (in french), 1996.

Hurd L. Formal language characterizations of cellular automata limit sets. *Complex Systems* **Vol. 1 no. 1**: 69–80, 1987.

Ishii S. Measure theoretic approach to the classification of cellular automata. *Discrete Applied Mathematics* **Vol. no. 39**: 125–136, 1992.

Jen E. Linear cellular automata and recurrence systems in finite fields. *Comm. Math. Physics* **Vol. no. 119**: 13–28, 1988.

Kari J. Reversibility and surjectivity problems of cellular automata. *Journal of Computer and System Sciences*, **Vol. no. 48**: 149–182, 1994.

Langton C. Self-Reproduction in Cellular Automata. *Physica*, **Vol. no. 100**: 135–144, 1984.

Lindgren K. and Nordahl M. Universal computation in simple one dimensional cellular automata. *Complex Systems* **Vol. no. 4**: 299–318, 1990.

Machí A. and Mignosi F. Garden of Eden configurations for cellular automata on Cayley graphs of groups. *SIAM Journal on Discrete Mathematics*, **Vol. no. 6**: 44–56, 1993.

Martin O., Odlyzko A. and Wolfram S. Algebraic properties of cellular automata. *Comm. Math. Phys.*, **Vol. no. 93**: 219–258, 1984.

Martin B. A universal automaton in quasi-linear time with its s-n-m form. *Theoretical Computer Science* **Vol. no. 124**: 199–237, 1994.

Mazoyer J. Computations on one-dimensional cellular automaton. *Annals of Mathematics and Artificial Intelligence* **Vol. no. 16**: 285–309, 1996.

Mazoyer J. and Rapaport I. Inducing an order on cellular automata by a grouping oper-

ation. in *STACS'98*, LNCS **Vol. 1373**: 116–127, 1998.
Mazoyer J. and Terrier V. Signals on one dimensional cellular automata. To appear in *Theoretical Computer Science.*, 1998.
McCullough W. and Pitts W. A logical calculus of the ideas immanent in nervous activity. *Bull. Math. Biophys.*, **Vol. no. 5**: 115–133, 1943.
Minsky M. *Finite and infinite machines.* Prentice Hall, 1967.
Morita K. A Simple Construction Method of a Reversible Finite Automaton out of Fredkin Gates, and its Related Model. *Transactions of the IEICE*, **Vol. no. E**: 978–984, 1990.
Morita K. and Imai K. A Simple Self-Reproducing Cellular Automaton with Shape-Encoding Mechanism. *Artificial Life V*, 1996.
Moore E. F. Machine models of self-reproduction. *Proc. Symp. Appl. Math.* **Vol. no. 14**: 17–33, 1962.
Mosconi J. La constitution de la théorie des automates. *Thèse d'Etat*, Université Paris 1, 1989.
Myhill J. The converse of Moore's garden-of-eden theorem. *Proc. AMS* **Vol. no. 14**: 685–686, 1963.
Peitgen H.-O., Rodenhausen A. and Skordev G. Self-similar Functions Generated by Cellular Automata. Research Report 426, Bremen University, 1998.
Reimen N. Superposable Trellis Automata. *LNCS*,**Vol. no. 629**: 472–482, 1992.
Richardson D.Tesselations with local transformations. *Journal of Computer and System Sciences*, **Vol. no. 5**: 373–388, 1972.
Rogers H. *Theory of Recursive Functions and Effective Computability*, MIT Press, 1967.
Róka Zs. Simulations between Cellular Automata on Cayley Graphs. To appear in *Theoretical Computer Science*, 1998.
Smith III A. Real-time language recognition by one-dimensional cellular automata. *Journal of Computer and System Sciences* **Vol. no. 6**: 233–253, 1972.
Smith III A. Simple computation-universal spaces. *Journal of ACM.* **Vol. no. 18**: 339-353, 1971.
Smith III A. Simple Non-trivial Self-Reproducing Machines. *Proceedings of the Second Artificial Life Workshop*, Santa Fe: 709–725, 1991.
Sutner K. De Bruijn graphs and linear cellular automata. *Complex Systems* **Vol.5 no. 1**: 19–30, 1991.
Thatcher J. Universality in the von Neumann cellular model. Tech. Report 03105-30-T, University of Michigan, 1964.
Turing A. On computable numbers, with an application to the Entscheidungsproblem. *P. London Math. Soc.* **Vol. no. 42**: 230–265, 1936.
von Neumann J. The General and Logical Theory of Automata. in *Collected Works, Vol. 5*, Taub. A., Eds, New York: Pergamon Press: 288–328, 1963.
von Neumann J. *The Computer and the Brain*, Yale University Press, New Haven, 1958.
von Neumann J. *Theory of Self-reproducing automata*, University of Illinois Press, Chicago, 1966.
von Neumann J. *Probabilistic logic and the synthesis of reliable organisms from unreliable components*, Shannon and McCarthy Eds, Princeton University Press, Princeton, 1956.
Wagner K. and Wechsung G. *Computational Complexity.* Reidel, 1986.
Wiener N. *Cybernetics, or control and communication in the animal and the machine*, M.I.T. Press, New Haven, 1961.
Willson S. Cellular automata can generate fractals. *Discrete Applied Mathematics*,**Vol. no. 8**: 91–99, 1984.
Wolfram S. Twenty problems in the theory of cellular automata. *Physica Scripta* **Vol. no. 79**: 170–183, 1985.
Wolfram S. *Theory and Applications of Cellular Automata*, World Scientific, 1986.

THE GAME OF LIFE: UNIVERSALITY REVISITED

BRUNO DURAND
LIP ENS-Lyon,
46 allée d'Italie,
69364 Lyon Cedex 07, France.

AND

ZSUZSANNA RÓKA
L.R.I.M, IUT de Metz,
Ile du Saulcy,
57045 Metz Cedex 01, France.

Abstract. The Game of Life was created by J.H. Conway. One of the main features of this game is its *universality*. We prove in this paper this universality with respect to several computational models: boolean circuits, Turing machines, and two-dimensional cellular automata. These different points of view on Life's universality are chosen in order to clarify the situation and to simplify the original proof. We also present precise definitions of these 3 universality properties and explain the relations between them.

1. Introduction

The *Game of Life* (Life for short) is a cellular automaton introduced by J.H. Conway in 1970: it can be seen as a game played on an infinite grid. At any time some of the cells are alive and the others dead. Which cells are alive at time 0 is up to the player who chooses the initial configuration. Then the player observes the evolution of the configuration: the state of each cell at any time follows inexorably the previous one according to the rules of the game (see Section 2). Conway proved in (Berlekamp *et al.*, 1982) the computational universality (let us say Turing-universality to avoid confusion in the remainder) of the game by defining some special patterns able to simulate the components of a real computer. Independently and simultaneously, it seems that Gosper obtained the same result but we could not have access to this — unpublished as far as we know — proof.

We focus on three aspects of Conway's proof of Turing-universality. First, one point is very complicated and difficult to check: the construction of a memory stack. Second, as far as we could understand it, it contains an error discussed in Section 3.3. In our opinion, it explains why none of the attempts (as far as we know) to build effectively a universal computer within the Game of Life have succeeded. Third, the author did not explain precisely what he means by "universality". It could be at first sight considered as a minor problem since it is also the case for instance in the works of Banks (Banks, 1971), Nourai and Kashef (Nourai and Kashef, 1975) and Smith (Smith, 1972; Smith, 1971). As a matter of fact, all authors we could read on this topic did not defined what they meant by Turing-universality although there exist some partial proposals by Codd (Codd, 1968), Albert and Čulik (Albert and Čulik, 1987). This point may seem of low interest since the definition is usually straightforward, but in the case of simulation by cellular automata, it is a *very delicate* point because cellular automata transform infinite objects — versus finite in standard models. One can find in literature *false* results because of definition problems. Such a definition can even be extremely difficult to establish when combined with self-reproduction which is also hard to define (von Neumann, 1966; Burks, 1972). Thus, we give first a precise definition of this notion.

In this paper, we also prove a stronger universality property: we prove that life is *intrinsically* universal, *i.e.* that it can simulate any 2-dimensional cellular automaton. There is also a problem of definition for this notion of simulation, even if it is broadly used in literature. We base our work on the definition of simulation presented in (Róka, 1998). The key point is that this definition is compatible with Turing-universality: if a Turing-universal cellular automaton is simulated by another cellular automaton, the latter is also Turing-universal. This intrinsic universality is implicitly used by Banks in (Banks, 1971) but it seems that he did not get the relationship with Turing-universality — which he also proves in a different way. This author focuses on the problem of the difference between finite and infinite initial configurations, which is indeed the key point of the definition of Turing-universality as explained in Section 4. The intrinsic universality allows to simplify proofs and, in our case, to prove directly the Turing-universality of Life without the difficult construction of the memory stack. This approach is also used in (Gajardo , 1998).

More precisely, the universality is proved in 3 different contexts: first for boolean circuits, then we prove it for Minsky machines (it corresponds to Turing-universality), and *in fine* we prove this universality for 2D cellular automata (intrinsic universality).

The paper is organized as follows: first in Section 2, we recall the rules of the game and show some basic configurations that are used as basic construction tools. In Section 3 we prove that any boolean circuit can be simulated by these basic patterns (our first kind of universality). In Section 4 we give and discuss a precise definition of Turing-universality for cellular automata. We give a hint of Conway's proof without entering the technical details of the construction of the memory stack. Section 5 is devoted to intrinsic universality: we propose adequate definitions and prove that Life is intrinsically universal.

2. A basic survey of Life

In this section we present a few important patterns and some reactions between them (without modifying Conway's constructions of (Berlekamp *et al.*, 1982)). Then, in Section 5, we shall use them for constructing some special configurations to prove universality.

Life is a game played on an infinite grid. At any time, some cells are alive and the others are dead. Each cell is surrounded by eight neighbors, forming the so-called Moore neighborhood (see Figure 1).

Figure 1. The Moore neighborhood

At every time step, each cell can change its state, in a parallel and synchronous way according to the following local rules:

Birth: a cell that is dead at time t becomes alive at time $t+1$ if and only if three of its eight neighbors were alive at time t (for an example, see Figure 2a).

Survival: a cell that was alive at time t will remain alive if and only if it had just 2 or 3 alive neighbors at time t (see Figure 2b).

Death

- **by overcrowding:** a cell that is alive at time t and has 4 or more of its eight neighbors alive at t will be dead by time $t+1$ (see Figure 2c).
- **by exposure:** a cell that has only one alive neighbor, or none at all, at time t, will be dead at time $t+1$ (see Figure 2d).

Alternative definition

Let us now consider that the center cell belongs to the set of neighbors (*i.e.* that the neighborhood of a cell consists of nine cells). Let us denote by n

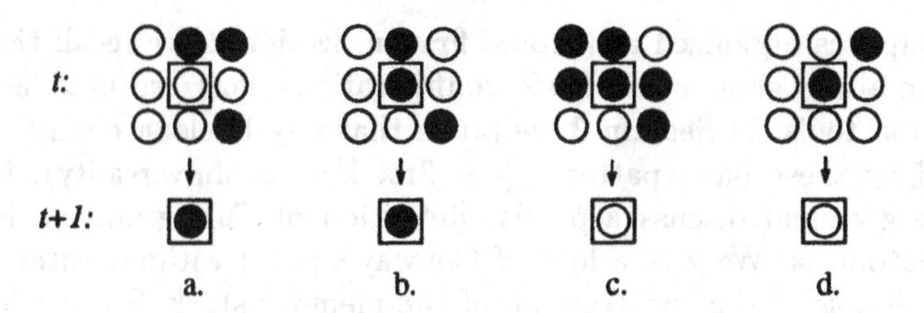

Figure 2. Examples for rules

the number of alive cells in this neighborhood. Then the rules of the game are:

- $n \leq 2$: the cell dies,
- $n = 3$: the cell lives,
- $n = 4$: the cell stays in the same state,
- $n \geq 5$: the cell dies.

We can remark that Life can be considered as a two-dimensional cellular automaton over the alphabet $\{0, 1\}$.

Definition 1

We call pattern *a configuration that contains only a finite number of alive cells.*

There can be many different types of evolutions for a given pattern. The pattern can turn into a stable configuration (which does not change) after a finite number of time steps, or its evolution can be dynamical. Some stable patterns can be seen in Figure 3. In the patterns of the present paper, we feature alive cells by black bullets.

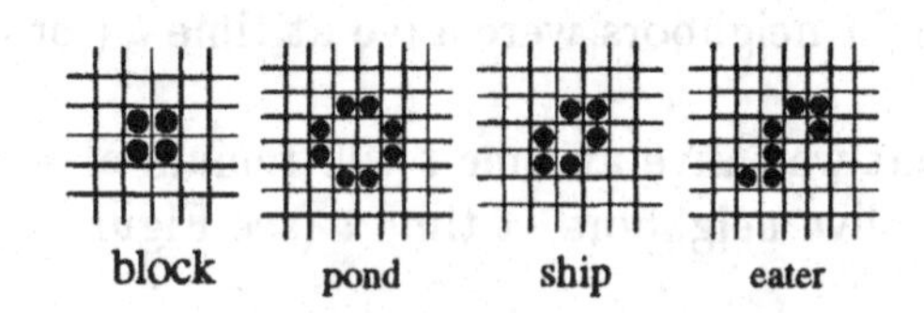

Figure 3. Stable patterns

Among patterns having dynamical evolution, we distinguish two cases:

- a pattern can have a periodical behavior: it repeats itself with period greater than one. Then, either its whole evolution needs a bounded room on the board (see an example in Figure 4), or it travels across the plane like the glider shown in Figure 5.
- a pattern can grow without limit as, for instance, Gosper's gun (see Figure 6) which emits a new glider every 30 generations (see Figure 7).

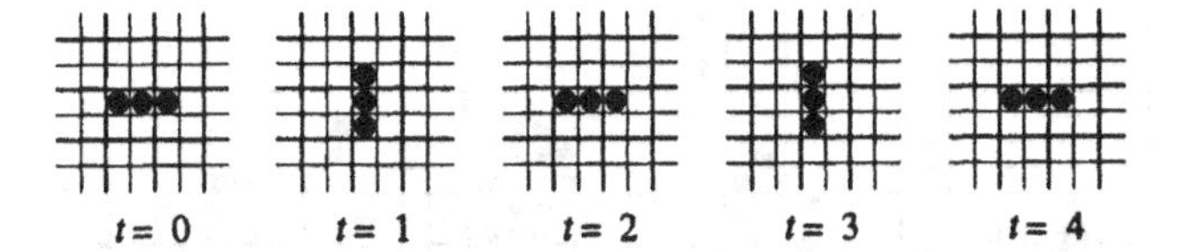

Figure 4. The blinker: a 2-period configuration

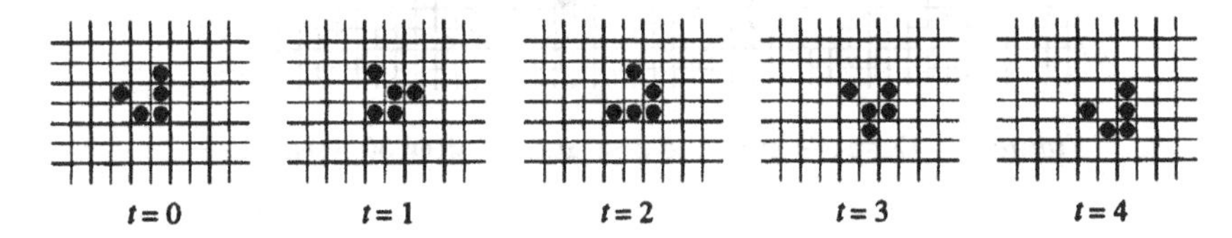

Figure 5. The glider: a 4-period South-East moving configuration

Other interesting patterns such as spaceships, guns etc. can be found in literature. Here, we only present configurations used to prove the computational universality of the game.

Let us now study pattern evolution in general. A natural question arises:

Can one find an algorithm predicting the kind of evolution of a given pattern?

In (Berlekamp *et al.*, 1982), it is showed that even for the simple example of a straight line of k alive cells, the evolution seems unpredictable. As far as we know this problem is still open; the answer to the more general question above is "no": the unpredictability of Life's evolutions is a consequence of its Turing-universality.

One can think about another natural question:

Given a configuration, does another configuration exist that precedes in time the first one?

The answer to this question is again "no": a configuration without predecessor is called a *Garden of Eden* configuration. There exist Garden of Eden

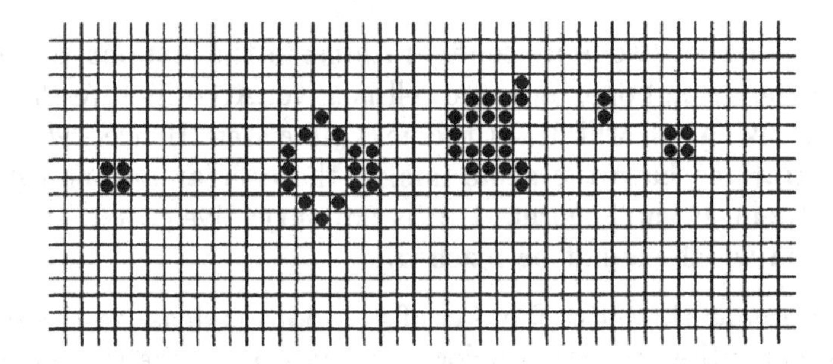

Figure 6. Gosper's gun

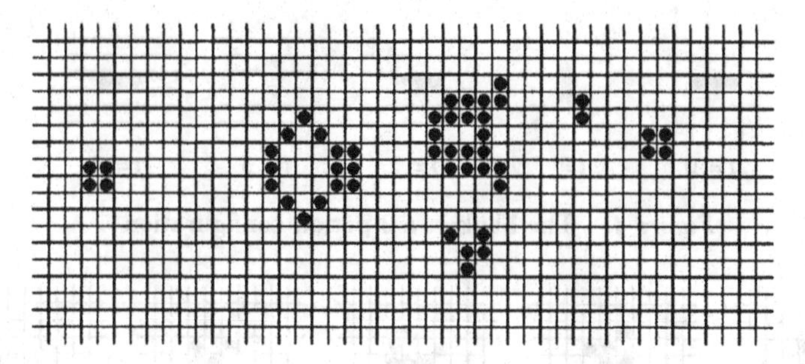

Figure 7. Gosper's gun: the first glider (after 30 iterations)

configurations for Life. One of them (shown in Figure 8) has been found by an M.I.T. group (E. Banks, M. Beeler, R. Schroeppel *et al.*)

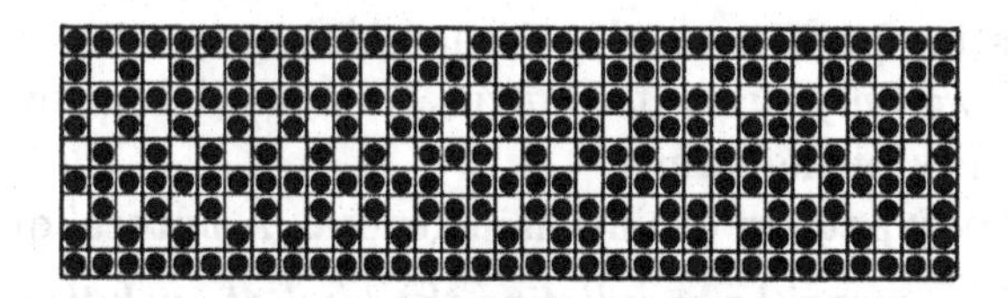

Figure 8. A Garden of Eden pattern

To prove the existence of such a Garden of Eden configuration, we use the general theory of cellular automata:

Theorem 1 (Moore-Myhill (Moore, 1962; Myhill, 1963))
A cellular automaton is injective (one-to-one) on finite configurations if and only if it is surjective (onto).

We do not give here the proof of this theorem. It is based on a combinatorial argument.

Theorem 2
Life has a Garden of Eden configuration. Equivalently, Life is not surjective.

Proof. Let us consider two finite configurations; in the first one, every cell is dead, and in the second one, only one cell is alive. According to the rules of the game, the living cell will die at the next time step. Hence, two different finite configurations have the same image: the cellular automaton is not injective on them. From Theorem 1, it is not surjective either: there exists a configuration which cannot be reached. □

Furthermore, as the set of finite configurations is dense in the set of all configurations for the product topology, one can prove the following: there exists a finite part of configuration (beware that it is sometimes called a pattern) that cannot be found in a configuration after one iteration of Life (see also Figure 8).

For those readers who are interested in global properties of cellular automaton, see the review paper (Durand, 1998).

This study of Garden of Eden configurations for life was motivated by the following notion (see (von Neumann, 1966; Burks, 1972)):

Definition 2 (von Neumann) *A cellular automaton is* construction- universal *for a set of configurations* $\mathcal{C}$ *if and only if any configuration of* $\mathcal{C}$ *is the image of a configuration.*

Remark that when no set is mentioned, the considered set of configurations is the set of *all* configurations. In this case the construction-universality is equivalent to surjectivity.

Corollary 1 (of Theorem 2)
Life is not construction-universal.

3. Boolean circuits simulation

Boolean circuits are crucial in simulations because they constitute the "heart" of computing devices. Beware that a boolean circuit alone is not universal: to obtain a universal device one can either add a memory — thus obtain a Turing (or a Minsky) machine, or consider recursive infinite families of boolean circuits (in this case, the code of the family is taken as a program). A formal definition of simulation of boolean circuits by cellular automata is rather tedious and not very useful since no example we could read in literature was ambiguous. Nevertheless, we decided to present a definition of this kind of simulation in order to formulate our meta-theorem 1.

Let us summarize the situation: we are given as input a boolean circuit and a boolean vector. Our aim is to construct the associated initial configuration of the cellular automaton that will perform the simulation. We think that it is always possible to restrict the initial configuration to finite configurations or eventually periodic ones. More precisely the initial configuration is defined by a finite support on which there is no restriction, and a periodic or constant media outside the support. This media should depend only on the cellular automaton and not on the circuit/vector to be simulated otherwise we could encode the result in it. For the same reason, the support should depend on the circuit to be simulated but not on the boolean vector. Now a problem remains: it is *a priori* possible to encode the result of the circuit straight into the initial configuration. We should restrict the definition in order that this is impossible, otherwise the computation of the boolean function would not be performed by the cellular automaton but by the encoding function. We perform this task using the following trick: each cell in the support may depend only on the value of at most one coordinate of the boolean vector.

Now the question is to observe the evolution of the cellular automaton and after some time, observe the result (*i.e.* the value of the boolean circuit in the chosen value). The question is: when and where should we observe the evolution? In order to allow complex simulation we decide to observe the states in a window of constant size located at a position that depends only on the boolean circuit (not on the vector so that the position does not encode the result of the computation). After a given number of steps (that depends only on the circuit for the same reason) the window is observed and the result 0 or 1 is given as output.

This definition can be further formalized: we only gave here an informal version. One can be convinced that the notion of simulation of a boolean circuit is well defined. But recall that we would like to take into account a family of boolean circuits and not an isolated one. This is why we think that one should require that the definition is constructive.

Given a configuration p, we call *p-finite* any configuration which is equal to p except on a finite number of cells. A *support* of such a p-finite configuration is a finite pattern such that outside this pattern, the configuration is equal to p. Beware that this pattern is not shift invariant, thus when a program outputs a pattern it should give not only it size[1] but also its position in the plane.

Definition 3

Let $\mathcal{A}$ denote a cellular automaton. $\mathcal{A}$ is boolean-circuit- universal iff there exist a periodic (may be constant) configuration p, a window size s, a decoding finite function d from windows of size s into $\{0,1\}$, and two programs P and Q such that P depends on a variable representing a boolean circuit denoted by B, Q depends on two variables representing a boolean circuit B and a boolean vector V; $P(B)$ outputs a support for an eventually p-finite configuration, an observation time t and a window position x; $Q(B,V)$ outputs a finite part of a configuration that fills the support such that the value of all its cells depend at most on one coordinate of V. The simulation corresponds to the requirement that after t steps of $\mathcal{A}$ on the initial configuration obtained via P and Q, if d is applied on the window placed on the cell x, we obtain the same result than B would perform on V.

Intuitively Q computes the initial configuration using B and V. As the support of this initial configuration (*i.e.* its non-periodic part) should not depend on V, it is determined independently by $P(B)$ which also computes the observation time and the window position. P computes all that does not depend on V but only on B.

[1] The shape is not needed since any pattern can be embedded into a square.

Concerning the Game of Life, the construction imitating a "real" machine is described in details in (Berlekamp *et al.*, 1982). Here, we only present the ideas and patterns used to show Life's universality for Boolean circuits.

3.1. WIRES AND PULSES

In a real machine, there are

1. *pulses* of electricity,
2. *wires* along which pulses of electricity go,
3. a *clock* generating pulses at regular intervals,
4. *logical gates* AND, OR and NOT,
5. auxiliary storage *registers*, each of which will store an arbitrarily large number.

In Life,

1. *pulses* are represented by *gliders*,
2. *wires* are represented by lines in the plane along which gliders travel. They travel diagonally, so for better understanding, we turn the plane through 45° in Figures 12, 13 and 14,
3. the rule of the *clock* is played by a *Glider Gun*,
4. patterns imitating *logical gates* are presented in Section 3.2.

3.2. LOGICAL GATES

Let us explain now how logical gates can be simulated. As there are patterns (such as gliders or spaceships) crossing the plane, they can meet each other. Depending on their relative positions, many things can happen. Here, we present three reactions that will take part in the construction of the gates.

Gliders' crash. When two gliders meet each other, they can have, for instance, a vanishing reaction. Here, we only show one (the fastest) of the many possible ways to get such a reaction (see Figure 9) that we call a *crash*.

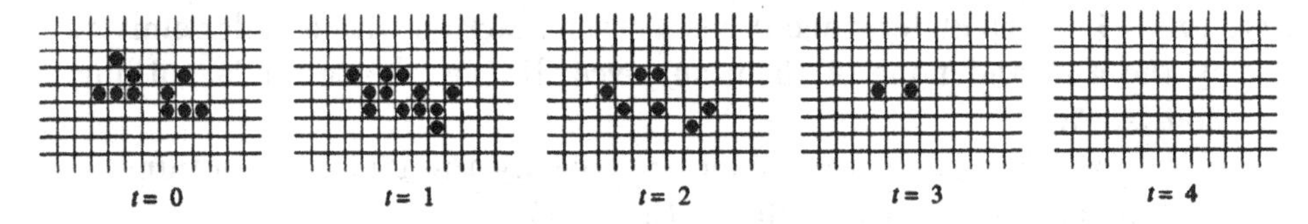

Figure 9. Gliders crash

Glider meets eater. Two gliders can also give birth to an eater —defined in Figure 3. Then, an eater, as its name shows, can eat many other patterns. For instance, a glider is eaten in Figure 10.

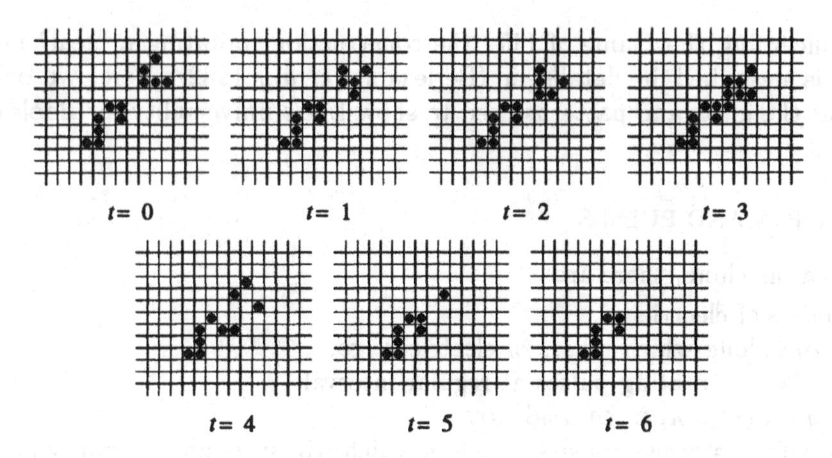

Figure 10. Glider is consummated by an eater

The kickback reaction. Another interesting reaction is the kickback shown in Figure 11. A glider can meet another one in such a way that one of them dies and the other one continues his way but in the opposite direction.

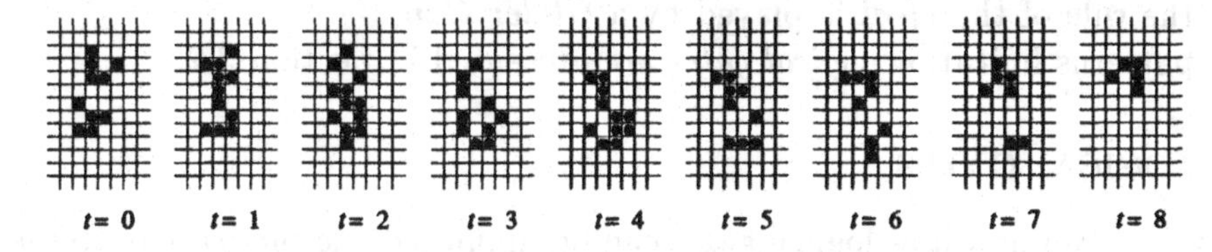

Figure 11. Kickback

With the help of these reactions, the logical gates NOT, AND and OR are constructed as shown in Figures 12a, b and c, respectively. The crash reaction is featured with a grey hexagon. A and B represent the input information: they are represented by a *glider* (for the value 1) or by *nothing* (for the value 0).

NOT **gate.** If A is a glider, then it has a crash reaction with a glider emitted by the gun and none of them survives. Hence, the result is nothing, that is, $\overline{A}$.
If A is nothing, then the glider emitted by the gun can continue its way, hence the result is a glider, that is $\overline{A}$.

AND **gate.** The result is a glider if and only if, after the second crash, there is a glider continuing its way from B's direction: B has to be a glider and nothing can arrive from the Gun's direction. It is possible if and only if the glider emitted by the Gun is killed in a crash reaction by a glider: A has to be a glider ((A and B) $= 1 \Longleftrightarrow A = 1$ and $B = 1$). If none of A and B is a glider, then the glider emitted by the Gun can walk on the plane even infinitely. An eater is placed after the second crash in order to avoid this problem.

OR **gate.** The result is not a glider if and only if, the glider emitted by the top gun is killed by a glider coming from the direction of the bottom gun. It is possible if and only if this glider survives the two crash reactions, that is, if and only if neither A nor B is a glider ($(A$ or $B)$ $= 0 \Longleftrightarrow A = 0$ and $B = 0$). For the same reason as above, an eater is placed at the issue of the first crash reaction.

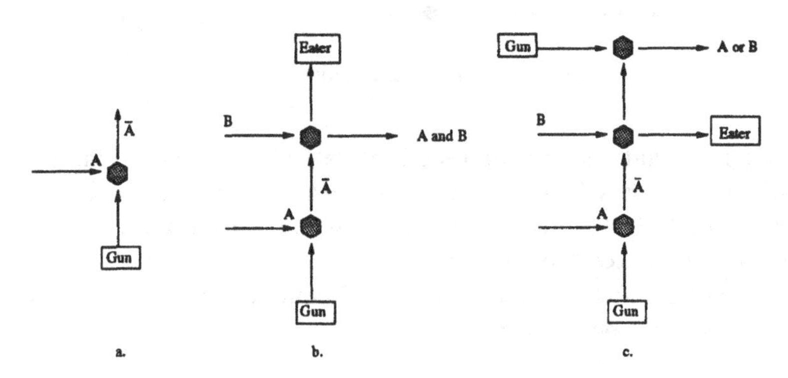

Figure 12. Logical gates NOT, AND and OR

Unfortunately, there is a problem: (A AND B) and (A OR B) are parallel to the input information A and B, while the NOT gate turns the information through 90°.

We have now to find a way to represent a turn without negation, or alternatively to perform a negation without turning the information. We show below how to realize the second operation.

The NOT *gate without turning information.* The glider streams that emerge from normal guns are so dense that they cannot collide without interfering with the following gliders. Before constructing a NOT gate without turning information, we show how to sparse a glider stream (see Figure 13).

Two guns emit gliders in parallel paths but in opposite directions in such a way that if we introduce a glider walking perpendicularly to these paths has a kickback reaction with gliders emitted by these guns. If the distance between these paths is $n/2$, then every nth glider is killed in both streams. Then, as shown in Figure 13, through these holes, every nth glider emitted by a third gun in a parallel path to the "walking" glider can cross without interacting. To get the phasing right, n must be divisible by 4, but it can be arbitrarily large and so we can make an arbitrarily thin stream.

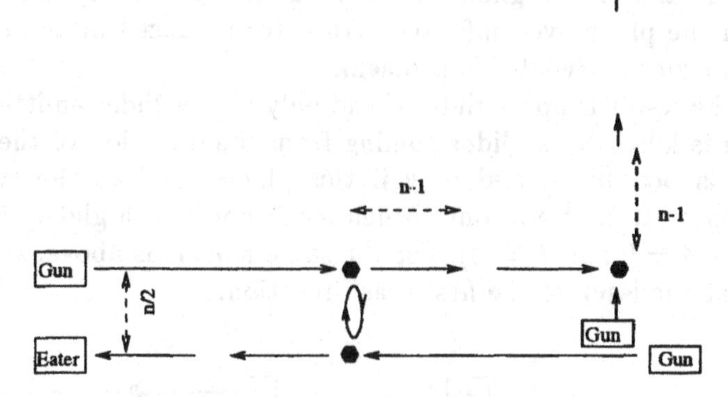

Figure 13. Sparsing a glider stream

We suppose that the gun emits gliders every 120 generations ($n = 4$). Then, when making a kickback reaction with the first glider, it is kicked back, the second glider crashes into the first one, forming a block and the third glider annihilates the block.

Let us see now how to make copies and complement of an information without turning it. See Figure 14.

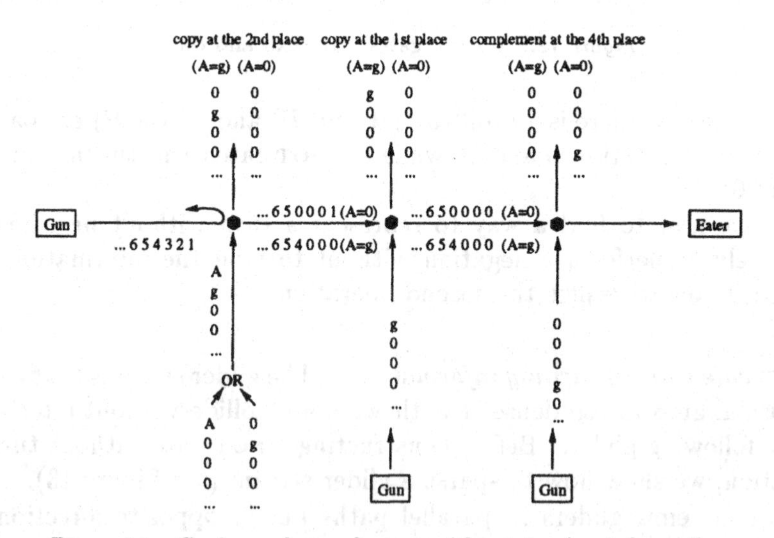

Figure 14. Copies and complement without turning information

We suppose that an information A is present every ten place (000000000A): A can be a glider ($A = g$) or nothing ($A = 0$). We consider that A is the first "emitted" information, followed by nine 0. When a stream 000000000A feeds into an OR gate with a stream being empty except at the second place where there is a glider g (00000000g0), the stream leaving the OR gate is the information A followed by a glider ($00000000gA$). Let us denote by X987654321 a stream of 10 gliders emitted by a gun (as described above), that we call a *full stream*. When the stream $00000000gA$ has a kickback reaction with a full stream, a copy of the information is reserved at the second place ($00000000A0$):

- if A was a glider ($A = g$), it has a kickback reaction with the first glider of the full stream. As described above, gliders 1-3 of the full stream are killed and X987654000 continues its way (as it is presented under the horizontal flash starting from the first crash reaction). As the second glider of the full stream is killed, $A = g$ can continue its way and the stream leaving the crash is $00000000g0$: in Figure 14, this case is represented at the left-hand-side of the flash ascending from the first crash.
- if A was not a glider ($A = 0$), the first glider of the full stream continues its way, and gliders 2-4 are dead, killed by the glider arriving at the second place: X987650001 continues its way (as it is presented on the horizontal flash starting from the first crash reaction). Then, after the crash reaction, the result is a stream without any glider, that is, 0000000000, as it is shown at the right-hand-side of the flash ascending from the first crash.

Using other crash reactions, we can reproduce another copy of A (crashing the stream ...654000/...650001 to a stream with only one glider at the first place $000000000g$) or the complement of A (crashing the stream ...654000/...650000 to a stream with only one glider at the fourth place $000000g000$), as shown in the figure.

3.3. THE CIRCUIT

Now let us organize all these devices to obtain a boolean circuit. First we use wires but we also need that the pertinent information arrives at the same time on the input of the boolean circuit. Thus we need to introduce delays which is not difficult by turning the wires (see Figure 15).

Thus we obtain a circuit. It is enough for our simulation and we can say that Life is universal for boolean circuits.

Now remark that because of the small period of the glider guns, only some information received at the end of the circuit is pertinent: the outputs at times corresponding to the traversal length of the circuit. This is not a problem in our case: we can ignore them! It becomes a serious problem for Turing-universality since in this case a 2 Registers Machine is simulated: these parasitic outputs can induce some extra modifications on registers.

In order to avoid these parasitic outputs, it is sufficient to increase the period of the glider guns and to adjust this period to the traversal length

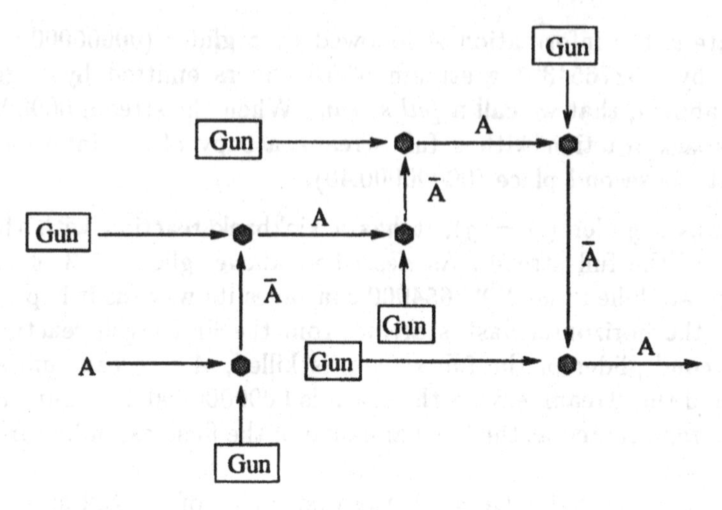

Figure 15. Delays

of the circuit. Alas this transformation by itself increases traversal length of the circuit because these new glider guns are much larger. Furthermore, in Conway's construction the size of such glider guns is proportional to the period (see Figure 13). In this particular point lies the problematic point of Conway's construction — as far as we understood it.

This problem can be solved by 3 methods. The first one and the most general would be to find an infinite family of recursively defined glider guns so that their size is little "*o*" of their period. We were told that this construction exists but could not get any reference of it. If some reader has some information about on such a family, we would be glad if he could pass it onto us. With the help of this family, it would be possible to use glider guns of period larger than the traversal length of the circuit, hence correct the original proof.

The second method is to remark that there exists a small universal 2 Registers Machine machine and that it is sufficient to simulate it (and not any 2 Registers Machine). In this case, we think that among all known glider guns (many of them can be found on the WEB), one can find an adequate glider gun for using in this universal 2 Registers Machine simulation. We agree that this does not constitute a formal proof!

The last method that we propose is to use intrinsic universality of Section 5 because, in this case, this problem of modifying a register does not arise; thus, it is not needed to suppress parasitic outputs. The simulation can be realized using a fixed period glider gun. The period of this gun should not be too small not to create jams in logical gates, but it is the only constraint that should be observed — we can use a *large* gun.

4. Turing universality

4.1. DEFINITION AND DISCUSSION

Let us first explain what "Turing universality" means. There are several approaches to handle this notion: first, a *constructive* one in which one should explain how a machine (e.g. a Turing machine, a Minsky machine, or any Kolmogorov-Uspensky machine on a given aggregate (see (Uspensky and Semenov, 1993)) is transposed into our cellular automata model. Second, an axiomatic one in which we define an acceptable programming system (see (Machtley and Young, 1978)). Although these approaches differ fundamentally, what is to be done in both cases is more or less the same: proper encodings of words, decodings, and halting conditions are to be defined.

Concerning the notion of simulation, there a 3 different possibilities:

1. an infinite family of cellular automata is considered and each of them is supposed to simulate a single machine. In this context, a word should be properly encoded into an initial configuration of the proper cellular automaton. This encoding should be uniform for all the family (a recursive dependency from both the word and the cellular automaton) The family of concerned cellular automata should be enumerable, and a cellular automaton that simulates a machine μ should be recursively obtained from the code of μ.
   Clearly, we are not in this context; nevertheless the conditions we give below for our context are also adequate for this kind of simulation with the above requirements;
2. an isolated cellular automaton is considered. It is supposed to simulate a chosen universal machine. In terms of simulations, it is the most simple situation. Nevertheless this case is not standard in literature because it is often much simpler to explain how to make a generic simulation of any machine than to give precisely the construction corresponding to a particular universal machine.
   The *ad hoc* conditions are the same that those that we propose for the 3rd case below where the variable representing the machine is assumed constant.
3. an isolated cellular automaton is considered. It is supposed to simulate any machine of a Turing-universal computational model (e.g. any Turing machine, any Minsky machine, or any Kolmogorov-Uspensky machine on a given aggregate (see (Uspensky and Semenov, 1993)). The goal is the following: given 2 variables representing respectively an input word and a machine, construct an initial configuration (encoding) so that the considered cellular automaton simulates the computation of the machine on the given word.

Thus we need now to explain what kinds of encoding functions, halting conditions, and decoding functions are allowed.

In the case of standard computation models an input word is transformed into an initial configuration which is also a finite object (see (Uspensky and Semenov, 1993) for an interesting discussion on the fundamentals of the notion of algorithmic machines). In this case, the natural requirement is that this transformation is recursive. In our case we consider a cellular automaton: the initial configuration is infinite and the cellular automaton acts on the whole configuration; thus we have to explain what kinds of transformations from finite words to infinite (initial) configurations are allowed. The standard restriction to recursive transformations is not meaningful since such a transformation maps a word onto an infinite object.

Recall that in order to define completely the computation model associated to a cellular automaton, we need to define

- the already mentioned transformation from a word into an initial configuration (*the encoding*),
- *the halting condition,*
- *the decoding*, *i.e.* the transformation forms the configuration to the output word. (As often, this is not a problematic point).

The first idea is to define the encoding as a transformation from words to finite configurations of the cellular automaton *i.e.* to almost everywhere quiescent configurations). The first problem is that all cellular automata do not have quiescent states; second and more important: in many simulations a periodic media —*i.e.* a periodic background configuration— is needed at infinity.

Thus the second idea is to transform an input word w into a program of a total recursive function f_w that defines the initial configuration. This total recursive function would map the set of cells to the states. Unfortunately, this elegant definition is not convenient for us: consider the following encoding function:

$$f_w(x) = \begin{cases} 0 & \text{if } x \leq 0, \\ w_i & \text{if } 1 \leq x \leq |w|, \\ 0 & \text{if } x = |w| + k \text{ where } k \in \mathbf{N}, \text{ and not } U(w,k), \\ 1 & \text{if } x = |w| + k \text{ where } k \in \mathbf{N}, \text{ and } U(w,k). \end{cases}$$

where $U(w,k)$ is true if and only if the universal Turing machine halts on the input word w in exactly k steps. The mapping $w \mapsto f_w$ is recursive but if we accept this encoding it contains in itself all the "computation" of the model and in this case, a trivial cellular automaton as the *shift* is universal. This is not acceptable thus our definition should be more restrictive.

We propose to consider the following definition of acceptable encodings. Such an encoding should map a word w and a machine μ to an almost everywhere periodic configuration. A periodic configuration is formed by the repetition of a finite pattern under the "natural" transformations in the space of cells. For instance in 3D the configuration should be the repetition of a cube *i.e.* invariant under 3 independent translations. The periodic part of the configuration should not depend on the input word w but may depend of the machine μ. The non-periodic part of the initial configuration should be obtained recursively from w, μ (and thus its size is bounded by a recursive function of w, μ).

We are not sure that this definition is the most general. But all the simulations we met in literature fit these requirements.

The decoding function can be very similar: take into account the non-periodic part of the configuration when halting is observed (see below for the restrictions on the halting condition); perform a recursive transformation of this part to a word.

Concerning the halting condition, the first idea is that a "special" state should appear somewhere in the configuration (see (Smith, 1972)). This is not good because a new specially dedicated state should be used. For instance in Life the halting condition is different. We could also ask that the dynamics of the cellular automaton becomes stable (*i.e.* enters a fixed point). This definition seems also too restrictive as it excludes the billiard models (see (Serizawa, 1987)). In some constructions the halting corresponds to the appearance of a special pattern but considering only a number of iterations multiple of an integer p. Intuitively this integer corresponds to the number of steps needed to simulate one time step of the machine.

We propose a more general notion: the halting condition is performed by the appearance of a finite pattern at precise observation time steps. More precisely, a function from a window of fixed size to the pair $\{\text{halt}, \text{continue}\}$ is performed everywhere and halting occurs if somewhere its image is "halt". Of course "everywhere" is not a problem since it is sufficient to check on the non-periodic part of the configuration. The general idea is that the halting condition is a local phenomena. A priori, this kind of definition would not include halting by entering fixed points. In all constructions that we know, we can see that this fixed point is obtained when a special pattern appears, thus it enters our definition.

We propose that the size of the observation window and the (finite) function from this window to the pair $\{\text{halt}, \text{continue}\}$ may depend only of the machine μ and not of the word w. The goal of this requirement is to impede the halting condition to make by itself the computation. The observation time steps should also be independent of w: we propose that it is performed by a recursive function that takes as inputs the machine μ

and a time t and outputs the observation time. The most complicated kind of functions that we saw in literature is in the form

$$(\mu, t) \mapsto f(\mu)t^2$$

where f is a recursive function.

4.2. CONSTRUCTION OF A TWO REGISTER MACHINE

In order to define a pattern imitating a real machine, in addition to the devices constructed for boolean circuit's simulation, we also need some auxiliary storage units. In (Berlekamp *et al.*, 1982), Conway chooses to simulate 2-Registers Machines (2RM for short) that are Turing-universal (Minsky, 1967). The construction of registers —and operations on them— are also described. In brief, a block is used to represent a static piece of information, two gliders are used to pull a block back three diagonal places and ten gliders are used to move up a block just one diagonal place. The precise construction is very complex.

5. Intrinsic universality

5.1. DEFINITIONS AND DISCUSSION

Although the intuitive notion of simulation between cellular automata is broadly used, it does not always correspond to the same definition. For instance, in some cases the number of steps used to simulate one step of the considered cellular automaton may vary. Another difference is that sometimes it is not the same function that is used to encode and decode the configurations.

We present below a rather strict definition of this intuitive notion inspired by the notion of simulation in (Róka, 1998). We do not know what is the most general definition that could be accepted.

Definition 4

Let $\mathcal{A}$ be a cellular automaton and $C_\mathcal{A}$ the set of its configurations. Let $\mathcal{B}$ be a cellular automaton and $C_\mathcal{B}$ the set of its configurations. We say that $\mathcal{B}$ simulates $\mathcal{A}$, *if there exist a one-to-one (injective) application $f : C_\mathcal{A} \to C_\mathcal{B}$ UL-defined (see below) and a constant T in $\mathbb{N}$ such that for all c in $C_\mathcal{A}$*

$$f(\mathcal{A}(c)) = \mathcal{B}^T(f(c)).$$

T is the simulation time factor, that is, the time which is necessary for $\mathcal{B}$ to simulate one iteration of $\mathcal{A}$. It depends on f but not on c. f is called the *encoding function;* "f is UL-defined" means that f is defined uniformly and locally. This notion is not mathematic but rather philosophical: our aim is that it is true in any space on which we could imagine a definition of cellular automaton [2]. This notion depends on the definition of cellular automaton we use. In the simple case of 2D cellular automaton, f is UL-defined means that it commutes with 2 independent shifts (uniformity) and that it is continuous (locality). This definition can be illustrated by the following diagram:

$$\begin{array}{ccc} c & \xrightarrow{f} & f(c) \\ \mathcal{A}\Big\Downarrow & & \Big\Downarrow\mathcal{B}^T \\ \mathcal{A}(c) & \xrightarrow{f} & \mathcal{B}^T(f(c)) \end{array}$$

Remark that if we endow "naturally" $C_{\mathcal{A}}$ and $C_{\mathcal{B}}$ with the product topology, f is a continuous function that is invariant under n independent shifts where n is the dimension of the space. From (Hedlund, 1969; Richardson, 1972) we deduce that, basically, f is a cellular automaton with, in addition, the possibility to take into account a finite information concerning the position of the cell. This finite information must be periodically distributed on the space.

The differences between our definition and the definition of (Róka, 1998) is that we add the requirement that f is UL-defined. Our goal is to make definition of simulation compatible with Turing universality: if a cellular automaton $\mathcal{A}$ is simulated by a Turing-universal cellular automaton $\mathcal{B}$, then $\mathcal{A}$ is universal and conversely.

But, this definition does not cover all simulations. In (Róka, 1994) the problem of simulating cellular automata with only one-way communication between cells has been studied: "given a Cayley graph, can all bidirectional cellular automata be simulated by a one-way cellular automaton on this graph?". Sometimes, such a simulation is possible, but after the simulation of each iteration of the bidirectional cellular automaton, the obtained configuration is shifted in the one-way cellular automaton.

Beware that a strict interpretation of the term "intrinsic universality" would mean that we simulate any other cellular automaton defined on the 2D grid, but not on an hexagonal lattice. Our result is a little stronger since we simulate any 2D cellular automaton. It is stronger because any cellular automaton—e.g. on a hexagonal lattice— can be simulated by a 2D cellular automaton (and even with Von Neumann neighborhood). This is also true for any "regular" lattice of the plane, as proved in (Róka, 1998) using Cayley graphs.

[2]This formal definition is not possible to formulate for classical paradoxes of set theory: there does not exist a set containing all other sets.

Let us present now a very useful meta-theorem:

Meta-theorem 1

Consider a cellular automaton $\mathcal{A}$ universal for boolean circuits. If one can simulate in $\mathcal{A}$ wires, crossings of wires, delays, duplicators, and all needed plugs, then $\mathcal{A}$ is intrinsically universal for all cellular automata defined on the same underlying architecture, and therefore $\mathcal{A}$ is Turing-universal.

The proof of this theorem is straightforward and consists only of the intuitive simulation of an infinite "electronic" circuit that performs the computation of the considered cellular automaton. The construction of the following Section illustrates this proof.

This theorem is not completely formalized: wires, delays duplicators and plugs are not formally defined. They could be, but the obtained definitions would be extremely complicated and of no use at all. We prefer not to give them, but even if meta-theorem 1 is not completely formalized in this context, it gives the basic ideas for constructing correct simulations.

5.2. OUR CONSTRUCTION

We recall that one can define patterns imitating

- pulses and
- logical gates.

As the NOT gate can be imitated with and without turning information, one can also turn information as often as one wishes. Hence,

- wires of arbitrary length can be constructed.

Let us consider any 2D cellular automaton $\mathcal{A}$ and let us define first our encoding function f. In the context of our definition, the cellular automaton $\mathcal{B}$ is Life. We shall explain what is the simulation time T at the end of this section.

Let us first consider the local transition rule of $\mathcal{A}$ that we call δ. If we encode states (for instance in binary) by words on the alphabet $\{0, 1\}$, δ can be seen as a boolean function. We saw in Section 3 that we can simulate with our cellular automaton any boolean function via the simulation of wires, logic gates, crossovers, fan-outs, delays, etc. Thus we can construct the circuit corresponding to δ in a rectangle (see Figure 16).

The idea of our simulation is to transform each cell into large squares containing this rectangle and the neighborhood connections between cells (see Figure 18). For simulating 1D cellular automata, we obtain the simple Figure 17 (output arrows represent copies of the same information).

The state of each cell is naturally encoded by the corresponding word (in binary) on the output wires of the associated rectangle.

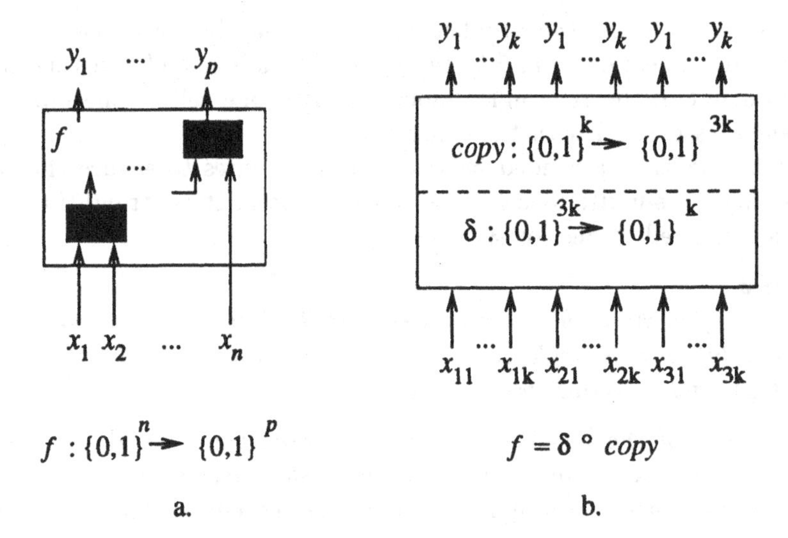

Figure 16. The simulation of δ

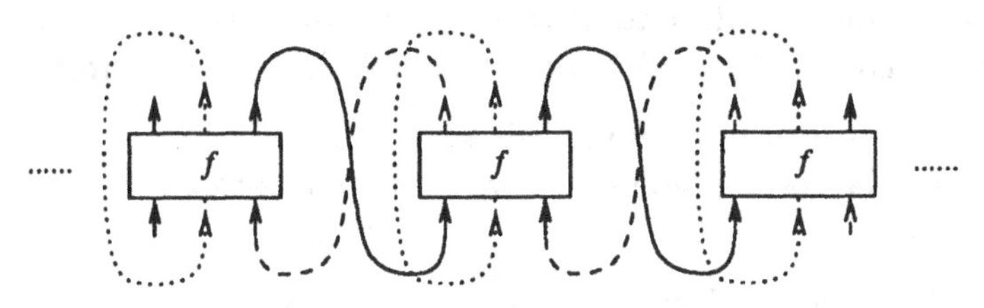

Figure 17. Simulation of 1D cellular automata

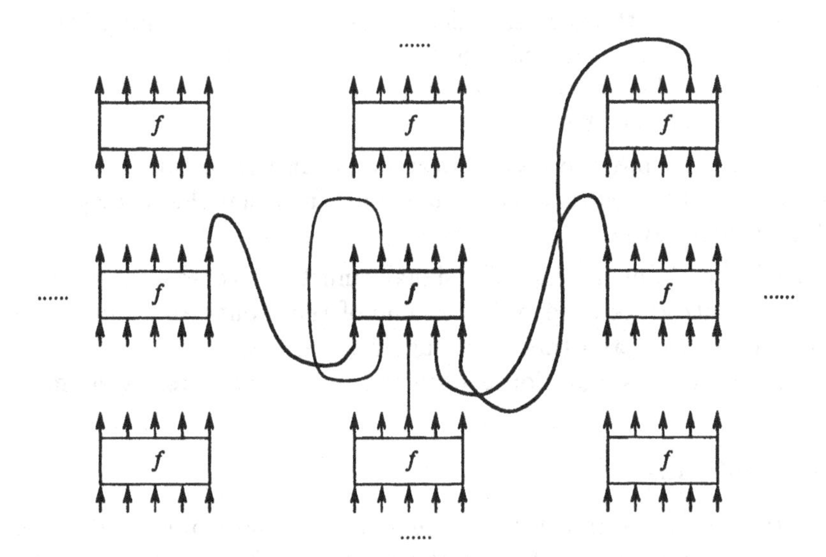

Figure 18. The general scheme

Now, let us examine the problem of connections between the cells. A neighborhood connection should be represented by a bunch of wires starting from the output of the rectangle associated with one cell to the input of the rectangle associated with the other one.

A synchronization is needed between all these wires to ensure that information arrives simultaneously from all neighbors. It is straightforward using delays defined in Section 3.

Theorem 3

For all neighborhood there exists a sufficiently large size of square such that the simulation is possible, i.e. *such that it is possible to draw all needed wires, delays, and crossover between cells.*

Proof. The idea of the proof is that we can zoom the general scheme of computations in order to introduce those necessary delays.

First, let us consider the general scheme of the construction in which wires are represented by curves (see Figure 18).

It is not difficult to see that we can embed this scheme into a configuration of our cellular automaton; the only problem is how to solve the synchronization constraints. We propose the following procedure:

- compute the length of all input wires,
- consider the largest difference between these lengths. Let n be such a difference.

Then we need a square of size $\mathcal{O}(\sqrt{n})$. If we have no space to put the necessary delay, we know that there exist a scale factor M such that we can include in our scheme the necessary delays making a $\times M$ zooming because, by making this zooming, the delay that we introduce is proportional to M^2, while the needed delay n grows linearly in M.

Let us now continue the procedure:

- zoom $\times M$, and put the needed even delays in the rectangle.
- if a delay is odd: move that cell of one unit, such that the corresponding delay becomes even.

To avoid the problem that the signals must arrive separated to the crossovers, we introduce a delay before one of the inputs and another one after the other output and then zoom again accordingly.

The time T of the simulation is obtained by these successive zooms. □

6. Open problems

The definition of simulation between cellular automata presented in Section 4 is maybe not the most general one. We think that this notion is worth studying.

Another problematics is to find the smallest Turing-universal or intrinsically universal cellular automaton: it is interesting to know whether small cellular automata can be universal in order to construct them with very elementary natural devices and observe complicated behaviors. Contrarily to Turing Machines, there are not many varieties of cellular automata: no problem to know whether heads of Turing machines could be allowed to stay or just go left or right, no problem of semi-infinite or bi-infinite tape, no problem to know whether the initial state can or cannot be used after the first step ... This problem of finding the smallest universal cellular automata is much more interesting than for Turing Machines.

Even if it is conjectured that 2 states and 3 neighbors are sufficient (see (Wolfram, 1983)) for Turing universality, the smallest universal cellular automaton that we know (both Turing and intrinsic universalities) has 3 states and 3 neighbors on an hexagonal lattice (see (Gajardo , 1998)).

References

Albert A. and Čulik H. A simple universal cellular automaton and its one-way and totalistic version. *Complex Systems*, **Vol. no. 1**: 1–16, 1987.

Banks E. Information processing and transmission in cellular automata. *PhD thesis*, Mass. Inst. of Tech., Cambridge, Mass, 1971.

Berlekamp E., Conway V., Elwyn R. and Guy R. *Winning way for your mathematical plays, volume 2* Academic Press, 1982.

Burks E. *Essays on Cellular Automata*, University of Illinois Press, 1972.

Codd E. *Cellular Automata*, Academic Press, 1968.

Durand B. Global properties of cellular automata. in *Cellular Automata and Complex Systems*, Goles E. and Martinez S., Eds, Kluver, 1998.

Gajardo A. Universality in a 2-dimensional cellular space with a neighborhood of cardinality 3. *Preprint*, 1998.

Hedlund G. Endomorphism and automorphism of the shift dynamical system. *Mathematical System Theory*, **Vol. no. 3**: 320–375, 1969.

Machtley M. and Young P. *An introduction to the general theory of algorithms*, Elsevier, 1978.

Minsky M. *Finite and infinite machines*, Prentice-Hall, 1967.

Moore E. Machine models of self-reproduction. *Proc. Symp. Apl. Math.*, **Vol. no. 14**: 13–33, 1962.

Myhill J. The converse to Moore's garden-of-eden theorem. *Proc. Symp. Apl. Math.*, **Vol. no. 14**: 685–686, 1963.

Nourai F. and Kashef S. A universal four states cellular computer. *IEEE Transaction on Computers*, **Vol.C24 no. 8**: 766–776, 1975.

Richardson D. Tesselations with local transformations. *Journal of Computer and System Sciences*, **Vol. no. 6**: 373–388, 1972.

Róka Zs. One-way cellular automata on Cayley graphs. *Theoretical Computer Science*, **Vol. no. 132**: 259–290, 1994.

Róka Zs. Simulations between cellular automata on Cayley graphs. To appear *Theoretical Computer Science*,1998 .

Serizawa T. Three state neumann neighbor cellular capable of constructing self reproducing machines. *System and Computers in Japan*, **Vol. 18 no. 4**: 33–40, 1987.

Smith III A. Simple computation-universal cellular spaces. *Journal of Assoc. Comp. Mach.*, **Vol. 18 no. 3**: 233–253, 1971.

Smith III A. Real-time language recognition by one-dimensional cellular automata. *Journal of Computer and System Sciences*, Vol. no. 6: 233–253, 1972.
Uspensky V. and Semenov A. *Algorithms : main ideas and applications.* Kluwer, 1993.
von Neumann J. *Theory of Self-Reproducing Automata* University of Illinois Press, 1966.
Wolfram S. Universality and complexity in cellular automata. in *Cellular Automata*, Toffoli T., Farmer D. and Wolfram S., Eds, North-Holland , 1–36, 1983.

Part 2 Algorithmics

How to build cellular automata that produce wanted behavior?

First a few examples are presented: to cut a segment of cells according to some given ratio, to synchronize a segment of cells, to recognize palindromes. These examples give special prominence to the notion of signal and to a dependency graph along which information moves.

The second part is dedicated to this dependency graph. It is shown how it can be handled through space-time and used to perform the same computations in different parts of it. The induced techniques are applied to prove that cellular automata, synchronous by nature, are able to uniformly realize asynchronous computations.

COMPUTATIONS ON CELLULAR AUTOMATA

Some examples

J. MAZOYER
LIP ENS Lyon
46 Allée d'Italie,
69364 Lyon Cedex 07,FRANCE

Abstract. We show simple methods in order to construct cellular automata with a defined behavior. To achieve this goal, we explain how to move local information through the networks and to set up their meeting in order to get the wished global behavior. Some well-known examples are given: Fischer's prime construction, Firing Squad Synchronization Problem and an "ad hoc" example of filter.

1. Introduction

The computational power of cellular automata has been emphasized since their origin: the self-reproducing two-dimensional Von Neumann's automaton is computation universal. Very soon, it was shown that one-dimensional cellular automata could also be computational universal. Computational universality is obtained by simulating a Turing machine (Smith, 1972) , a cellular automaton (Albert and Čulik, 1987) and (Martin, 1994) or the "real" architecture of a computer (Berlekamp *et al.*, 1982) and other papers as *The Game of Life: universality revisited* in this book). To design algorithms on cellular automata comes later. In fact, it is important to distinguish the one dimension case from the other ones. If lines of cells are well studied, few things are known on two dimensional spaces. In fact, such algorithms may be found in papers on "mesh computers" or on "image processing". An original approach may be found in (Achasova *et al.*, 1994). The basic one dimensional notions developed below seem may be generalized to higher dimensions (see (Signorini, 1992) and papers on the Firing Squad). In the following, we only study lines of automata. Some reference papers are (Smith, 1972), (Wolfram, 1986) and (Čulik and Hurd, 1990).

To design an algorithm is to set up state transitions of a cellular automaton with a defined behavior on a set of initial configurations. This behavior may be of computational nature or intrinsically connected with parallelism (as Firing Squad Synchronization Problem) without any counterpart in any sequential model of computation.

The interest of CA for massive parallel computation is due to the fact that they are a natural model of parallelism. They give birth to local algorithms, independent of the number of cells, with a complex global behavior even with very simple cell-behavior (see (Berlekamp *et al.*, 1982)).

When designing an algorithm on a line of automata, we have to

- i - define atoms of information,
- ii - set up their moves in the space time diagram,
- iii - avoid (or resolve) any problem of meetings (local synchronization).

To define the nature of the atoms of information depends on the chosen algorithm. But, avoiding any question of optimization, we simply attribute states to atoms. Then we define moves of these atoms in the space-time diagram. Trajectories of such atoms will be called "signals". An exact definition of signal is not simple. We only need some elementary notions on this notion: a signal appears at the meeting of other signals, it moves giving (possibly) birth to new information and so on. These signals are useful to construct cellular automata with a given behavior. To know if cellular automata have in their evolution some significant information that we may call signal is a difficult question, viewed in terms of "particles" (Fokas *et al.*, 1989). In the following we give examples of such signals.

2. Simple signals

The simpler example of use of signal is to design 1D-cellular automata which cut a line in a rational ratio. We study this example in great detail. First of all, two notions must be precise:

- a - Given n cells, what is the cell numbered $\frac{pn}{q}$ $(p, q \in \mathrm{N})$? As usual, in the discrete field, $\frac{pn}{q}$ means $\lfloor \frac{pn}{q} \rfloor$.
- b - This cell is distinguished (enters a special state) at some time $\theta(n)$ depending on n. We choose this time to be "as soon as possible" and we compute it later.

We set up formally this question as follows:

Problem 1

Given two integers p and q, design an 1D-CA (Q, δ) with a distinguished state F and 2 quiescent states, A (active state) and ! (outside state) with the transitions indicating that state ! corresponds to the inactive part of the *configuration*

$$\forall Q_1, Q_2 \in Q, \delta(Q_1, !, Q_2) = !$$

such that ($\forall n \in \mathbf{N}$):

- *i - Starting from any initial configuration*

$$C(n) = {}^{\omega}!\overbrace{A \ldots A}^{n \text{ times}}!^{\omega}$$

- *ii - the system evolves such that at some time* $t(n)$*, the reached configuration is*

$$^{\omega}! \quad \overbrace{A \ldots A}^{\lfloor \frac{pn}{q} \rfloor - 1 \text{ times}} \quad F \quad \overbrace{A \ldots A}^{n - \lfloor \frac{pn}{q} \rfloor \text{ times}} \quad !^{\omega}$$

Now, we are able to define $\theta(n)$ such that the state F appears as soon as possible. At initial time, two cells play special roles:

- cell 0 with its left neighbor in state !; itself and its right neighbor in state A (we call it the "left-end" cell).
- cell $n-1$, similarly defined (we call it the "right-end" cell).

When at time $\theta(n)$ a cell enters state F, it "must" know its distance to the left-end and the right-end cells. Thus, the time $\theta(n)$ depends on the relative position of the cell $\lfloor \frac{pn}{q} \rfloor$ with the middle of the segment. When this cell is near the left-end cell, we have:

$$\theta(n) = n - \lfloor \frac{pn}{q} \rfloor$$

and, if not

$$\theta(n) = \lfloor \frac{pn}{q} \rfloor - 1 \, .$$

A formal proof of the value of $\theta(n)$ is in two points: first we define cellular automata which cut the line at the previous time $\theta(n)$ (it will be constructed later), second we prove the impossibility to cut the line in a time less than $\theta(n)$. In the last case we place ourselves in the case when $\lfloor \frac{pn}{q} \rfloor \leq \lfloor \frac{n}{2} \rfloor$. In order to obtain a contradiction, we assume that, for a fixed value of n, large enough, denoted n_0, a segment of length n_0 is cut at time $t_{n_0} < \theta(n_0)$.

Let $\langle h, t \rangle$ denotes the state of cell h at time t. The value of $\langle \lfloor \frac{pn_0}{q} \rfloor, t_{n_0} \rangle$ depends:

- at step 1, only on values of $\langle \lfloor \frac{pn_0}{q} \rfloor - 1, t_{n_0} - 1 \rangle$, $\langle \lfloor \frac{pn_0}{q} \rfloor, t_{n_0} - 1 \rangle$, $\langle \lfloor \frac{pn_0}{q} \rfloor + 1, t_{n_0} - 1 \rangle$;
- at step 2 only on values of $\left\{ \langle \lfloor \frac{pn_0}{q} \rfloor + i, t_{n_0} - 2 \rangle \mid i \in \{-2, \ldots, +2\} \right\}$;

– and finally at any step j, only on values of
$\left\{\langle\lfloor\frac{pn_0}{q}\rfloor+i, t_{n_0}-j\rangle \mid i\in\{-j,\ldots,+j\}\right\}$.

At step t_{n_0}, we obtain that the value of $\langle\lfloor\frac{pn_0}{q}\rfloor, t_{n_0}\rangle$ only depends on $\left\{\langle\lfloor\frac{pn_0}{q}\rfloor+i, 0\rangle \mid i\in\{-t_{n_0},\ldots,+t_{n_0}\}\right\}$.
These values correspond to the initial configuration and, thus, are known:
$\forall i\in\{-t_{n_0},\ldots,-\lfloor\frac{pn_0}{q}\rfloor-1\}$, $\langle\lfloor\frac{pn_0}{q}\rfloor+i,0\rangle = !$
$\forall i\in\{\lfloor\frac{pn_0}{q}\rfloor,\ldots,t_{n_0}\}$, $\langle\lfloor\frac{pn_0}{q}\rfloor+i,0\rangle = A$ because we have $\lfloor\frac{pn_0}{q}\rfloor+i \leq \lfloor\frac{pn_0}{q}\rfloor+\lfloor\frac{pn_0}{q}\rfloor < n_0$ by the hypothesis $\lfloor\frac{pn}{q}\rfloor\leq\lfloor\frac{n}{2}\rfloor$.
Let us now consider a segment of length $n_1 = n_0+q$, the value of $\langle\lfloor\frac{pn_1}{q}\rfloor, t_{n_0}\rangle$ is the same as in the case of a segment of length n_0 because the involved initial values are the same. Thus $\langle\lfloor\frac{pn_1}{q}\rfloor, t_{n_0}\rangle = F$ also, but $\frac{p(n_0+q)}{q} = \frac{pn_0}{q} + p \neq \frac{pn_0}{q}$. We get the contradiction.

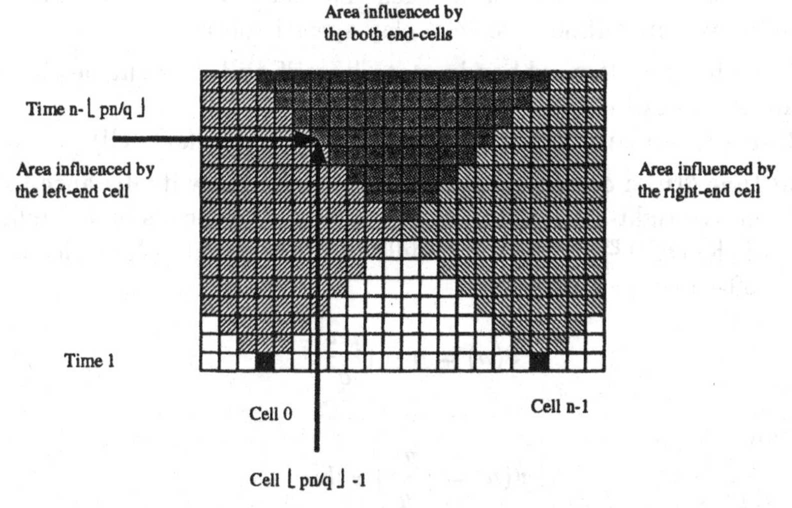

Figure 1. Cutting a segment: areas of influence: n even

Another (direct but unformal) way to see the previous proof is:

- i - To consider the important data of an initial segment: namely the left and right ends of the segment, indicated by the values $\langle 0,1\rangle$ and $\langle n-1,1\rangle$.
- ii - To see how that may influence the values of $\langle h,t\rangle$. The values influenced by $\langle i,j\rangle$ are:
  $I_{\langle i,j\rangle} = \{\langle i+i^\star, j+j^\star \mid j^\star\geq 0, i^\star\in\{-j^\star,\ldots,j^\star\}\}$.
- iii - The cutting point must be influenced by $I_{\langle 0,1\rangle}$ and $I_{\langle n-1,1\rangle}$ as proved previously.

The Figures 1 and 2 show $I_{\langle 0,1\rangle}$ and $I_{\langle n-1,1\rangle}$ and we also get the wanted optimal value of $\theta(n)$.

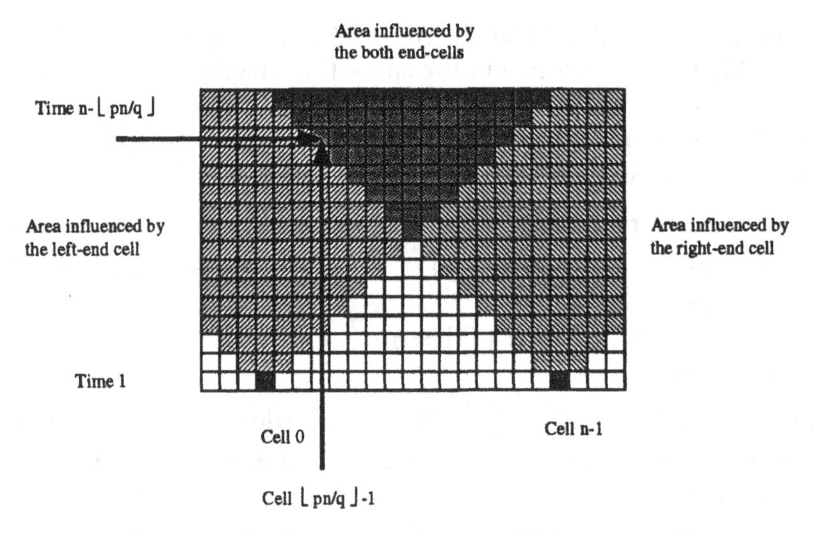

Figure 2. Cutting a segment: areas of influence: n odd

Now we indicate how to construct an automaton cutting any segment at its $\frac{pn}{q}$, we choose the special case when the cell to be distinguished is "near" the left-end cell. On the Figure 3, we mark all the sites which become distinguished when the length n increases with the values $p = 2$ and $q = 7$. These sites belong to a broken line. The equation of this line depends on the values of p and q. This line is always ultimatly periodic. In the following of this section we suppose that p and q are relatively prime. In this case, the period of the line has length q.

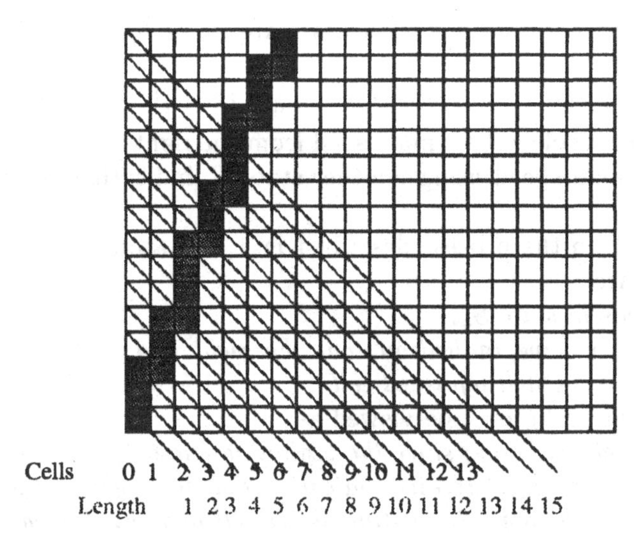

Figure 3. Cutting a segment: the distinguished line

We use $2q$ new states (denoted S_h, $h \in \{0, \ldots, q-1\}$ and $S_h^\star$, $h \in \{0, \ldots, q-1\}$) to draw this line in the space-time diagram. And we assume that state F appears instead of one of the S_h (and not $S_h^\star$).

Suppose that S_0 is replaced by F when the length of the initial line is mq, then $\lfloor \frac{p(mq+\ell)}{q} \rfloor$ is increased by 1 unit when $p\ell$ becomes greater than q. Thus, in the general case, $\lfloor \frac{p(mq+\ell)}{q} \rfloor$ is increased by 1 unit every time $p\ell$ becomes greater than αq (with $\alpha \in \{1, \ldots, p\}$). It is to say that $\lfloor \frac{p(mq+\ell)}{q} \rfloor$ increases of 1 unit for:
$\ell = \lceil \frac{q}{p} \rceil, \ldots, \ell = \lceil \frac{\alpha q}{p} \rceil, \ldots, \ell = \lceil \frac{pq}{p} \rceil = q$.

Now, let us denote by:

- $(x_\ell, t_\ell) = \left(\lfloor \frac{p\ell}{q} \rfloor, 1 + (\ell - \lfloor \frac{p\ell}{q} \rfloor)\right)$, the site on which is the cutting point along the antidiagonal born on the right end of the segment,
- $(x_\ell^\star, t_\ell^\star) = (x_\ell, t_\ell + 1)$.

Due to the periodicity of the cutting line, we need to define states of the previous points only for q values of ℓ. Our choice is the following:
If $\ell = mq + h$ then
$\langle x_\ell, t_\ell \rangle = S_h$ *and* $\langle x_\ell^\star, t_\ell^\star \rangle = S_h^\star$.

This choice may be set up by local transitions:

1. If $\langle k, t \rangle = S_h$, then
   - $\langle k, t+1 \rangle = S_h^\star$
   - The value of $\langle k+1, t+1 \rangle$ is:
     - $S_{h+1 \bmod q}$ if $h = \lceil \frac{\alpha q}{p} \rceil$ for $\alpha \in \{1, \ldots, p\}$,
     - state A else.
2. If $\langle k, t \rangle = S_h^\star$ and $\langle k+1, t \rangle$ is not a state S_h or $S_h^\star$, then $\langle k, t+1 \rangle = S_{h+1 \bmod q}$.
3. If $\langle k, t \rangle = S_h^\star$ and $\langle k+1, t \rangle$ is $S_{h+1 \bmod q}$, then $\langle k, t+1 \rangle = A$.

To set up the whole cut we must send from the right-end cell a signal to the left at maximal speed using another state R, indicating the right end of the segment.

Coming back to the process described in 1, the atoms of information, carried by $S_h^\star$, are:
Previously, I was in state S_h.
If, next time, I receive indication that the right end of the segment is reached, I become the distinguished cell.

Those carried by S_h may be viewed as:
I know that the elapsed time since the beginning (time 1) is $h \mod q$.
If, entering the state S_h, I receive indication that the right end of the segment is reached, I become the distinguished cell and, in this case I know that the remainder of the length of the segment by q is h.

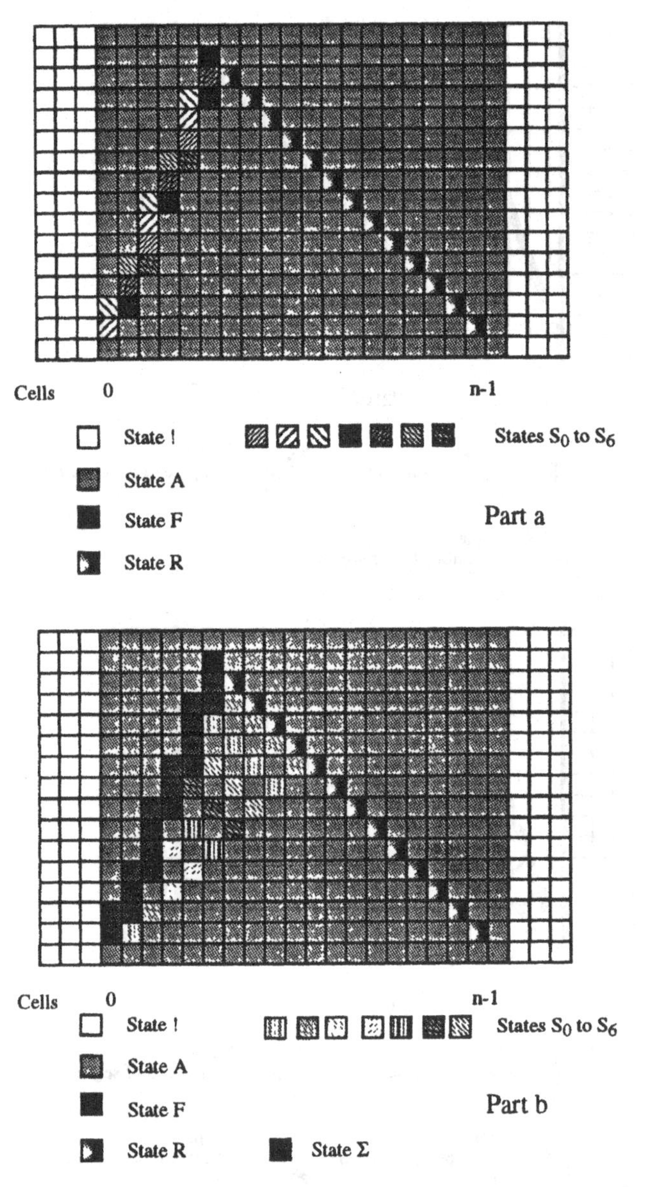

Figure 4. Cutting a segment: setting up the distinguished line with states

In order to consider the case where the cell to be distinguished is near of the right-end, we set up, in parallel, a symmetric process. By this way we obtain a cellular automaton with $(2q+1)^2+2$ states (left and right cases). We do not describe all its state transitions. In fact, it is easy to prove by induction that if a cellular automaton has the wished behavior on lines of length up to $2q$ (a period is achieved) then it is correct. Thus it is easy to

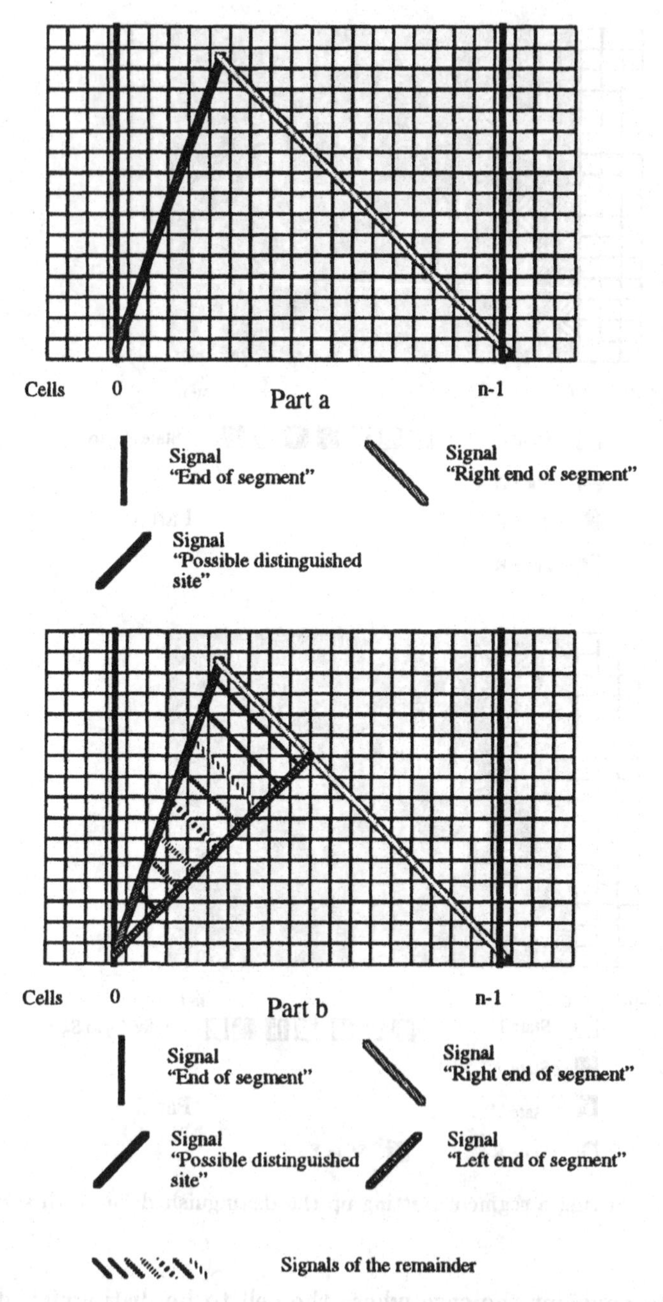

Figure 5. **Cutting a segment: setting up the geometric diagram**

construct it and to verify it with a computer.

Looking to the atoms of information given above, we feel that they are not atomic. Atomic ones will be:
I am a site to be distinguished
the remainder of the segment is h
and
I indicate that I am a left or right end of the segment.

It is possible to design another algorithm with these atoms of information, as soon as possible, the information on the remainder of the length of the segment is sent into the network as quickly as possible and the atom *I am a site to be distinguished* moves according to this information. This process is illustrated on the Figure 4.

This new algorithm is more efficient than the first one not only for aesthetic reasons (due to atoms of information), optimization reasons (its number of states is $(q+1)^2+2$) but also because it is more resistant to faults. For example if some delay between two cells is of k units of time (instead of 1), it works as well.

If it is easy to cut a segment in a rational ratio, what happens when the ratio will approximate some real number? This question is complex. First of all, we must restrict ourselves to constructive reals. In some case, it is impossible. For instance (Mazoyer and Terrier, 1998), if we choose a line which moves to the right with slope:

$$1.01\ldots1 \quad \underbrace{\underbrace{0\ldots0}_{log(log(\ldots))(i)}}_{i \text{ times}} \quad 1\ldots$$

In other cases, this signal may be constructed: for instance (Mazoyer and Terrier, 1998) with the movement:

$$1.01\ldots1\underbrace{0\ldots0}_{log_2(i)}1\ldots$$

We easily observe a figure such that the Figure 4 has a plenty of meaningful information but it becomes very quickly difficult to read (think at what becomes this figure with the needed symmetries). On other hand the spirit of the algorithms is to set up straight lines and to give sense to their presence and to their meetings. Putting these straight lines in the continuous plane we obtain a geometric description of algorithms. For instance the Figure 4 becomes the Figure 5. We call such a representation a "geometric diagram". But we must be cautious: translation of a geometric diagram to a real space-time diagram with states is not really straightforward. As shown by the previous remarks, lines with non rational ratio may be set up or not; thus, it is necessary to ensure that all the involved movements have a rational slope $\frac{p}{q}$ with q bounded. This is not sufficient, we must check that only a finite number of meetings occur in the evolution of the CA on the set of initial lines. In the following we shall use the two sorts of diagrams.

3. Fischer's recognition of primes

In the previous section, we have seen simple examples of linear signals. How can they interact in order to construct non linear signals? The first algorithm and the most known is due to P.C. Fischer (Fisher, 1965): he shows that 1D-CA are able to recognize in real time the language

$$\{1^p \mid p \text{ is a prime}\}$$

We shall explain this algorithm. But first let us start with some comments. In this paper, P.C. Fischer illustrates four points:

- i - Optimal parallel algorithms may be greedy and near classical ones.
- ii - Due to the definition of cellular automata, the difficulty to design some automata is the finite character of the automata located on cells.
- iii - There exists non linear signals (the" Fischer's parabola").
- iv - The number of states may be huge and we need special tricks in order to understand the process. In particular, Fischer designed an automaton which must be transformed ("grouping cells") to obtain the wished one.

First we indicate the wished behavior of our automaton:

Problem 2

Design a 1D-CA (Q, δ) with a distinguished set of states P and 2 quiescent states, A (active state) and ! (outside state) and 2 special states indicating the beginning and the end of the active part (G and D) with the transitions indicating that state ! corresponds to the inactive part of the configuration

$$\forall Q_1, Q_2 \in Q, \delta(Q_1, !, Q_2) = !$$

such that ($\forall n \in \mathbf{N}$):

- *i - Starting from any initial configuration*

$$C(n) = {}^{\omega}!\ G\ \overbrace{A \ldots A}^{n-2 \text{ times}}\ D\ !^{\omega}$$

- *ii - the system evolves such that at time n, the state of the first cell is in P if and only if n is prime.*

As indicated in point i), the used algorithm is the classical one, known as Eratosthene's jigger. We recall it: in order to know if n is prime, we mark all the multiples (less than n) of any integer i less than n. One may observe that the first integer to be crossed at level i is i^2.

3.1. MARKING MULTIPLES OF *I*

An integer is represented by a time on the first cell. To mark such an integer is obtained by giving a state belonging to P to the first cell at the corresponding time. According to the fact that the first cell indicates the prime nature of the elapsed time, to mark an integer is to give some "marking" information to the first cell. The right end cell sending a signal E_1 to the left marks (on the first cell) the time corresponding to the length of the active segment. The Figure 6 shows a way to mark all times multiple of 12. An information (signal M_7) remains on cell 7. Another signal zigzags between the cell 0 and the cell 7. It needs 6 units of time to reach cell 7 and then 6 other ones to come back. And thus marks times 12, 24, ... We distinguish the signal moving right R_1 from the signal moving left W_1.

Unfortunately, this method cannot be easily translated in order to mark odd times. The trick is to define "significant times". As an example, in the Figure 7, we consider that one time out of 5 is significant and we mark the significant times multiple of 6 (in fact, times multiple of 30). We describe this figure. The cell 5 is marked by M_5. The signal R_1 becomes a signal R_4 of slope 4 (only one unit of time out of 4 is goes to the right) and the signal W_1 becomes W_2 (only one unit of time out of 2 goes to the left). By this way, signals R_4 and W_2 take 30 times to achieve a zig-zag marking any multiples of 30 *or* marking any multiple of 6 if we restrict ourselves to the significant times. In addition the signal E_1, indicating the end of the active segment must reach only significant times. We have two choices: it goes to the left with slope 5, becoming E_5 or the active segment is supposed to have length $5n$. In order to set up the process described in 3.2, we choose the second solution assuming that the active segment is 5 times greater. By this way we recognize multiples of 6 in time 5n.

Our choice of the values of the significant times (multiples of 5), of the marked cell (M_5), and of the slopes (R_4 and W_2) is arbitrary. Fischer in his original paper chooses as significant times the multiples of 3, and signal R_2 and W_1. By this choice to mark significant multiples of i induces to mark the cell i by M_i. Any choice will give a solution but his choice "minimalizes" (at first glance) the number of states. In the following, we will follow his choice.

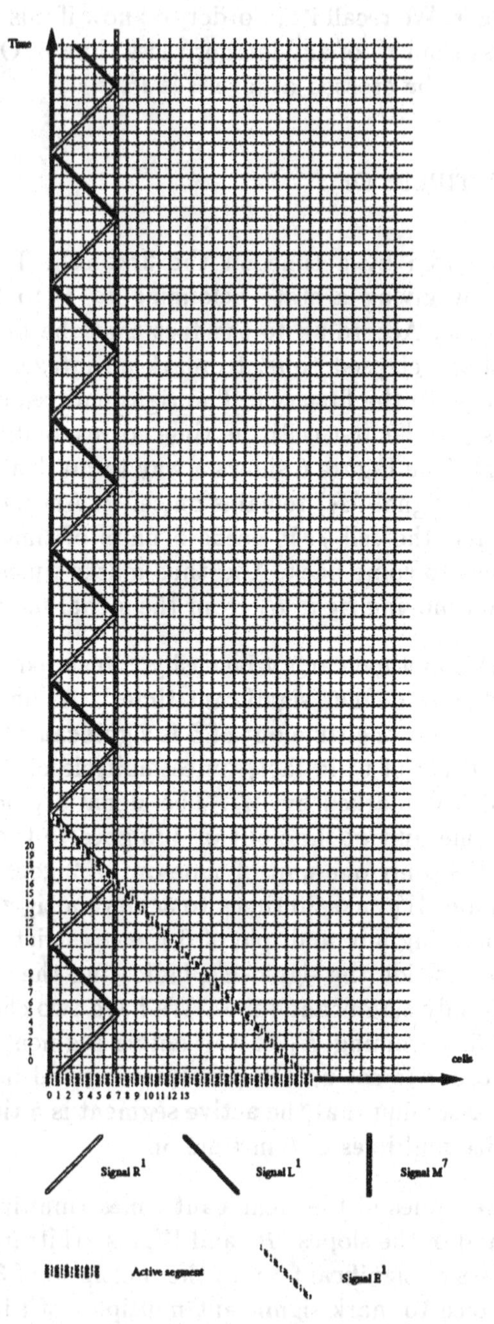

Figure 6. Marking the multiples of 12

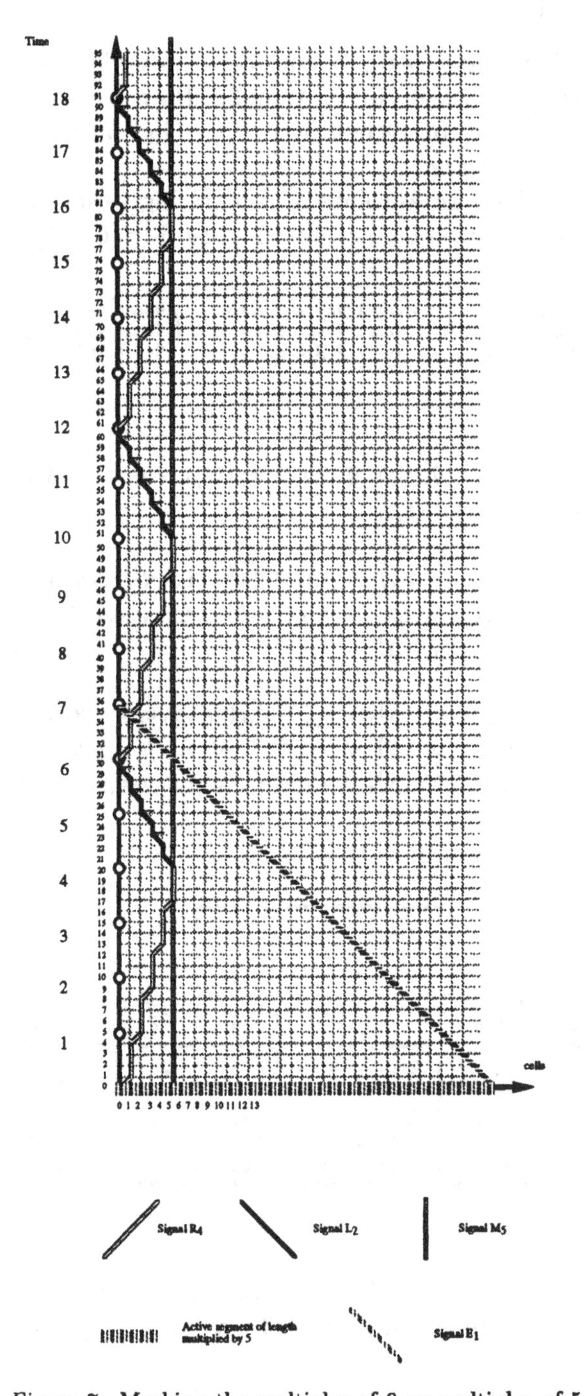

Figure 7. Marking the multiples of 6 on multiples of 5

3.2. GROUPING TIMES

Following Fischer's method, we only consider times which are multiples of 3. Thus the recognition will occurs in time 3n. How can we deduce of such an automaton an automaton recognizing primes in real time? The trick is to "group" times (and states).

Let $\mathbf{A} = (Q_{\mathbf{A}}, \delta_{\mathbf{A}})$ be an automaton recognizing primes in time $3n$, as indicated in 3.1, we construct a new automaton $\mathbb{B} = (Q_{\mathbb{B}}, \delta_{\mathbb{B}})$ by:

- a - $Q_{\mathbb{B}} = Q_{\mathbf{A}}^9$
- b - $\delta_{\mathbb{B}}\left(\begin{pmatrix} q_{\ell,3,1} & q_{\ell,3,2} & q_{\ell,3,3} \\ q_{\ell,2,1} & q_{\ell,2,2} & q_{\ell,2,3} \\ q_{\ell,1,1} & q_{\ell,1,2} & q_{\ell,1,3} \end{pmatrix}, \begin{pmatrix} q_{c,3,1} & q_{c,3,2} & q_{c,3,3} \\ q_{c,2,1} & q_{c,2,2} & q_{c,2,3} \\ q_{c,1,1} & q_{c,1,2} & q_{c,1,3} \end{pmatrix}, \begin{pmatrix} q_{r,3,1} & q_{r,3,2} & q_{r,3,3} \\ q_{r,2,1} & q_{r,2,2} & q_{r,2,3} \\ q_{r,1,1} & q_{r,1,2} & q_{r,1,3} \end{pmatrix}\right)$

$= \begin{pmatrix} q_{c,6,1} & q_{c,6,2} & q_{c,6,3} \\ q_{c,5,1} & q_{c,5,2} & q_{c,5,3} \\ q_{c,4,1} & q_{c,4,2} & q_{c,4,3} \end{pmatrix}$

where:

$q_{c,6,1} = \delta_{\mathbf{A}}(q_{\ell,5,3}, q_{c,5,1}, q_{c,5,2})$, $q_{c,6,2} = \delta_{\mathbf{A}}(q_{c,5,1}, q_{c,5,2}, q_{c,5,3})$,
$q_{c,6,3} = \delta_{\mathbf{A}}(q_{c,5,2}, q_{c,5,3}, q_{r,5,1})$

with:

$q_{\ell,5,3} = \delta_{\mathbf{A}}(q_{\ell,4,2}, q_{\ell,4,3}, q_{c,4,1})$, $q_{c,5,1} = \delta_{\mathbf{A}}(q_{\ell,4,3}, q_{c,4,1}, q_{c,4,2})$,
$q_{c,5,2} = \delta_{\mathbf{A}}(q_{c,4,1}, q_{c,4,2}, q_{c,4,3})$, $q_{c,5,3} = \delta_{\mathbf{A}}(q_{c,4,2}, q_{c,4,3}, q_{r,4,1})$,
$q_{r,5,1} = \delta_{\mathbf{A}}(q_{c,4,3}, q_{r,4,1}, q_{r,4,2})$

and, finally:

$q_{\ell,4,2} = \delta_{\mathbf{A}}(q_{\ell,3,1}, q_{\ell,3,2}, q_{\ell,3,3})$, $q_{\ell,4,3} = \delta_{\mathbf{A}}(q_{\ell,3,2}, q_{\ell,3,3}, q_{c,3,1})$,
$q_{c,4,1} = \delta_{\mathbf{A}}(q_{\ell,3,3}, q_{c,3,1}, q_{c,3,2})$, $q_{c,4,2} = \delta_{\mathbf{A}}(q_{c,3,1}, q_{c,3,2}, q_{c,3,3})$,
$q_{c,4,3} = \delta_{\mathbf{A}}(q_{c,3,2}, q_{c,3,3}, q_{r,3,1})$, $q_{r,4,1} = \delta_{\mathbf{A}}(q_{c,3,3}, q_{r,3,1}, q_{r,3,2})$,
$q_{r,4,2} = \delta_{\mathbf{A}}(q_{r,3,1}, q_{r,3,2}, q_{r,3,3})$

The function $\delta_{\mathbb{B}}$ is illustrated on the Figure 8.

			$q_{c,6,1}$	$q_{c,6,2}$	$q_{c,6,3}$			
		$q_{l,5,3}$	$q_{c,5,1}$	$q_{c,5,2}$	$q_{l,5,3}$	$q_{r,5,1}$		
	$q_{l,4,2}$	$q_{l,4,3}$	$q_{c,4,1}$	$q_{c,4,2}$	$q_{c,4,3}$	$q_{r,4,1}$	$q_{r,4,2}$	
$q_{l,3,1}$	$q_{l,3,2}$	$q_{l,3,3}$	$q_{c,3,1}$	$q_{c,3,2}$	$q_{c,3,3}$	$q_{r,3,1}$	$q_{r,3,2}$	$q_{r,3,3}$
$q_{l,2,1}$	$q_{l,2,2}$	$q_{l,2,3}$	$q_{c2,1}$	$q_{c,2,2}$	$q_{c,2,3}$	$q_{r,2,1}$	$q_{r,2,2}$	$q_{r,2,3}$
$q_{l,1,1}$	$q_{l,1,2}$	$q_{l,1,3}$	$q_{c,1,1}$	$q_{c,1,2}$	$q_{c,1,3}$	$q_{r,1,1}$	$q_{r,1,2}$	$q_{r,1,3}$

Figure 8. The function $\delta_{\mathbb{B}}$

Now, let us consider the evolution of automaton $\mathbf{A}$ on an initial valid configuration. We denote the state of cell i at time t by $q_{i,t}$. It is easy to observe that:

- $\forall i \in \{2, \ldots, n-3\}$, states $q_{i,1}$ are identical to $q_1 = \delta_{\mathbf{A}}(A, A, A)$.
- $\forall i \in \{3, \ldots, n-4\}$, states $q_{i,2}$ are identical to $q_2 = \delta_{\mathbf{A}}(q_1, q_1, q_1)$.

Now we may achieve the description of $\mathbb{B}$.

- c - $A_{\mathbb{B}} = \begin{pmatrix} q_2 & q_2 & q_2 \\ q_1 & q_1 & q_1 \\ A_{\mathbb{A}} & A_{\mathbb{A}} & A_{\mathbb{A}} \end{pmatrix}$, $!_{\mathbb{B}} = \begin{pmatrix} !_{\mathbb{A}} & !_{\mathbb{A}} & !_{\mathbb{A}} \\ !_{\mathbb{A}} & !_{\mathbb{A}} & !_{\mathbb{A}} \\ !_{\mathbb{A}} & !_{\mathbb{A}} & !_{\mathbb{A}} \end{pmatrix}$
- d - $G_{\mathbb{B}} = \begin{pmatrix} q_{0,2} & q_{1,2} & q_{2,2} \\ q_{0,1} & q_{1,1} & q_1 \\ G_{\mathbb{A}} & A_{\mathbb{A}} & A_{\mathbb{A}} \end{pmatrix}$, $D_{\mathbb{B}} = \begin{pmatrix} q_{n-3,2} & q_{n-2,2} & q_{n-1,2} \\ q_1 & q_{n-2,1} & q_{n-1,1} \\ A_{\mathbb{A}} & A_{\mathbb{A}} & D_{\mathbb{A}} \end{pmatrix}$
- e - $\begin{pmatrix} q_a & q_b & q_c \\ q_d & q_e & q_f \\ q_g & q_h & q_i \end{pmatrix} \in F_{\mathbb{B}}$ if and only if one of the states q_a, q_b, q_c, q_d, q_e, q_f, q_g, q_h, q_i is in $F_{\mathbb{A}}$.

It is easy to show that, evolving on the configuration

$$C(n)_{\mathbb{B}} = {}^{\omega}!_{\mathbb{B}}\ G_{\mathbb{B}}\ \overbrace{A_{\mathbb{B}} \ldots A_{\mathbb{B}}}^{n-2 \text{ times}}\ D_{\mathbb{B}}\ !_{\mathbb{B}}^{\omega}$$

the automaton $\mathbb{B}$ enters, at time n, a state of $F_{\mathbb{B}}$, if and only if the automaton $\mathbb{A}$ enters at time n a state of $F_{\mathbb{A}}$ starting from the configuration

$$C(n)_{\mathbb{A}} = {}^{\omega}!_{\mathbb{A}}\ G_{\mathbb{A}}\ \overbrace{A_{\mathbb{A}} \ldots A_{\mathbb{A}}}^{n-2 \text{ times}}\ D_{\mathbb{A}}\ !_{\mathbb{A}}^{\omega}.$$

3.3. AVOIDING AN INFINITE NUMBER OF STATES

Now, we construct the automaton $\mathbb{A}$ with the Fischer's values. If it is easy to cross all multiples of a fixed integer, crossing all multiples of all necessary integers is more difficult.
The first attempt (illustrated on the Figure 9) is to set up simultaneously all these processes. In this case some integer may be marked an arbitrary number of times. For example, we have $6 = 2 \times 3$. Thus, on cell 0, at significant time 6 (namely time 30 for $\mathbb{A}$), two signals R_2 are sent: one will be reflected on cell 2, the second on cell 3 and none will reach cell 4. To distinguish between these two signals, we aim to introduce 2 states, corresponding to the cells on which the reflection occurs. But, as there exist integers with an arbitrary large number of dividers, this leads us to introduce an infinite number of states which is forbidden by the finite nature of a cell.

We may see the marking process at level j (to mark multiples of j) as a cellular process occurring on some band of width j (introducing a new signal G indicating the first cell and the left border of the band).
Putting the band at another place of the space time diagram (its origin on cell $f(j)$ at time $-g(j)$ will mark all times $(j \times k) - g(j)$ $(k \in \mathbf{N})$ on cell $f(j)$. If we introduce a new signal D (as "delete"), running from cell $f(j)$ to the first cell at maximal speed, we may mark on cell 0 all times $(j \times k) - g(j) + f(j)$ $(k \in \mathbf{N})$. Obviously, we take $f = g$ in order to mark all multiples of j on cell 0. By this way, we translate the useful area of the

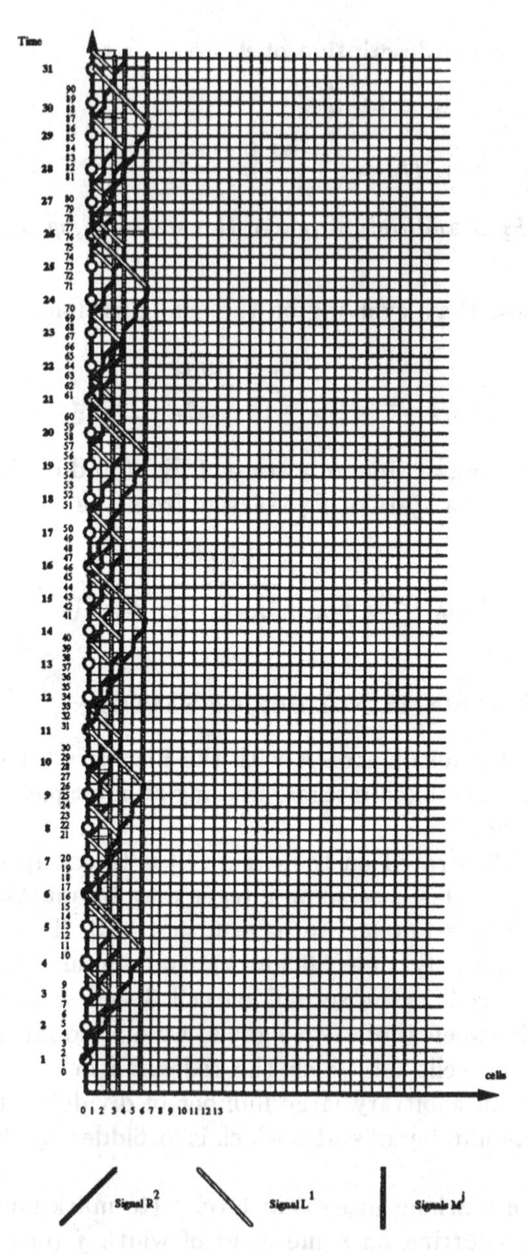

Figure 9. Crossing simultaneously all multiples of all integers

space-time diagram (the band of width j) in space (by $f(j)$) and also in time (by $-f(j)$).

We observe that, if a cell $f(j)$ must send a signal D corresponding to the j-th process when it receives a signal D created by some k-th process

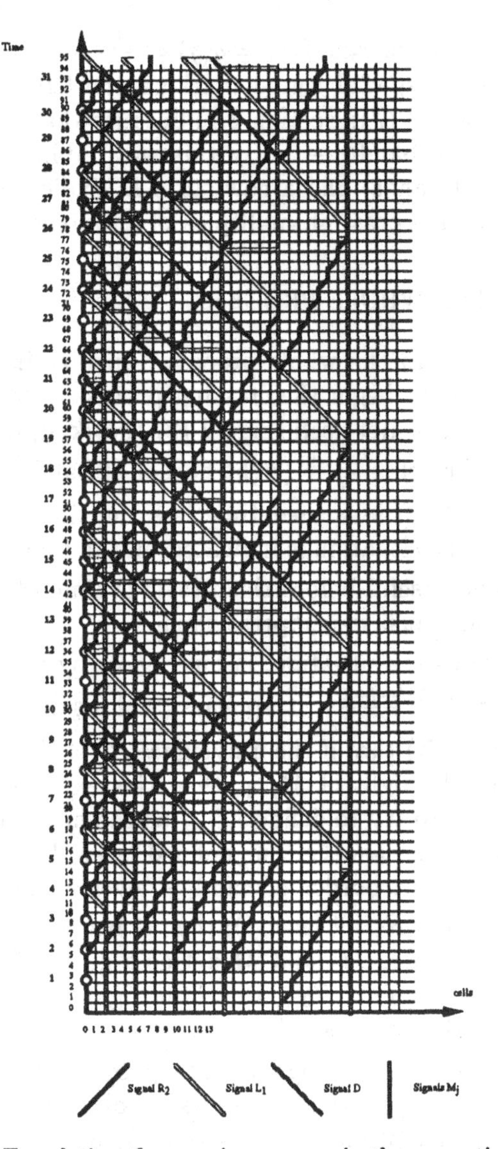

Figure 10. Translating the crossing process in the space-time diagram

$(k > j)$, it sends only one signal D (indicating to the first cell to mark the time of reception). It is not needed to distinguish them because it is sufficient to mark once.

The only remaining question is: How to place the working bands in the space-time diagram? We put them as soon as possible as shown on the Figure 10. We put the first band (level 2) from cell 0 to 2, the second one (level 3) from cell 2 to 5, and so on (level j from cell $2+3+\ldots+(j-1)) =$

$\frac{(j-1)j}{2}-1$ to $\frac{j^2-j-2}{2}$. We obtain $f(j)=g(j)=\frac{j^2-j-2}{2}$. Finally, we observe that we do not need a special signal marking the left border of a band (playing the role of signal G in section 3.1) because the left border of band j is the right border of band $j-1$. By this way, adding a new signal M (indicating borders of bands) and a new state D (indicating to mark time on the first cell), we may cross all multiples of all integers.

3.4. SETTING UP THE WHOLE PROCESS

In this section, we set up the whole process. First, as in the classical algorithm, it is sufficient to begin the crossing process at level j for integers greater than j^2. Thus the first useful signal W_1 (marking significant time j^2 on cell 0) starts on the cell $\frac{j^2-j-2}{2}$ (right border of band of level j) at time $3j^2-\frac{j^2-j-2}{2}=\frac{5j^2+j+2}{2}$ (the wished time less the time needed for W_1 to reach the first cell). The situation is indicated on the Figure 11. In addition the signals M_j are not useful before this previous time $\frac{5j^2+j+2}{2}$ (nothing is reflected or sent) and, thus, we suppress them.

The Figure 12 shows the situation. Let us denote by S_j the point of $\mathbb{Z}\times\mathbb{N}$ of coordinates $\left(\frac{j^2-j-2}{2},\frac{5j^2+j+2}{2}\right)$. It appears (and a simple computation or Thales' theorem proofs it) that if the first signal R_2 of the $(j-1)^{th}$ working band is extended to the right of M_{j-1}, then this extended signal reaches S_j. This fact allows us to try to define M_{j+1} as the meeting of two signals: the extended signal S_j and another one defined later. To extend S_j, we use 2 states marking the right border of band of level j: the first $M^\star$ marks cell $\frac{j^2-j-2}{2}$ until it meets R_2 (of level j); then it becomes M stopping new signals R_2. Thus, the states $M^\star$ and M act as a gate: $M^\star$ corresponds to the open state (the first signal R_2 goes to the right) and then the gate is clesed in state M (and later signal R_2 are only reflected).

It remains to set up another signal defining S_j. A simple computation shows that $S_j=S_{j-1}+(j,5j+2)$. A signal Π of slope 5 sent from S_{j-1} with a good initialization (it stays two time on S_j before to go to the right with slope 5) will reach S_j. We observe that points S_j occur on a discrete parabola (trace of signal Π), called the Fischer's parabola. The Figure 13 shows the whole process. The only thing it remains to do is to initialize the whole process. This is done on cells 1 and 2 at time 3, using new states.

We do not evaluate the needed number of states of $\mathbf{A}$ which is more important that it may seem: all possibilities of crossing must be taken into account. States of $\mathbb{B}$ are triplets of states of automaton $\mathbf{A}$ because:

- if a cell knows its own states $(q_{c,1,1},q_{c,1,2},q_{c,1,3})$ and the states $(q_{\ell,1,1},q_{\ell,1,2},q_{\ell,1,3})$ and $(q_{r,1,1},q_{r,1,2},q_{r,1,3})$ of its neigh-

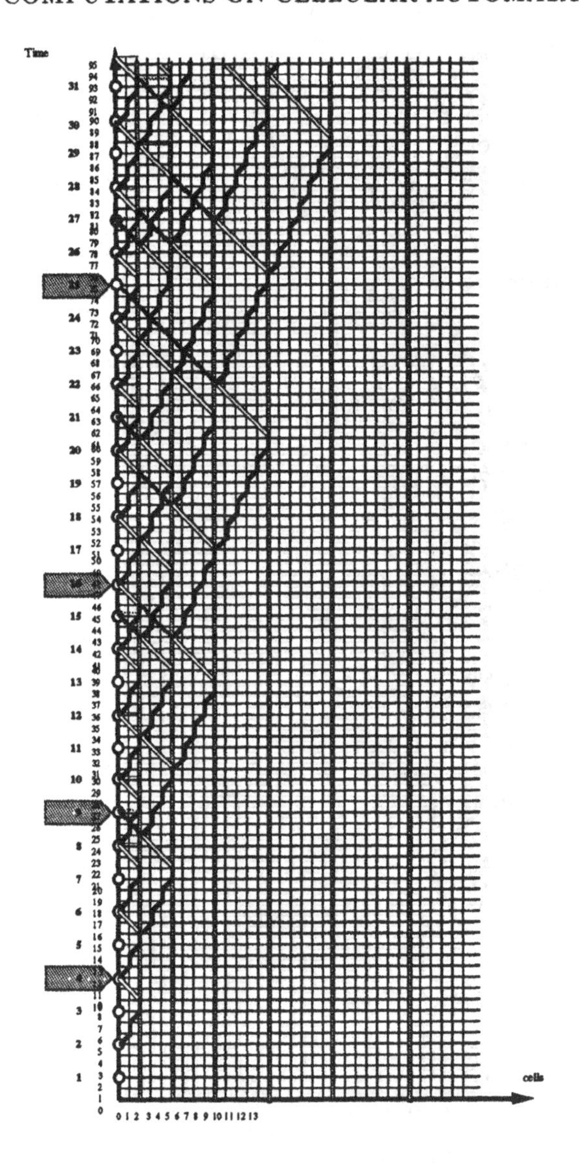

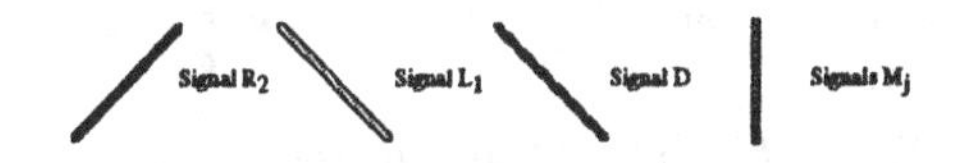

Figure 11. Initialization of working bands

bors,

it can deduce $\begin{pmatrix} q_{c,3,1} & q_{c,3,2} & q_{c,3,3} \\ q_{c,2,1} & q_{c,2,2} & q_{c,2,3} \\ q_{c,1,1} & q_{c,1,2} & q_{c,1,3} \end{pmatrix}$

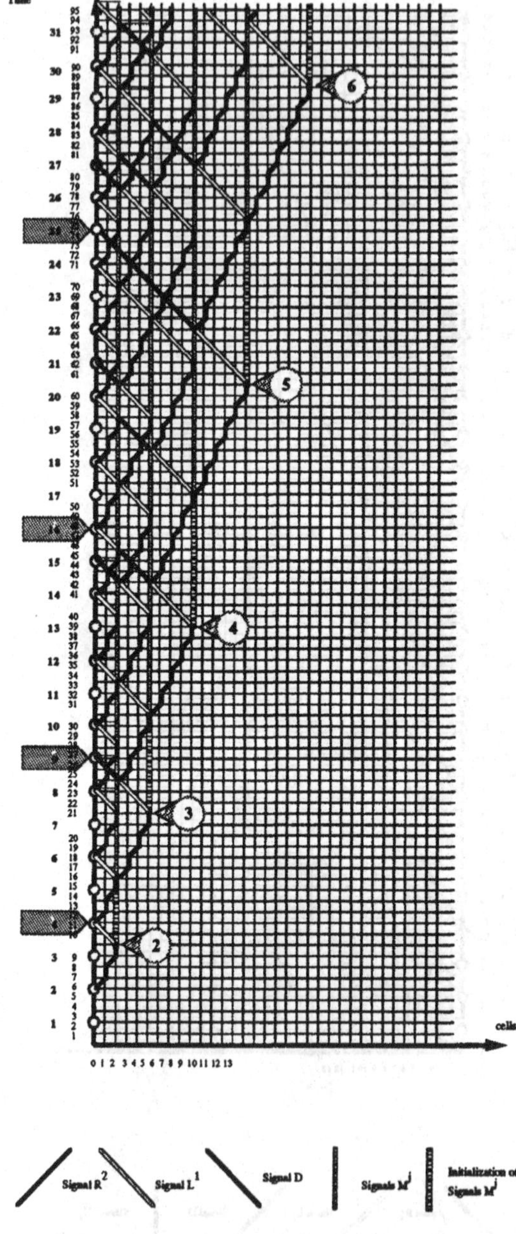

Figure 12. Introducing the signal $M_j^\star$

4. Firing Squad

Looking to the space-time diagram constructed in the section 3, we have a complex behavior but all involved signals are linear with a slope of 1,

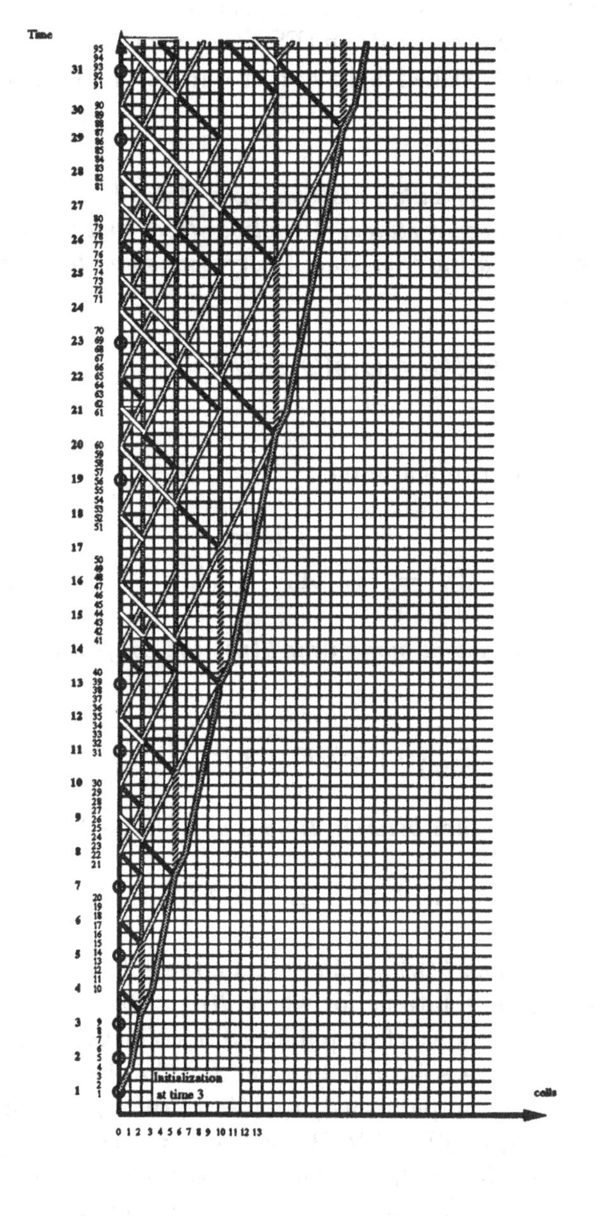

Figure 13. Initialization of the whole process

2 or 5. Is it possible to set up signals of arbitrary large slope in a space-time diagram? The answer is "yes" and the generic example is the so-called *Firing Squad Synchronization Problem* (FSSP for short).

First we indicate the wished behavior of our automaton, solving the FSSP:

Problem 3

Design a 1D-CA $((Q,\delta))$ *with 2 distinguished states F (as "fire") and G (as "general") and 2 quiescent states, A (active state) and ! (outside state) with the transitions indicating that state ! corresponds as the inactive part of the configuration*

$$\forall Q_1, Q_2 \in Q, \delta(Q_1, !, Q_2) = !$$

such that $(\forall n \in \mathbb{N})$*:*
Starting from any initial configuration

$$C_0(n) = {}^{\omega}!\ G\ \overbrace{A \ldots A}^{n-1 \text{ times}}\ !^{\omega},$$

the system evolves such that at some time $\theta(n)$*, the configuration is*

$$C_{\theta(n)}(n) = {}^{\omega}!\overbrace{F \ldots F}^{n \text{ times}}!^{\omega}$$

and the state F never appears between time 0 *and* $\theta(n) - 1$*.*

$$\forall t \in \{0, \ldots, \theta(n) - 1\}, \quad F \notin C_t(n).$$

It is easy to show that $\theta(n) \geq 2n - 2$. In the following, we restrict ourselves to minimal time solutions for which $\theta(n) = 2n - 2$.

Historically, this problem was first published in (Minsky, 1967). M. Minsky and M. McCarthy give solutions in time $3n$. The first minimal time solution is due to E. Goto (unpublished, see (Goto, 1962)). Simultaneously, A. Waksman (Waksman, 1966) and R. Balzer (Balzer, 1967) have published minimal time solutions. The Balzer's solution has 8 states. More recently, J. Mazoyer (Mazoyer, 1987) has published a 6 states minimal time solution. Since Balzer's studies (Balzer, 1966), and very recently Yunes' work (Yunes, 1993), it is known that there does not exist solution with 4 states, so the remaining problem if the existence of a 5 states solution. Here, we describe quickly the 6 states solution of (Mazoyer, 1987) in order to show how a family of signals of slope $\left(\frac{3}{2}\right)^i$ is introduced.

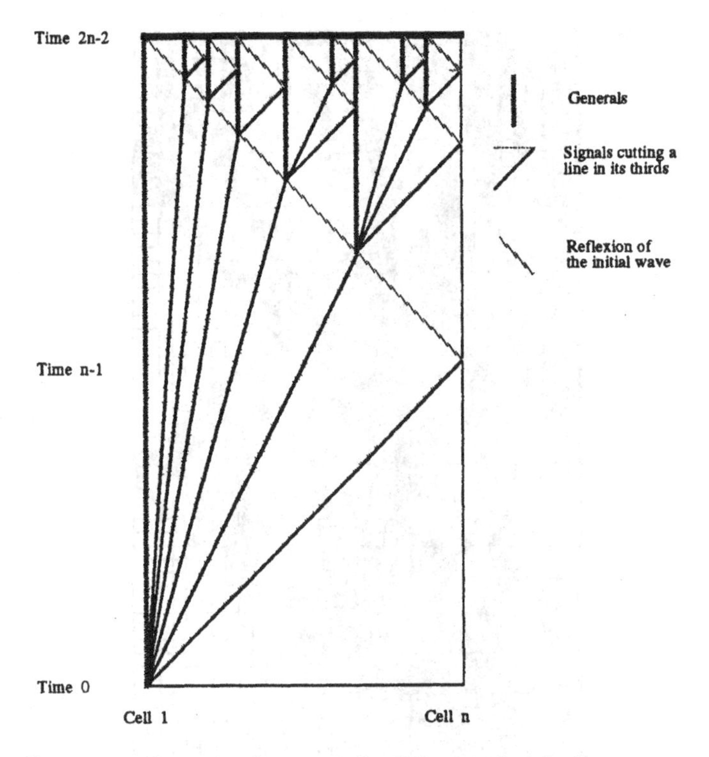

Figure 14. A geometric approach of the synchronization process.

4.1. HOW TO SYNCHRONIZE

The main idea to synchronize is given on Figure 14 using a geometric diagram defined in section 2 and already displayed in Figure 5. Forgetting the discrete nature of the segment $[1, n]$, we cut it at points $\frac{p}{q}n$, $\left(\frac{p}{q}\right)^2 n$, $\left(\frac{p}{q}\right)^3 n$, ..., $\left(\frac{p}{q}\right)^i n$, ... By this way, we obtain new segments of length $\frac{q-p}{q}n$, $\frac{q-p}{q}\frac{p}{q}n$, ..., $\frac{q-p}{q}\left(\frac{p}{q}\right)^{i-1} n$, ... We choose that the ith segment will be synchronized from its leftmost cell G_i: thus it consists of cells between G_i and $G_{i-1} - 1$ included (we define G_0 as $n+1$). The ith segment needs $2(\left(\frac{q}{q}\right)^i - \left(\frac{p}{q}\right)^{i-1})n - 2$ units of time to be synchronized. Its synchronization begins as soon as possible:

- The length of the initial segment must be known. This occurs (roughly speaking) at time n.
- This information must come back to the cell G_i. The amount of needed time is $2n - \frac{p^i}{q}n$.

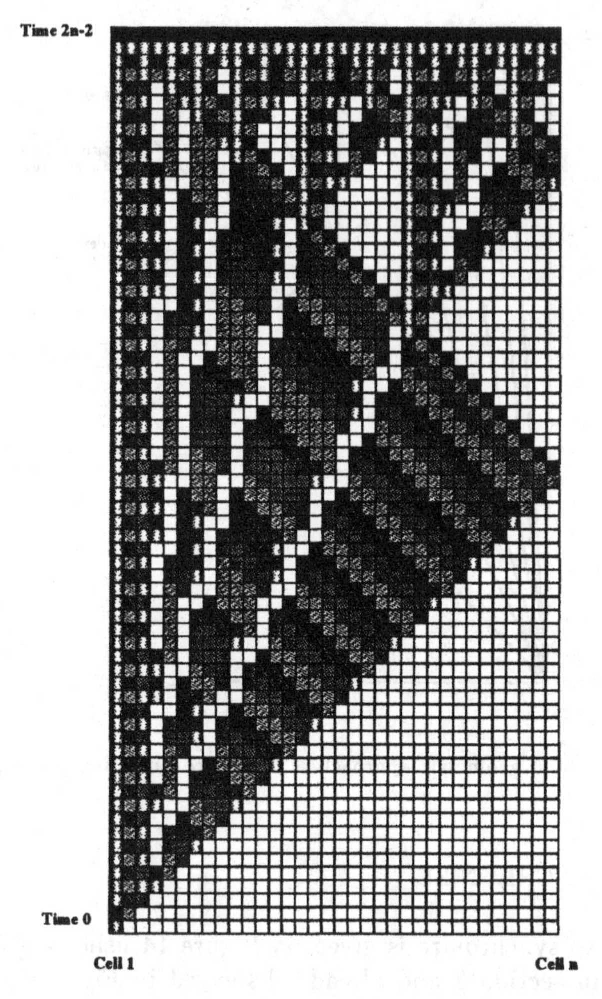

Figure 15. **Synchronization of a segment of 34 cells with a minimal time solution to the FSSP with 6 states.**

Thus the synchronization of the ith segment may be obtained at time:

$$2n - \left(\frac{p}{q}\right)^{i} n + 2\left(\left(\frac{q}{q}\right)^{i} - \left(\frac{p}{q}\right)^{i-1}\right)n - 2$$

and this time will be $2n - 2$. Thus, we get:

$$\left(\frac{p}{q}\right)^{i} n = 2\left(\left(\frac{q}{q}\right)^{i} - \left(\frac{p}{q}\right)^{i-1}\right)n$$

which may be written $2q - 3p = 0$ and is obtained for $p = 2$ and $q = 3$. In the following, we will choose these values.

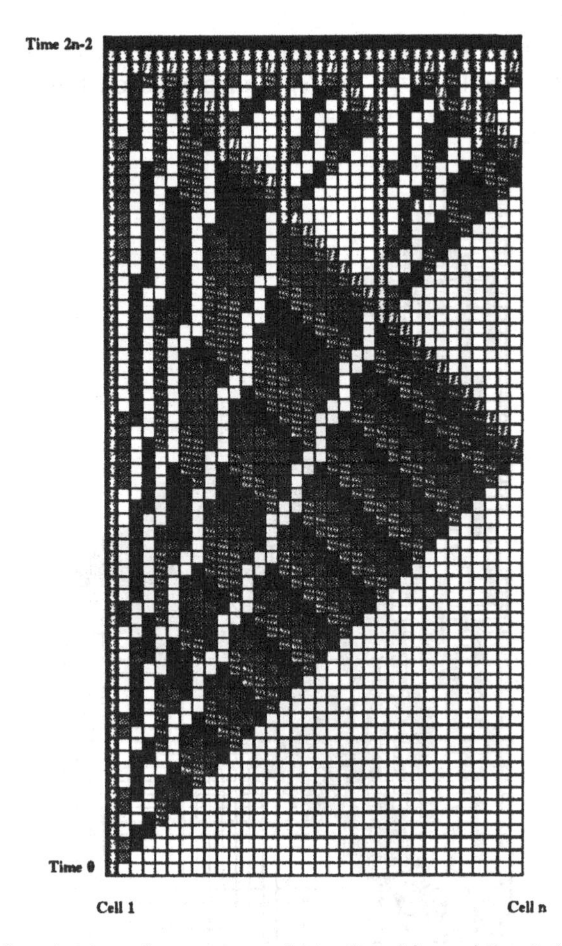

Figure 16. Synchronization of a segment of 34 cells with a minimal time solution to the FSSP with 8 states.

Due to the discrete nature of the line, the cell $\left(\frac{p}{q}\right)^i n$ will be chosen to be $\lfloor\left(\frac{p}{q}\right)^i n\rfloor + \alpha$ with $\alpha \in \{1, 2, 3\}$ and this choice will induce to begin the synchronization of new "small" segments with a delay of 1, 2 or 3 units of time. We do not describe this part and focus ourselves on the way to obtain points G_i. The idea is "to send" from cell 0 at time 0 a family of signals of slope $\left(\frac{p}{q}\right)^i$. Unfortunately, this is impossible. Thus, we choose that the signal W_i of slope $\left(\frac{p}{q}\right)^i$ is initialized on the cell 1 at the wished time.

The Figure 15 shows an example of synchronization with a 6 states solutions. In fact, this solution is obtained with some technical encodings. Thus we shall describe the construction of the family $(W_i \mid i \in \mathbf{N})$ from the same solution without encodings. Now we need 8 states and an example of synchronization may be found on Figure 16. The area concerned by the construction of the family $(W_i \mid i \in \mathbb{N})$ is displayed on Figure 17.

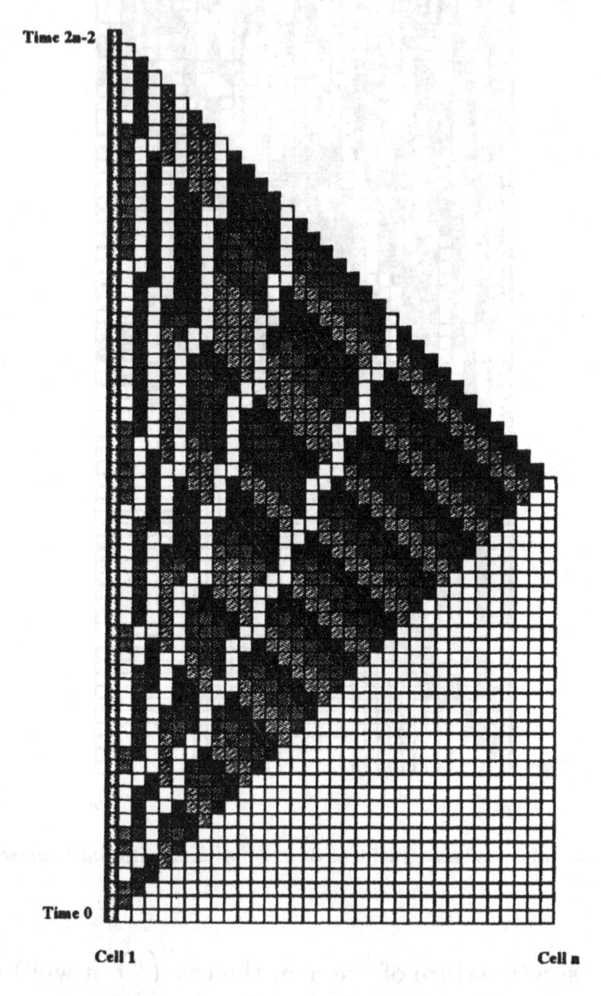

Figure 17. Construction of the family $(W_i \mid i \in \mathbb{N})$ with 6 states

4.2. CONSTRUCTION OF THE FAMILY $(W_I \mid I \in \mathbb{N})$

In order to simplify the construction, we first deal with the automaton depicted on Figure 18 which needs one more state.

We describe this process pointing out a succession of facts:

- a - As previously indicated, the fact that the leftmost cell begins the synchronization process, must be sent at maximal speed to the right. By definition of the active state A which is quiescent (see problem 3), this is done putting all sites (i, i) in a state different from A. By this

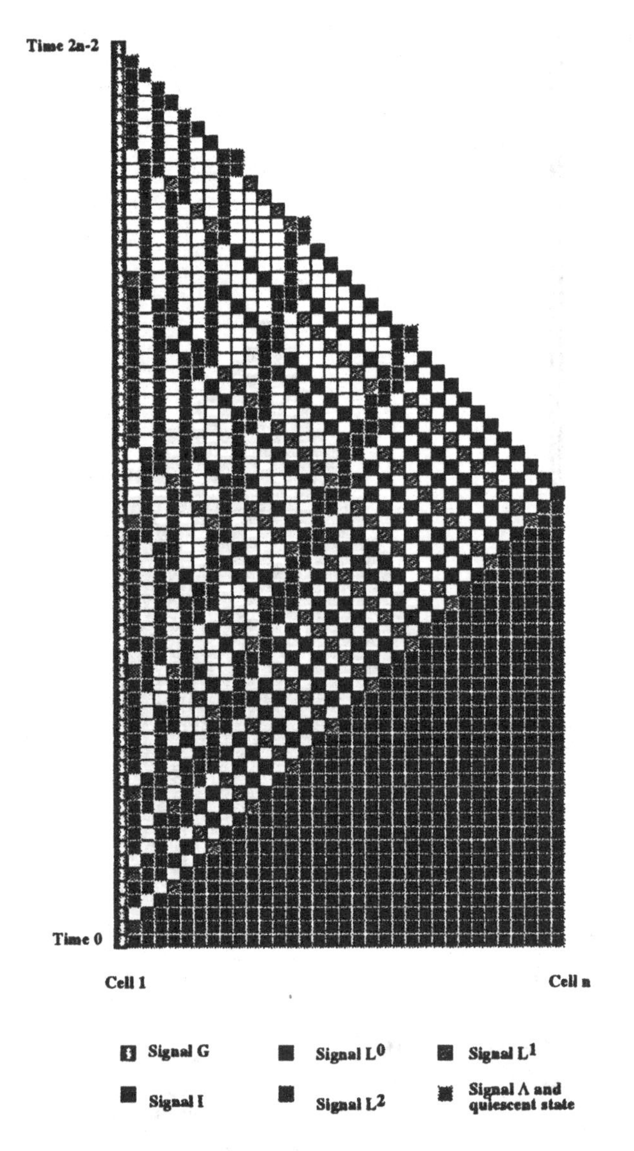

Figure 18. Construction of the family $(W_i \mid i \in \mathbb{N})$ with 7 states

way, we obtain a signal of slope 1 which is W_0. We observe that, when a cell leaves for the first time state A, it knows that it enters signal W_0.

- b - How to set up the signal W_1?

To set up a signal of slope $\frac{2}{3}$ has been indicated in section 2. Two choices are possible; using 3 states to represent the signal (Figure 4, part a)), using 4 states, one indicating the segment and the 3 others

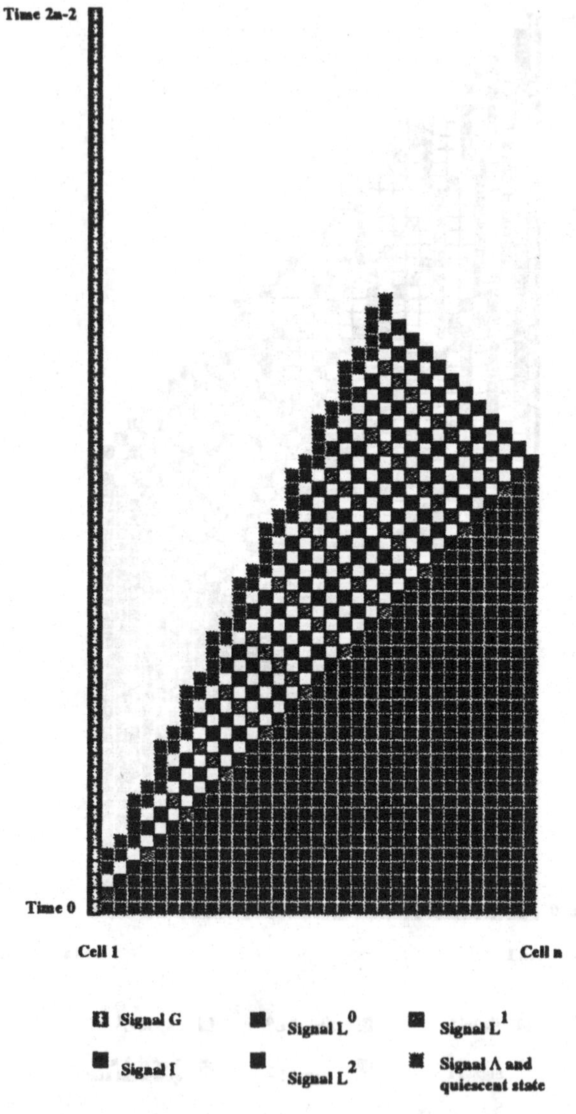

Figure 19. Construction of W_1

ones indicating moves of the signal (Figure 4, part b)). Here we cannot use a state S_i to represent each signal W_i; thus, we will use 4 states to set up W_1 according to Figure 4, part b). If we do not use state A to set up moves of signals W_i, we are allowed to use state A to indicate the position of W_i ($i \geq 1$). We make this choice and denote these 3 needed signals by L_0, L_1 and L_2. According to the nature of the FSSP, we choose to begin signals W_i on the second cell (the first cell must

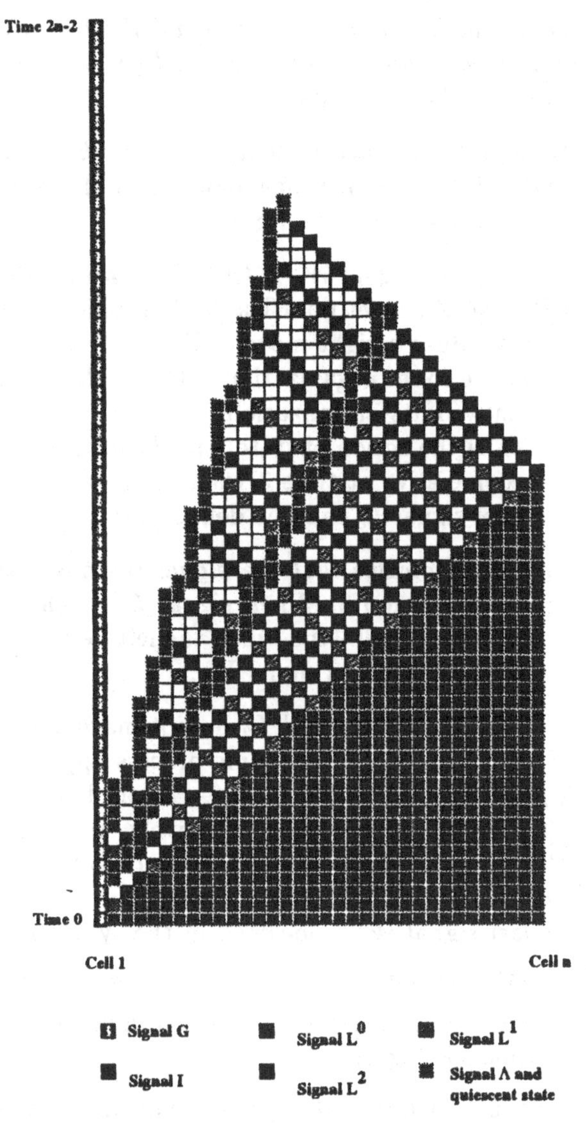

Figure 20. Construction of W_1 and W_2

remember that it is the left end of the segment), and, in addition, we mark cell 2 by some state I. The obtained process is shown on Figure 19. Let us indicate the meaning of the signals L_0, L_1 and L_2 in "human" words:

Signal L_0 *The signal* L_0 *indicates that signal* W_0 *has reached a cell whose number is* 0 mod 3. When signal L_0 reaches signal W_1, W_1 will stay on the same cell.

Signal L_1 *The signal* L_1 *indicates that signal* W_0 *has reached a cell whose number is* 1 mod 3. When signal L_1 will reach signal W_1, W_1 will move to the cell at its right.

Signal L_2 *The signal* L_2 *indicates that signal* W_0 *has reached a cell whose number is* 2 mod 3. When signal L_2 will reach signal W_1, W_1 will move to the cell at its right.

By this way, the slope of signal W_1 is the third two thirds $(\frac{2}{3})$ of the slope of signal W_0 and W_1 is set up. Clearly, it is sufficient that exactly 2 of these 3 signals induce a right move of W_1.
Our choice of L_0 is due to the organization of the whole synchronization process and especially of its initialization.
By a "good" choice of the action of state I (*indicating the second cell*), signal W_1 may become alive at time 3 when L_0 is on the second cell. Thus the behavior of signal L_0 may be defined as:

Signal L_0 *Signal* L_0 *indicates that the signal* W_0 *has reached a cell whose number is* 0 mod 3. When signal L_0 reaches signal W_1, W_1 stays on the same cell. And when the left neighbor is in state G (the first cell), L_0 creates signal W_1.

- c - The behaviors of signals L_0, L_1 and L_2 do not make any assumption on the slope of signal W_0. Thus, the same process applied on a signal of slope $\frac{\mu}{\nu}$ will give a new signal of slope $\frac{2\mu}{3\nu}$. Applied to signal W_1, we get signal W_2 (see Figure 20). Applied iteratively, we get for an initial configuration $^{\omega}!A^{\omega}$ all the family $(W_i \mid i \in \mathbb{N})$ as depicted on Figure 18.
- d - It remains to see when W_i is created. According to the FSSP point of view of 4.1, the last signal W_{i_0} appearing in the synchronization of a line of length n is the last integer such that: $\left(\frac{2}{3}\right)^{i_0} n \geq 2$ and it appears at time $2n - 3$. We obtain $4\lfloor\frac{3}{2}\rfloor^{i} - 3$. Thus the number of signals W_i appearing until time t is logarithmic in t.

In Figure 18, a not used white state appears, indicating that no signal pass through this site. To avoid this useless state, we change the meaning of states indicating L_0, L_1 and L_2 by: The last signal received is "L_0, L_1 or L_2". By this way, we obtain only the 6 states of Figure 17.

We observe that, if in section 3 we have obtained non linear signals by use of a finite number of basic linear ones, now we obtain a family of linear signals with growing slopes by use of a finite number of basic linear ones. Clearly, we may mix the two processes obtaining a family of non linear signals with growing slopes by use of a finite number of basic linear ones (as, for example, signals of slope $A(n,n)$ where $A(x,y)$ is the Ackermann's function).

5. Virtual filters

In the section 3 and 4, we have given two examples of algorithms using the notion of signal. In fact, when designing an algorithm, we have in mind some ideas (about what are the atoms of information and how they must meet -see 1), but we have some choices on the moves and where occur meetings. We shall illustrate this point on an example: the palindrome recognition. First, we set up our problem and then describe solutions with 3 possibilities of moves.

Problem 4

Design an 1D-CA $((Q, \delta))$ *with a distinguished set of states* $Y \cup N$ *(as "Yes" and "No"), another distinguished set of states* $(0, 1)$ *(the "input alphabet") and a quiescent state ! (outside state) and 4 special states indicating the beginning and the end of the active part* ($0G$, $1G$, $0D$, $1D$, *the active part begins (ends) with* 0 (1)) *with the transitions indicating that state ! corresponds as the inactive part of the configuration*

$$\forall Q_1, Q_2 \in Q, \delta(Q_1, !, Q_2) = !$$

such that ($\forall n \in \mathbb{N}$):

- *i - Starting from any initial configuration*

$$C(n) = ^{\omega}!\ XG\ x_2 \ldots x_{n-1}\ ZD\ !^{\omega}$$

where X *and* Z *are* 0 *or* 1 *and* x_i ($i \in \{2, \ldots, n-1\}$) *are in* $\{0, 1\}$. *We refer to* $x = x_1 \ldots x_n$ *as the "input word"* ($x_1 = X$ *and* $x_n = Z$).
- *ii - the system evolves such that at time* n, *the state of the first cell is in* Y *if and only if* x *is a palindrome and is in* N *else.*

The features of the natural algorithms are the following:

- i - We choose as atoms of information:
  - the values of the x_i ($i \in \{1, \ldots, n\}$),
  - the indication that the leftmost (rightmost) cell may have modified the information of a site,
  - the fact that a cell is the middle cell (if the length n is odd) or that is "between" the 2 cells, called middle cells (if n is even) - we shall define later "between"-,
  - the fact that some successive comparisons have been or not successful.
  - the input word is (not) a palindrome.
- ii - The links between these atoms are:

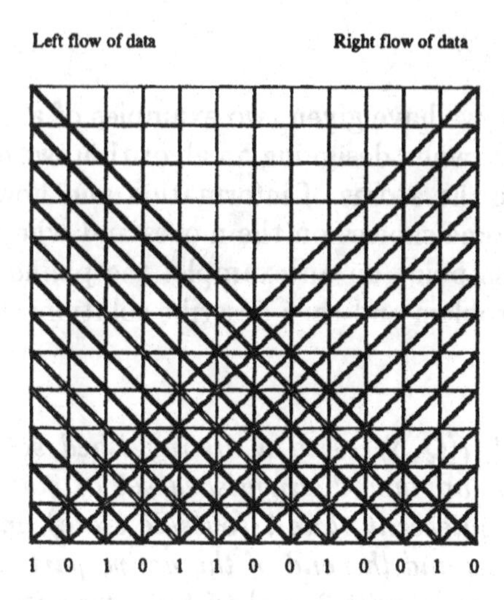

Figure 21. Moves of data: virtual filter 1.

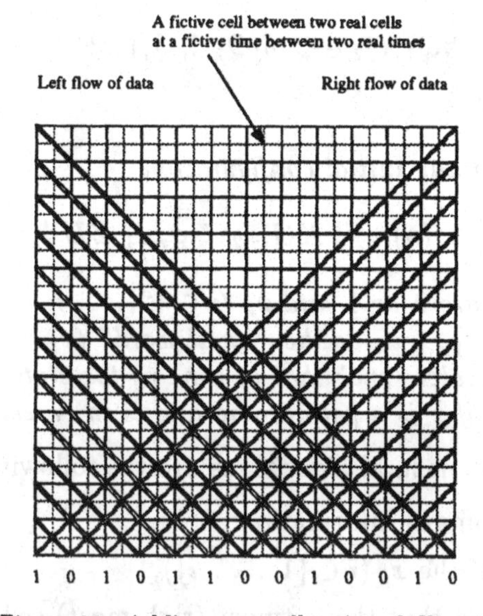

Figure 22. Adding new cells: virtual filter 1.

- The cell (or the fictive cell between two cells) on which the information *I am the first cell* and *I am the last cell* meet themselves for the first time gets the information *I am the middle.*
- When an information x_{a-i} meet an information x_{a+i} a comparison is achieved an its value is added to the results of the comparisons between x_{a-j} and x_{a+j} ($j \in \{1, \ldots, i-1\}$) giving the value of *successive comparisons around a up to i are (or not) successful.*

- when a cell gets *I am the middle*, looking to the information *successive comparisons around me up to now are (or not) successful*, it sends (at maximal speed to the left) the information *Success or failure*.

- iii - The moves of this information may be defined in various manner, but we must have:
The first meeting of values x_k and x_j encountered by *Successive comparisons around a up to i are (or not) successful* must be such that $k = a - i - 1$ and $j = a + i + 1$.

We describe 3 possible ways to set up the indicated moves. The key point is the moves of data (x_j). We call *Virtual filter* (see (Terrier, 1991)) the points where they meet themselves.

Virtual Filter 1 The simpler way is to send all the data both to the right and to the left at maximal speed. By this way, we obtain two flows of information as shown on Figure 21.
Where occurs their meeting? The first point is that it may occur between two cells and between two times. To avoid this difficulty, we use the grouping method of section 3.2. We introduce a new initial configuration:

$$C^\sharp(n) = ^\omega!\, XG\ L\ x_2 \dots Lx_{n-1}L\ ZD\ !^\omega$$

where L is a new state and replace n by $n + (n - 1)$ and wish to solve the problem 4 with these new settings. If we success we obtain an automaton $A^\sharp$, the automaton A obtained from $A^\sharp$ by grouping 2×2 states instead of 3×3, is a solution of our problem 4. We number cells and times of the space-time diagram of $A^\sharp$, using values in $\frac{\mathbb{N}}{2\mathbb{N}}$. We refer to the cell with a not integer number as *fictive cells* (versus "real cells"). We also observe that state L is not significant for our problem and that, **now**, all our palindromes are odd. Figure 22 illustrates this point.
Our virtual filter is now made of points:

$$\{(k, \ell) \mid k + \ell \in \mathbb{N}, k \geq 0, k \leq \ell, n - k \leq \ell\}$$

We renumber these points by $\{(i, j) \mid i, j \in \{1, \dots, n\}\}$ where (i, j) is the meeting point of x_i during its right move and of x_j during its left move. We have $k = \frac{i+j}{2}$ and $\ell = \frac{j-i}{2}$.

Virtual Filter 2 The second way is to send the data to the left at maximal speed, when they meet the first cell (with a left neighbor in state !), this left flow is reflected and goes to the right at maximal speed becoming the reflected flow (see Figure 23). As previously, meetings may occur

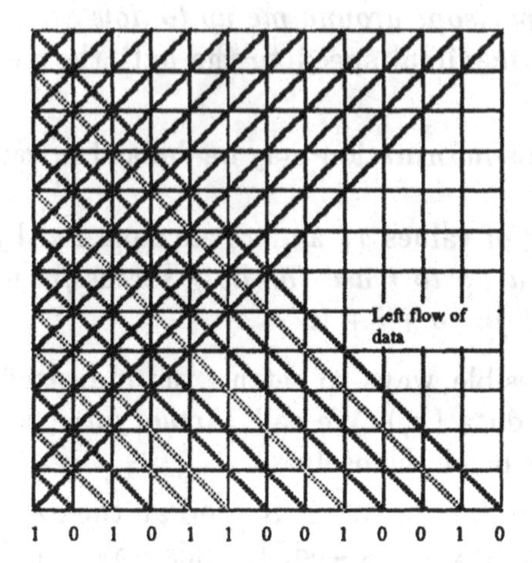

Figure 23. Moves of data: virtual filter 2.

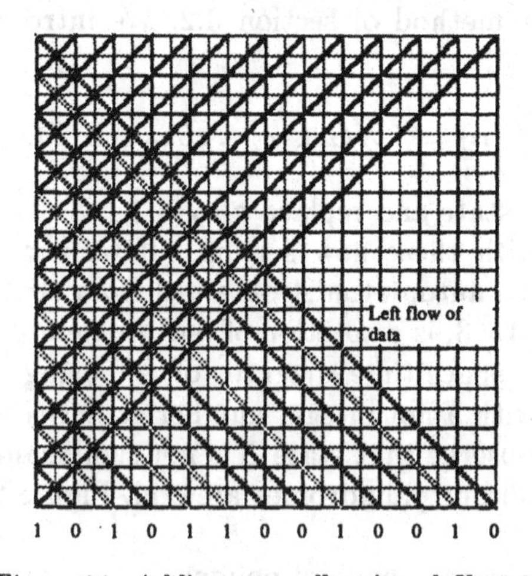

Figure 24. Adding new cells: virtual filter 2.

between cells and we introduce the same automaton as in the preceding case and solve also this question by grouping 2×2 (see Figure 24). Now the intersections between the left flow and the reflected flow are points:

$$\{(k,\ell) \mid k+\ell \in \mathbf{N}, k \geq 0, k \geq \ell, k+\ell \leq 2n\}$$

We also renumber these points by $\{(i,j) \mid i,j \in \{1,\ldots,n\}\}$ where (i,j) is the meeting point of x_i during its left move and of x_j during its reflected move. We have $k = 1 + \frac{i-j}{2}$ and $\ell = (j-1)\frac{i-j}{2}$ with $i \geq j$.

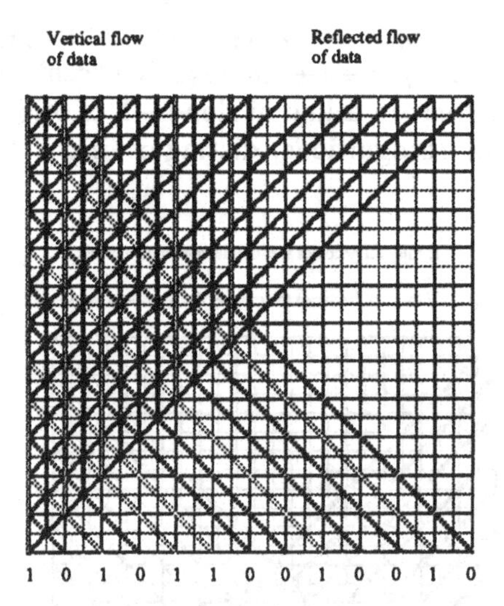

Figure 25. Moves of data: virtual filter 3.

Virtual Filter 3 More complex moves may be set up. For example, when the value x_i of the previous reflected flow meets x_j (on the first diagonal of the active part of the space-time diagram), the value x_j remains also on the meeting cell, creating a third vertical flow. We consider the intersections between the reflected and the vertical flow (see Figure 25). Now the intersections between the reflected flow and the vertical flow are the same as previously:

$$\{(k,\ell) \mid k+\ell \in \mathbf{N}, k \geq 0, k \geq \ell, k+\ell \leq 2n\}$$

But when renumbering these points by $\{(i,j) \mid i,j \in \{1,\ldots,n\}\}$. where (i,j) is the meeting point of x_i during its reflected move and of x_j during its vertical move. We have $k = \frac{i+j}{2}$ and $\ell = 2i$.

Now, we deal with the point iii) of the general features of our algorithms: comparisons must occur around a central letter. With our 3 filters, we obtain:

Virtual Filter 1 Values x_{a-j} and x_{a+j} when j is in $\{1,\ldots,a-1\}$, meet themselves on the line $x = a$. The "central cell" of number a generates comparisons on a vertical line. Figure 26 shows where occur recognitions, distinguishing if the initial input is odd or even.

Virtual Filter 2 Values x_{a-j} and x_{a+j} when j is in $\{1,\ldots,a-1\}$, meet themselves on the line $y = 3a - 2x$. The "central cell" of number a generates comparisons on a line of slope -2 of origin (a,a). Figure 27 shows where occur recognitions, distinguishing if the initial input is odd or even.

Virtual Filter 3 Values x_{a-j} and x_{a+j} when j is in $\{1, \ldots, a-1\}$, meet themselves on the line $y = -x+2(a+1)$. The "central cell" of number a generates comparisons on a line of slope -1 of origin (a, a). Figure 28 shows where occur recognitions, distinguishing if the initial input is odd or even.

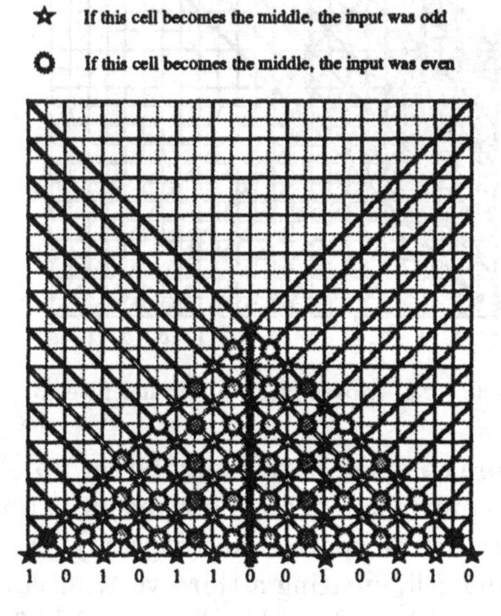

Figure 26. Palindromes recognition on the virtual filter 1.

It remains to define moves of the not studied information in the 3 previous cases.

Virtual Filter 1 We must select as central cell the middle of the line: so the information *I am the middle* will appear on point $(\frac{n+1}{2}, \frac{n+1}{2})$. The available line of comparisons is shown on Figure 26. Thus, *I am the first cell* (*I am the last cell*) is sent at maximal speed to the right (the left) from point $(1, 0)$ $((n, 0))$. The information *I am or not a palindrome* is sent at maximal speed to the left from $(\frac{n+1}{2}, \frac{n+1}{2})$.

Virtual Filter 2 As previously, we must select as central cell the middle of the line: so the information *I am the middle* must appear on point $(1, n)$. It cannot appears before due to the number of comparisons to do and the fact that the information *I am the last cell*, sent at maximal speed to the left, cannot reach the line $y = 3\frac{n}{2} - 2x$ before time n. The available line of comparisons is shown on Figure 27. In this case, *I am the last cell* is sent at maximal speed to the left from $(n, 0)$; and *I am the first cell* stays on cell 1. Information *I am or not a palindrome* is not needed.

★ If this cell becomes the middle, the input was odd

○ If this cell becomes the middle, the input was even

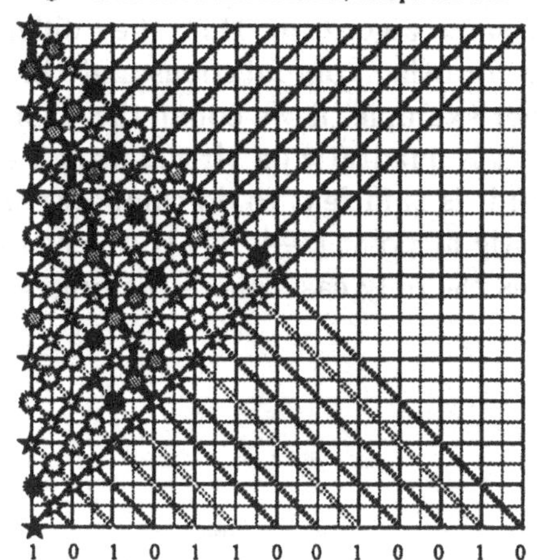

Figure 27. Palindromes recognition on the virtual filter 2.

★ If this cell becomes the middle, the input was odd

○ If this cell becomes the middle, the input was even

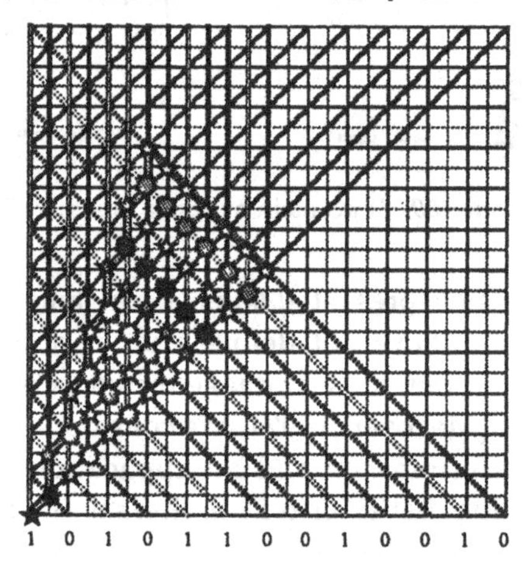

Figure 28. Palindromes recognition on the virtual filter 3.

Virtual Filter 3 Comparisons occurs on antidiagonals and comparisons around the central cell on the line $y = -x + 2(n+1)$ starting from point $(\frac{n+1}{2}, \frac{n+1}{2})$. We sent *I am the last cell* at maximal speed to the left from $(n, 0)$. *I am the first cell* runs to the right at speed $\frac{1}{2}$ from $(1, 0)$ (see Figure 28). Their meeting on $(\lfloor\frac{n}{4}\rfloor, \lceil\frac{3n}{4}\rceil)$ generates *I am the middle*. Information *I am or not a palindrome* runs at maximal speed to the left from this last point.

Thus, the same basic algorithmic ideas may conduct to 3 different cellular automata. Are they really different? We look at them as 3 different implementation of the same algorithm. But, they differ by the remaining area that may be used for other computations and their place around the chosen virtual filter. For example, if we count the number of comparisons to achieve, we see that it is quadratic on virtual filter 1 and 2, but becomes linear in the third case: it is sufficient to do comparisons on the useful line of which we know the origin (point $(\frac{n+1}{2}, \frac{n+1}{2})$), the end (point $(\lfloor\frac{n}{4}\rfloor, \lceil\frac{3n}{4}\rceil)$) and the slope (-1). It is the only case where the origin of the useful line is known.

6. Input-Output

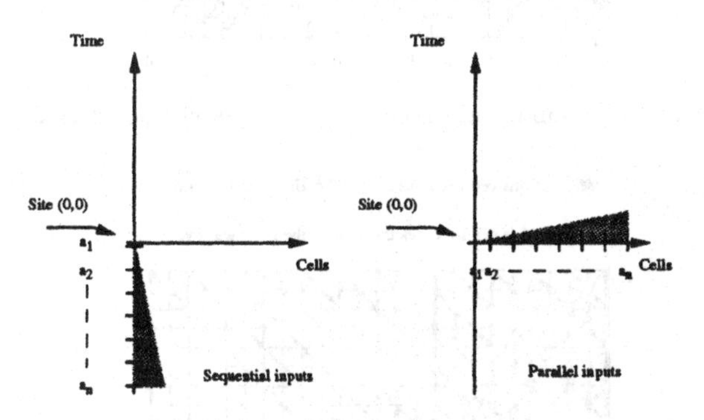

Figure 29. Two main types of inputs.

In previous sections, we have presented some algorithms on one dimensional cellular automata, but we have not define what is a computation. When cellular automata are considered as language recognizers, various notions of input have been defined (Mazoyer, 1992). We consider a subset $\Gamma^\star$ of the set of states of the cellular automaton and an application ψ (non necessarily injective) of the input alphabet $\mathbb{S}$ onto Γ. In literature, two main inputs are defined (see Figure 29):

1. *Sequential inputs* (iterative arrays) where inputs appear on the first cell time after time. The last one corresponding to the first letter of the input word is set up at time 0.
2. *Parallel inputs* (cellular automata) where inputs appear simultaneously, at time 0, on the first cells of the line.

We may consider that the most general input is such that a state of Γ, with indication of x_{i+1} (the $(i+1)^{th}$ letter of the input word) appears anywhere in the space time diagram in an area which does not interact with the point on which has appeared the state of x_i. If x_i appears on site (ξ_i, η_i), the condition is written $\xi_{i+1} + \eta_{i+1} > \xi_i + \eta_i$. We observe that the previous definition of iterative arrays contradicts this new one: we must consider the reverse input word. In (Mazoyer, 1992), it is shown that given a cellular automaton with parallel (sequential) inputs we may syntactically construct another one with sequential (parallel) inputs having the same states on the first diagonal. Thus, an interesting input is $\xi_i = 1$ and $\eta_i = 1$.

In the case of computations of functions things are less clear. First of all, now the output cannot appear on one cell at one time. Observe that the output is a word on some alphabet $\mathbb{S}$. We distinguish:

- i - A subset Γ of set of states of our automaton **A** and an application $\phi : \Gamma \longrightarrow \mathbb{S}$, possibly non injective: a state in Γ represents a letter of $\mathbb{S}$. But we do not know that this value in $\mathbb{S}$ corresponds to a piece of the output.
- ii - A subset of Γ, Θ and an application, maybe non injective, $\phi^{\sharp} : \Theta \longrightarrow \mathbb{S}$ representing letters of $\mathbb{S}$. A state of Θ indicates to one (and only one cell) that the computation is achieved.

The classical notions of outputs are the following (see (Mazoyer, 1992)) and are illustrated on Figure 30:

- i - *Sequential outputs* Output letters appear on the first cell at some times. In this case, the output does not need to appear time after time and set of states Θ is not useful. One may decide that any letter of Γ is significant: the output begins the first time that a state of Γ appears on the first cell.
- ii - *Parallel synchronized outputs* At some time θ depending on the input, some cells are in a state of Θ. The output is $x_1 \ldots x_m$ where x_i is the ith cell in a state of Θ. In some way, all cells with output know that the computation is achieved. In this case, all cells with output froze their evolution. In the case where output appears on cells 1 to m, all the output remains unchanged in the time and may clean up the remaining part of the active segment.
- iii - *Parallel unsynchronized outputs* At some time θ depending on the input, the first cell is in a state of Θ and some cells are in a state of $\Gamma \setminus \Theta$. The output is $x_1 \ldots x_m$ where x_i is the $(i-1)^{th}$ cell in a state of Γ $(i > 1)$. x_1 is given on the first cell via Θ. Only, the first cell knows that the computation is achieved.
- iv - *Parallel unknown outputs* At some time θ depending on the input, some cells are in a state of Γ. No cell knows that the computation is achieved. Thus the system evolves forgetting the result of the computation. It is not clear that this notion is really an output.

The three first notions of outputs are equivalent: there exist linear time syntactical simulations between themselves using the "Firing Squad " of

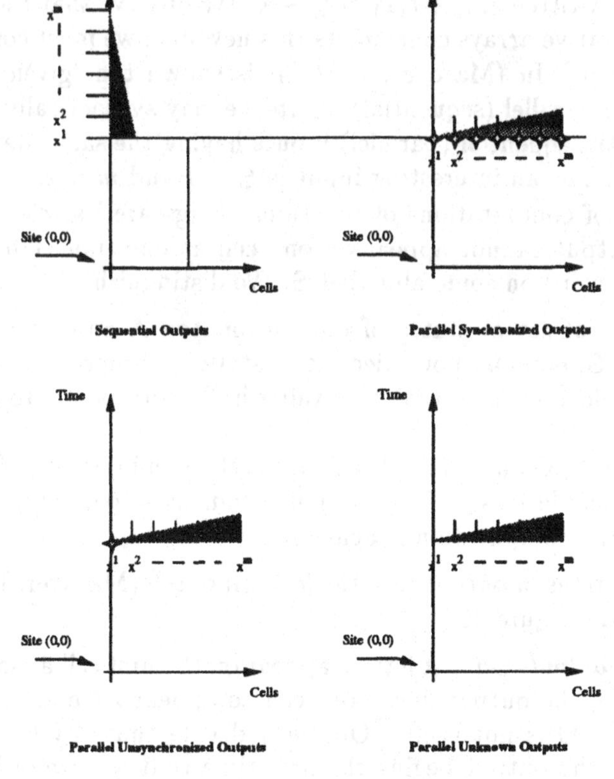

Figure 30. Four main types of outputs.

section 4 (see (Mazoyer, 1992)).

Various other notions of output may be defined. We cite some of them:

- v - Starting from one configuration containing the input, all the system evolves. Two possibilities occurs, it reaches a periodic configuration or not. If it reaches a periodic configuration, the computation is achieved and its result depends on the period (its length for example).
- vi - Starting from one configuration containing the input, all the system evolves. The computation is achieved if and only if the system reaches a fixed point with all its states in Γ and the output is written on this fixed point. Using again FSSP, this definition is equivalent in quadratic time to definitions 1, 2, 3.

Thus, we have many notions of outputs. In fact, to define such a notion is not easy. Input and output are "how a system communicates with the outside". In the case of a sequential machine (Turing machine for example), it is well known that the input is a sequence of letters (with an internal order) and that to give the letter in the natural order of the tape or in the natural order of the time is equivalent. In the one dimensional cellular case the situation is very similar. But in the two dimensional case, the input may be a picture (without natural order) and not a word; and, now to define input becomes troublesome. Looking to output in the sequential case appears to be: the single machine stops and the output word is written somewhere. But, even, in the one dimensional cellular case, the first point becomes more complex:

- Does only one cell stops (definitions 1, 3) or do all cells stop simultaneously (definitions 2, 5, 6)?
- Where does the output appear? on one cell (definitions 1, 5)? on many cells (definitions 1, 2, 3, 6)? In other words, is the outside sequential or parallel?
- If only one machine stops, do other machines pursue their computation (definition 3)?

Clearly, the situation in two dimensional case is more complex.

We conclude this section with a general remark. The previous points explain why cellular automata are not presented as the basic model of computations. A notable exception is (Levin, 1991). But we must not forget that all results concerning the theory of computability remain true. For example, it is undecidable if two cellular automata compute the same function (with fixed definitions for inputs and outputs).

7. Concluding remarks

Looking to constructed cellular automata, one may think that they have a very great number of states. This contradicts our idea that very simple cellular automata with few states (less than fifty) may have very complex behavior. This idea is comforted by the fact that there exists an universal cellular automaton with 8 states (Lindgren and Nordahl, 1990). How to reduce the number of states of our constructed automata? This optimization is very difficult as shown by the fact that it is not known if the FSSP has a 5 states solution (see 4). We do not know powerful methods.

Another interesting point is that parallelism speeds up computations. On the examples, this fact is true but more astonishing is the fact that the best cellular algorithms are the simpler ones (the "human" ones).

The main features of cellular algorithms are their "locality" and for those designed the fact that the virtual construction (built in the space-time diagram) is made of lines (signals).

What happens in the two dimensional case? Now, the basic lines are not only set up by the programmer but must also be found in the configurations themselves. So, two dimensional algorithms are closely connected with discrete geometry (Delorme *et al.*, 1997).

References

Achasova S., Bandman O., Markova V. and Piskunov S. *Parallel substitution algorithm.* World Scientific, 1994.

Albert J. and Čulik K. A simple universal cellular automaton and its one-way and totalistic version. *Complex Systems* **Vol. no.1**: 1–16, 1987.

Balzer R. Studies concerning minimal time solutions to the firing squad synchronization problem. Ph.D. thesis, Carnegie Institute of Technology, 1966.

Balzer R. An 8-states minimal time solution to the firing squad synchronization problem. *Information and Control* **Vol. no.10**: 22–42, 1967.

Berlekamp E., Conway V., Elwyn R. and Guy R. *Winning way for your mathematical plays, volume 2* Academic Press, 1982.

Čulik K. and Hurd L. Computation theoretic aspects of cellular automata. *Physica D* **Vol. no.32**: 357–378, 1990.

Delorme M., Mazoyer J. and Tougne L. Discrete parabolas and circles on 2D cellular automata. To appear in *Theoretical Computer Science.*

Fischer P.C. Generation of primes by a one dimensional real time iterative array. *Journal of A.C.M.* **Vol. no. 12**: 388–394, 1965.

Fokas A., Papadopoulou E. and Saridakis V. Particles in soliton cellular automata. *Complex Systems* **Vol. no. 3**: 615–633, 1989.

Goto E. A minimal time solution to the firing squad synchonization problem. Course notes for applied mathematics Harvard University, 1962.

Levin L. Theory of computation: How to start. *SIGACT News* **Vol. 22 no.1**: 47–56, 1991.

Lindgren K. and Nordahl M. Universal computation in simple one dimensional cellular automata. *Complex Systems* **Vol. no. 4**: 299–318, 1990.

Martin B. A universal automaton in quasi-linear time with its s-n-m form. *Theoretical Computer Science* **Vol. no. 124**: 199–237, 1994.

MazoyerJ. A six states minimal time solution to the firing squad synchronization problem. *Theoretical Computer Science* **Vol. no. 50**: 183–238, 1987.

Mazoyer J. Entrées et sorties sur lignes d'automates. *Algorithmique Parallèle*, Masson, Cosnard C., Nivat M. and Robert Y., Eds, 47–61, 1992.

Mazoyer J. and Terrier V. Signals on one dimensional cellular automata. To appear in *Theoretical Computer Science.*, 1998.

Minsky M. *Finite and infinite machines* Prentice Hall, 28–29, 1967.

Signorini J. Programmation par configurations des ordinateurs cellulaires à très grande échelle. Ph.D. thesis, Université Paris 8, 1992.

Smith A. Real time languages by one-dimensional cellular automata. *Journal of Computer and System Sciences* **Vol. no. 6**: 233–253, 1972.

Terrier V. Temps réel sur automates cellulaires. Ph.D. thesis, Université Lyon 1, 1991.

Waksman A. An optimum solution to the firing squad synchronization problem. *Information and Control* **Vol. no. 9**: 66–78, 1966.

Wolfram S. *Theory and Applications of Cellular Automata*, World Scientific, Singapore, 1986.

Yunes J.B. Seven states solutions to the Firing Squad Synchronization Problem. *Theoretical Computer Science* **Vol. 127 no. 2**: 313–332, 1993.

COMPUTATIONS ON GRIDS

J. MAZOYER
LIP ENS Lyon
46 Allée d'Italie,
69364 Lyon Cedex 07,FRANCE

Abstract. We study how the underlying graph of dependencies of one dimensional cellular automaton may be used in order to move and compose areas of computations. This allows us to define complex cellular automata, relaxing in some way the inherent synchronism of such networks.

1. Introduction

We only consider one-dimensional cellular automata (1D-CA for short) and wonder about what is a computation?

- First of all, a computation needs some interaction between the network of cells and the outside (human or other sequential or parallel machine).
- Second, calculi occurs both on cells and on connections between these cells.

The first point is the input-output issue in 1D-CA. In the literature, many studies have been done on this point (see, for instance, (Ibarra *et al.*, 1985) (Mazoyer, 1992)). In fact, we may summarize inputs used in general by:

- *Parallel inputs* (PCA in this book).
  At initial time, the input $(a_0, \ldots, a_{n-1})$ is encoded in the states of cells $0, \ldots, n-1$ and all other cells are in a quiescent state. All the cells have the information that the time is the initial one (the network is synchronized).
- *Sequential inputs* (SCA in this book).
  At initial time all cells are quiescent and the input is encoded in the state of cell 0 at times $0, \ldots, n-1$.

In fact, it is possible to show (Mazoyer, 1992), that in both cases it is possible to obtain the same states on the diagonal (cell m at time m). In the following we will choose as inputs:

- $(a_0, \dots, a_{n-1})$ is encoded in the states of cells $0, \dots, n-1$ at times $0, \dots, n-1$. Denoting by $\langle k, t \rangle$ the state of cell k at time t, called *site* $\langle k, t \rangle$, the input a_i in encoded in $\langle i, i \rangle$.

The second point is the study of the graph of dependencies. We observe that the natural graph (see Figure 1) may be viewed as redundant. The non

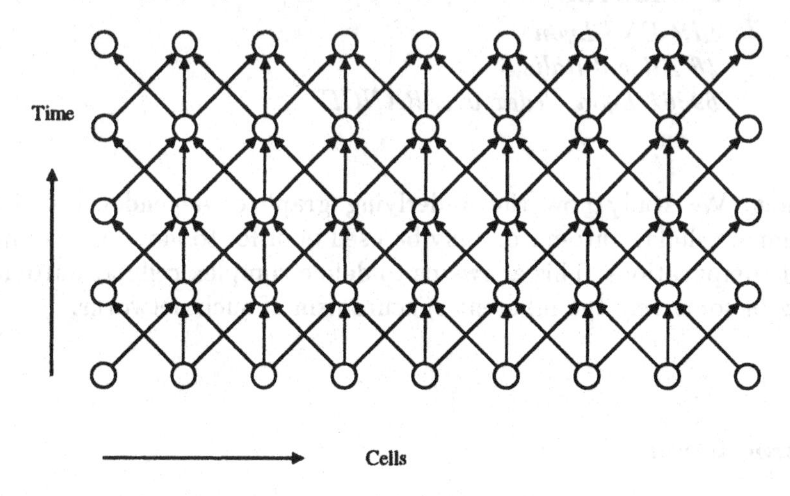

Figure 1. Graph of dependencies of an 1D-CA.

redundant notion linked with is the trellis notion of (Čulik II *et al.*, 1981) and (Choffrut and Čulik II, 1984). To explain this fact, from a given 1D-CA $\mathbf{A} = (Q_\mathbf{A}, \delta_\mathbf{A})$ with $\delta_\mathbf{A} : Q_\mathbf{A}^3 \longrightarrow Q_\mathbf{A}$ and neighborhood $(-1, 0, +1)$, we construct a new 1D-CA $\mathbf{B} = (Q_\mathbf{B}, \delta_\mathbf{B})$ with $\delta_\mathbf{B} : Q_\mathbf{B}^2 \longrightarrow Q_\mathbf{B}$ and neighborhood $(-1, +1)$ as follows:

- $Q_\mathbf{B} = Q_\mathbf{A} \cup Q_\mathbf{A}^2$
- $\delta_\mathbf{B}((q_\mathbf{A}, q'_\mathbf{A}), (q'_\mathbf{A}, q^\star_\mathbf{A})) = \delta_A(q_\mathbf{A}, q'_\mathbf{A}, q^\star_\mathbf{A})$
  and
  $\delta_\mathbf{B}(q_A, q'_\mathbf{A}) = (q_A, q'_\mathbf{A})$

To any configuration $C_\mathbf{A}$ for $\mathbf{A}$, we associate by Φ the configuration $C_\mathbf{B} = \Phi(C_\mathbf{A})$ of $\mathbf{B}$ defined by $C_\mathbf{B}(2z) = C_\mathbf{A}(z)$ and $C_\mathbf{B}(2z+1) = q_{\mathbf{A}_0}$ where $q_{\mathbf{A}_0}$ is the quiescent state of $\mathbf{A}$. It is easy to see ((Choffrut and Čulik II, 1984)), that starting from a configuration $C_\mathbf{A}$ for $\mathbf{A}$, we have

$$\Phi(C_\mathbf{A}^t) = (\Phi(C_\mathbf{A}))^{2t}$$

The graph of dependencies of $\mathbf{B}$, called trellis automaton is depicted on the Figure 2 on which we have indicated the sites of $\mathbf{B}$ on which values

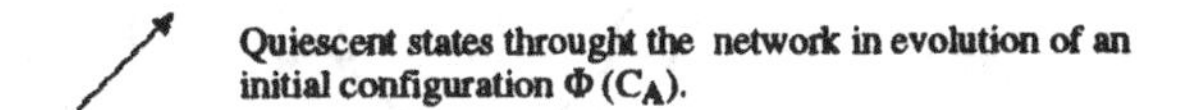

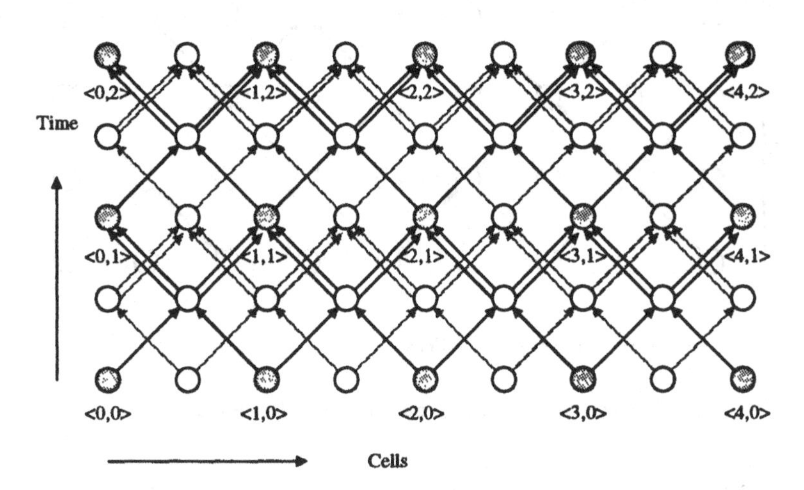

Figure 2. Graph of dependencies of a trellis automaton. The grid corresponding to computing nodes as its dependencies drawn in black. The computations of **A** occur on gray nodes. The grid in which are only transitting quiescent states is drawn in light gray.

of sites of **A** are computed. We observe that the graph of dependencies of **B** is now two twisted grids in the space-time diagram (they are rotated by 90 degrees) which never interact. Starting from an initial configuration $\Phi(C_{\mathbf{A}})$, only quiescent states transit on one of the two grids, thus we skip this grid and view the graph of dependencies as only one twisted grid. In the following we will consider this new graph (made of a twisted grid) as the new dependencies graph (see Figure 3). In other words, we forget odd cells. In this case, a cell in network **B** computes a state transition of **A** one time out of two (the other time is used to distribute information).

2. Grids

Looking to the graph of Figure 3, we have now only two elementary moves of information: *to the right* or *to the left*, denoted $\nearrow$ (of vector $(+1,1)$ on a space time diagram) and $\nwarrow$ (of vector $(-1,1)$). We observe that an information cannot stay on a cell: it is dispatched to its two neighbors, reflected by them and then, thus, received and available the next time. Thus, the notion of stationary move is not an elementary move. We define a temporal move, denoted $\uparrow$ (of vector $(0,2)$ on the space time diagram), as the succession of previous moves $\nearrow$ and $\nwarrow$ or $\nwarrow$ and $\nearrow$. Every time we would be able to define two notions of elementary moves, denoted $\nearrow^{\sharp}$ and $\nwarrow^{\sharp}$, called to the right and to the left, we shall define a new graph of

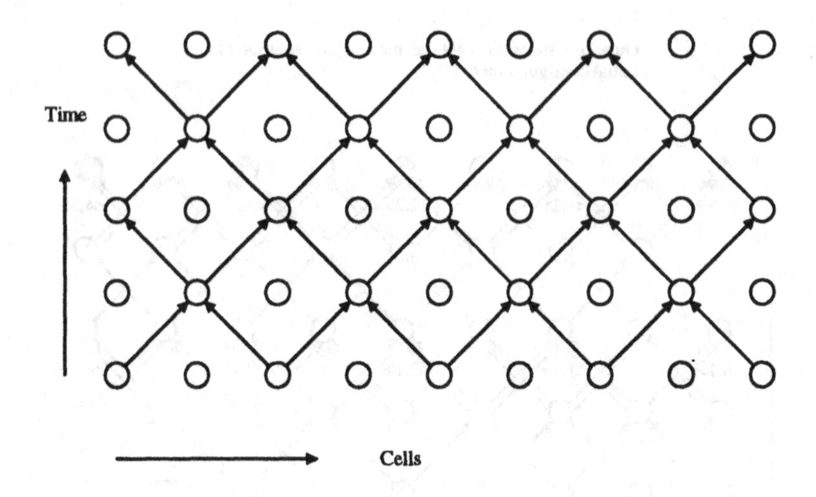

Figure 3. The basic graph of dependencies, Γ_0.

dependencies. Every behavior on the first one may be translated to the new one.

How to define a new graph? Intuitively we use 2 signals setting up all elementary moves $\nearrow^{\sharp}$ and $\nwarrow^{\sharp}$. The formal definition needs to consider states with at least two components in order to distinguish between $\nearrow^{\sharp}$, $\nwarrow^{\sharp}$ and computations. This leads us to the following definition in which:

- A grid Γ may be viewed as a wire netting distorded and put in the space time diagram. Its wires are made of elementary moves of the underlying grid (for instance, the treillis of the Figure 3).
- In order to define a cellular automaton with significant computations occurring on vertices (sites in the space-time diagram) of a grid, we distinguish two component to its states. The first one, denoted by Q_G is used to set up wires between vertices. States of $Q_{G,r}$ set up wires between points $\Gamma(x,y)$ and $\Gamma(x+1,y)$. States of $Q_{G,\ell}$ set up wires between points $\Gamma(x,y)$ and $\Gamma(x,y+1)$. The second component is used to achieve computations when the first component belongs to $Q_{G,v}$.
- Clearly, only some grids can be set up by cellular automata.

Definition 1

Let $\mathbf{A} = (Q_G \times Q_C, \delta)$, *with 3 distinguished subsets of* Q_G, *denoted* $Q_{G,\ell}$, $Q_{G,r}$ *and* $Q_{G,v}$ *and* $q_0 = (q_{0,G}, q_{0,C}) \notin (Q_{G,\ell} \cup Q_{G,r} \cup Q_{G,v}) \times Q_C$ *as quiescent state. We say that* $\mathbf{A}$ sets up a grid Γ *on the basic trellis if:*

1. $\Gamma = (\Gamma_x, \Gamma_y)$ *is an application of* $\mathbf{N}^2$ *into* $\mathbf{Z} \times \mathbf{N}$ *such that:*
   $\Gamma_x(x+1,y) > \Gamma_x(x,y)$, $\Gamma_y(x+1,y) > \Gamma_y(x,y)$
   $\Gamma_x(x,y+1) > \Gamma_x(x,y)$, $\Gamma_y(x,y+1) > \Gamma_y(x,y)$.

2. *Starting from a configuration* $^{\infty}q_0(M_G, M_C)q_0^{\infty}$ *with* $M_G \in Q_{G,v}$, *the evolution of* $\mathbf{A}$ *is such that:*

   - *i* - *For any point* (ξ, θ) *of* $\mathbb{Z} \times \mathbf{N}$, $\langle \xi, \theta \rangle \in Q_{G,v} \times Q_C \iff \exists x, y \in \mathbf{N}$ *such that* $(\xi, \theta) = \Gamma(x, y)$.
     If the first component of a site (a state associated to a point) is in $Q_{G,v}$, then the corresponding point is a node of Γ.
   - *ii* - *For any point* (x, y) *of* $\mathbb{Z} \times \mathbf{N}$, *there exists a finite sequence of points* $R(x,y) = ((\xi_i^\star, \theta_i^\star) \mid i \in \{1, \ldots, n(x,y)\})$ *of* $\mathbf{N}^2$ *such that:*
     - *a* - $(\xi_1^\star, \theta_1^\star) = \Gamma(x, y)$
     - *b* - $(\xi_{n(x,y)}^\star, \theta_{n(x,y)}^\star) = \Gamma(x+1, y)$
     - *c* - $\forall i \{2, \ldots, n(x,y)-1\}$,
       $\langle \xi_i^\star, \theta_i^\star \rangle$ *is in* $(Q_{G,r} \cup Q_{G,v}) \times Q_C$.
     - *d* - $\forall i \{2, \ldots, n(x,y)-1\}$,
       $(\xi_{i+1}^\star, \theta_{i+1}^\star) - (\xi_i^\star, \theta_i^\star) = (1, 1)$ *or*
       $(\xi_{i+1}^\star, \theta_{i+1}^\star) - (\xi_i^\star, \theta_i^\star) = (-1, 1)$
   - *iii* - *For any point* (x, y) *of* $\mathbb{Z} \times \mathbf{N}$, *there exists a finite sequence of points* $L(x,y) = ((\xi_i^\star, \theta_i^\star) \mid i \in \{1, \ldots, n(x,y)\})$ *of* $\mathbf{N}^2$ *such that:*
     - *a* - $(\xi_1^\star, \theta_1^\star) = \Gamma(x, y)$
     - *b* - $(\xi_{n(x,y)}^\star, \theta_{n(x,y)}^\star) = \Gamma(x, y+1)$
     - *c* - $\forall i \{2, \ldots, n(x,y)-1\}$,
       $\langle \xi_i^\star, \theta_i^\star \rangle$ *is in* $(Q_{G,r} \cup Q_{G,v}) \times Q_C$.
     - *d* - $\forall i \{2, \ldots, n(x,y)-1\}$,
       $(\xi_{i+1}^\star, \theta_{i+1}^\star) - (\xi_i^\star, \theta_i^\star) = (1, 1)$ *or*
       $(\xi_{i+1}^\star, \theta_{i+1}^\star) - (\xi_i^\star, \theta_i^\star) = (-1, 1)$

   The vectors $\Gamma(x+1, y) - \Gamma(x, y)$ *and* $\Gamma(x, y+1) - \Gamma(x, y)$ *are called* the elementary right (left) move at node $\Gamma(x, y)$ *an denoted by* $\nearrow_\Gamma (x, y)$, $\nwarrow_\Gamma (x, y)$.

The figure 4 gives an example of a grid the elementary moves of which are uniform and equal to $\nearrow_\Gamma (x, y) = \nearrow + \nearrow$ (vector $(2, 2)$) and $\nwarrow_\Gamma (x, y) = \nwarrow$ (vector $(-1, 1)$). In the following this grid will be denoted by $\Gamma_{2,2;-1,1}$. Such a grid may be set up by the automaton $\mathbf{A}_{2,2;-1,1} = (Q_G \times \{q_{0,C}\}, \delta)$ with $Q_G = Q_{G,\ell} \cup Q_{G,r} \cup Q_{G,v}$ and $Q_{G,\ell} = \emptyset$, $Q_{G,r} = \{N\}$ and $Q_{G,v} = \{M_G, q_{1,G}\}$ and:
$\delta((M_G, q_{0,C}), q_0, q_0) = (N, q_{0,C})$, $\delta(q_0, q_0, M_G) = ((q_{1,G}, q_{0,C}), q_{0,C})$,
$\delta((N, q_{0,C}), q_0, q_0) = (q_{1,G}, q_{0,C})$, $\delta((q_{1,G}, q_{0,C}), q_{0,C}, q_0, q_0) = (N, q_{0,C})$,
$\delta((q_0, q_0, (q_{1,G}, q_{0,C})) = (q_{1,G}, q_{0,C})$,
$\delta((q_{1,G}, q_{0,C}), q_{0,C}), q_0, (N, q_{0,C})) = (N, q_{0,C})$,
$\delta((N, q_{0,C}), q_{0,C}), q_0, (q_{1,G}, q_{0,C})) = (q_{1,G}, q_{0,C})$,

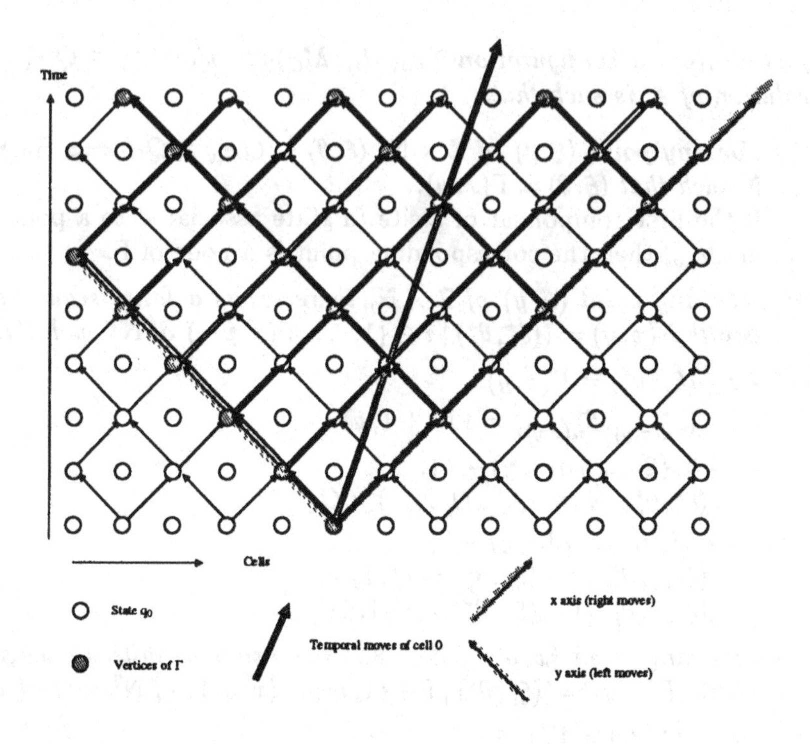

Figure 4. The grid $\Gamma_{2,2;-1,1}$.

the other transitions giving state q_0. State $(q_{1,G}, q_{0,C})$ corresponds to vertices labelled by (x, y). State $(N, q_{0,C})$ is used to set up the move to the right. The corresponding "temporal moves" of the grid of Figure 4 are always the same corresponding to vector $(1, 3)$. The "temporal" trajectory corresponding to cell 0 is indicated on figure 4.

In the definition 1, we have only consider grids constructed on the basic trellis (uniform with rigth and left moves always of vectors $(1, 1)$ and $(-1, 1)$). We may clearly iterate the process. Given a grid $\widetilde{\Gamma}$ constructed on the basic trellis, we consider it as a trellis and construct a new grid as in definition 1. The figure 5 illustrates this point. This leads us to an inductive definition.

Definition 2

For any application Γ *of* $\mathbf{N}^2$ *into* $\mathbb{Z} \times \mathbf{N}$, *we define the vectors* $\nearrow_{\Gamma,(x,y)}$ $(\nwarrow_{\Gamma,(x,y)})$ *as* $\Gamma(x+1, y) - \Gamma(x, y)$ $(\Gamma(x, y+1) - \Gamma(x, y))$.

For an application $\widetilde{\Gamma}$ *of* $\mathbf{N}^2$ *into* $\mathbb{Z} \times \mathbf{N}$ *and an automaton* $\mathbf{A} = (Q_G \times Q_C, \delta)$ *with the same properties as in the definition 1, another application* Γ *of* $\mathbf{N}^2$ *into* $\mathbb{Z} \times \mathbf{N}$ *is called set up by* $\mathbf{A}$ *on* $\widetilde{\Gamma}$ *if:*

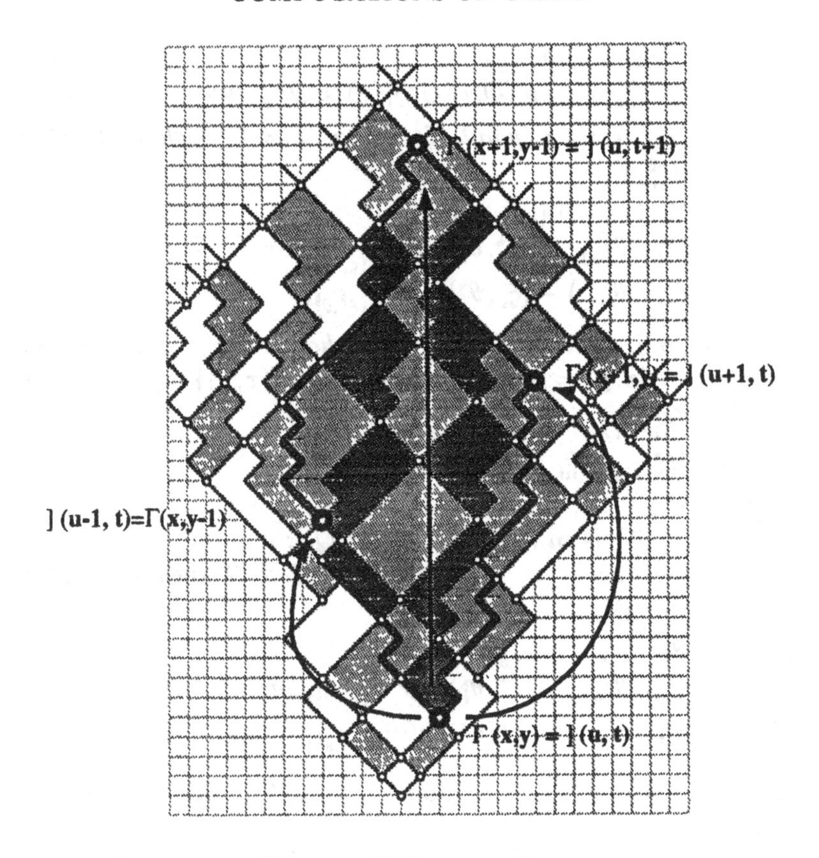

Figure 5. Grid construction.

1. $\Gamma = (\Gamma_x, \Gamma_y)$ *is an application of* $\mathbf{N}^2$ *into* $\mathbb{Z} \times \mathbf{N}$ *such that:*
   $\Gamma_x(x+1, y) > \Gamma_x(x, y)$, $\Gamma_y(x+1, y) > \Gamma_y(x, y)$
   $\Gamma_x(x, y+1) > \Gamma_x(x, y)$, $\Gamma_y(x, y+1) > \Gamma_y(x, y)$.
2. *Starting from a configuration* ${}^{\infty}q_0(M_G, M_C)q_0^{\infty}$ *with* $M_G \in Q_{G,v}$, *the evolution of* $\mathbf{A}$ *is such that:*

   - *i* - *For any point* (ξ, θ) *of* $\mathbb{Z} \times \mathbf{N}$, $\langle \xi, \theta \rangle \in Q_{G,v} \times Q_C \iff \exists x, y \in \mathbf{N}$ *such that* $(\xi, \theta) = \Gamma(x, y)$*and* *there exists a point* $(\xi^\natural, \theta^\natural)$ *of* $\mathbb{Z} \times \mathbf{N}$ *such that* $(\xi, \theta) = \widetilde{\Gamma}(\xi^\natural, \theta^\natural)$. If the first component of a site (a state associated to a point) is in $Q_{G,v}$, then the corresponding point is a node of Γ.

   - *ii* - *For any point* (x, y) *of* $\mathbb{Z} \times \mathbf{N}$, *there exists a finite sequence of points* $R(x, y) = ((\xi_i^\star, \theta_i^\star) \mid i \in \{1, \ldots, n(x, y)\})$ *of* $\mathbf{N}^2$ *such that:*

     - *a* - $(\xi_1^\star, \theta_1^\star) = \Gamma(x, y)$
     - *b* - $(\xi_{n(x,y)}^\star, \theta_{n(x,y)}^\star) = \Gamma(x+1, y)$

- *c* - $\forall i\{1, \ldots, n(x,y)-1\}$,
 $\langle \xi_i^\star, \theta_{i+1}^\star \rangle$ *is in* $(Q_{G,r} \cup Q_{G,v}) \times Q_C$; *and*
 there exist a point $(\xi_i^\sharp, \theta_i^\sharp)$ *of* $\mathbb{Z} \times \mathbf{N}$ *such that* $(\xi_i^\star, \theta_i^\star) = \widetilde{\Gamma}(\xi_i^\sharp, \theta_i^\sharp)$.
- *d* - $\forall i\{1, \ldots, n(x,y)-1\}$,
 $(\xi_{i+1}^\star, \theta_{i+1}^\star) - (\xi_i^\star, \theta_i^\star) = \nearrow_{\widetilde{\Gamma},(\xi_i^\sharp,\theta_i^\sharp)}$ *or*
 $(\xi_{i+1}^\star, \theta_{i+1}^\star) - (\xi_i^\star, \theta_i^\star) = \nwarrow_{\widetilde{\Gamma},(\xi_i^\sharp,\theta_i^\sharp)}$

- *iii* - *For any point* (x,y) *of* $\mathbb{Z} \times \mathbf{N}$, *there exists a finite sequence of points* $L(x,y) = ((\xi_i^\star, \theta_i^\star) \mid i \in \{1, \ldots, n(x,y)\})$ *of* $\mathbf{N}^2$ *such that:*
 - *a* - $(\xi_1^\star, \theta_1^\star) = \Gamma(x,y)$
 - *b* - $(\xi_{n(x,y)}^\star, \theta_{n(x,y)}^\star) = \Gamma(x, y+1)$
 - *c* - $\forall i\{1, \ldots, n(x,y)-1\}$,
 $\langle \xi_i^\star, \theta_i^\star \rangle$ *is in* $(Q_{G,r} \cup Q_{G,v}) \times Q_C$; *and*
 there exists a point $(\xi^\sharp, \theta^\sharp)$ *of* $\mathbb{Z} \times \mathbf{N}$ *such that* $(\xi_i^\star, \theta^\star) = \widetilde{\Gamma}(\xi_i^\sharp, \theta_i^\sharp)$.
 - *d* - $\forall i\{1, \ldots, n(x,y)-1\}$,
 $(\xi_{i+1}^\star, \theta_{i+1}^\star) - (\xi_i^\star, \theta_i^\star) = \nearrow_{\widetilde{\Gamma},(\xi_i^\sharp,\theta_i^\sharp)}$ *or*
 $(\xi_{i+1}^\star, \theta_{i+1}^\star) - (\xi_i^\star, \theta_i^\star) = \nwarrow_{\widetilde{\Gamma},(\xi_i^\sharp,\theta_i^\sharp)}$

Finally, let Φ *be the binary relation on the set of applications of* $\mathbf{N}^2$ *into* $\mathbb{Z} \times \mathbf{N}$ *defined by:*
$\Phi(\Gamma, \widetilde{\Gamma})$ *iff there exist* $\mathbf{A}$ *such that* Γ *is set up from* $\widetilde{\Gamma}$ *using* $\mathbf{A}$.

We obtain the following definition:

Definition 3

The set of grids is the smaller subset of the set of applications of $\mathbf{N}^2$ *into* $\mathbb{Z} \times \mathbf{N}$ *closed by* Φ *and containing* Γ_0 *defined by* $\Gamma_0(x,y) = (x-y, x+y)$ *(the basic trellis).*

On the Figure 5, we have represented meshes of the grid $\widetilde{\Gamma}$ with two level of gray. One mesh of Γ is shown. It is made of meshes of $\widetilde{\Gamma}$ in darker gray. The two elementary right and left moves are indicated. Viewed as a path on the grid $\widetilde{\Gamma}$, they are complex enough. If $\Gamma(x,y) = \widetilde{\Gamma}(x^\sharp, y^\sharp)$, then:
$\nearrow_{\Gamma,(x,y)} = \nearrow_{\widetilde{\Gamma},(x^\sharp,y^\sharp)} + \nearrow_{\widetilde{\Gamma},(x^\sharp+1,y^\sharp)} + \nwarrow_{\widetilde{\Gamma},(x^\sharp+2,y^\sharp))} +$
$\nearrow_{\widetilde{\Gamma},(x^\sharp+2,y^\sharp+1)} + \nwarrow_{\widetilde{\Gamma},(x^\sharp+3,y^\sharp+1)} + \nearrow_{\widetilde{\Gamma},(x^\sharp+3,y^\sharp+2)}$,
due to the values of moves of $\widetilde{\Gamma}$ the associated vector is $(5, 17)$; and
$\nwarrow_{\Gamma,(x,y)} = \nwarrow_{\widetilde{\Gamma},(x^\sharp,y^\sharp)} + \nwarrow_{\widetilde{\Gamma},(x^\sharp,y^\sharp+1)} + \nwarrow_{\widetilde{\Gamma},(x^\sharp+3,y^\sharp+2)} +$
$\nearrow_{\widetilde{\Gamma},(x^\sharp,y^\sharp+3)} + \nwarrow_{\widetilde{\Gamma},(x^\sharp+1,y^\sharp+3)}$, due to the values of moves of $\widetilde{\Gamma}$
due to the values of moves of $\widetilde{\Gamma}$ the associated vector is $(-6, 10)$. We define a temporal move $\uparrow_{\Gamma,(x,y)}$ given by the vector $\Gamma(x+1, y+1) - \Gamma(x,y) =$

(1, 27). We observe that, on $\widetilde{\Gamma}$, we have $\uparrow_{\Gamma,(x,y)} = \nearrow_{\widetilde{\Gamma},(x',y')} + bL_{\widetilde{\Gamma},(x'+1,y')}$ or $\uparrow_{\Gamma,(x,y)} = \nwarrow_{\widetilde{\Gamma},(x',y')} + bR_{\widetilde{\Gamma},(x',y'+1)}$, but the succession of moves on the basic trellis is different in the two cases since $bR_{\widetilde{\Gamma},(x',y')} \neq bR_{\widetilde{\Gamma},(x',y'+1)}$.

Given a recursive application γ of $\mathbf{N}^2$ into $\mathbb{Z} \times \mathbf{N}$, it may not belong to the set of grid. For instance, in the case where $\gamma(x,0) = x + \log_2(x)$ (see (Mazoyer and Terrier, 1997)).

Such a grid Γ may be understood as a "virtual" trellis of dependencies, set up by some automaton $\mathbf{A}_\Gamma$ in which the "computational part", state component in Q_C, does nothing as in the previous example. We recall that any behavior of a cellular automaton $\mathbf{A} = (Q_\mathbf{A}, \delta_\mathbf{A})$ may be translated to the behavior of some cellular automaton $\mathbf{B} = (Q_\mathbf{A} \cup Q_\mathbf{A}^2, \delta_\mathbf{B})$ as defined in the section 1 working only on a trellis. Then this last $\mathbf{B}$-evolution may be embedded on the grid Γ instead of the basic trellis as shown by the following theorem:

Theorem 1

For any grid Γ *set up by the automaton* $\mathbf{A}_\Gamma = (Q_{\Gamma,G} \times Q_{\Gamma,C}, \delta_\Gamma)$ *(in the sense of the definition 1) and any automaton* $\mathbf{A} = (Q_\mathbf{A}, \delta_\mathbf{A})$*, there exists an automaton*
$\mathbf{B}_{\longrightarrow\Gamma}(((Q_{\Gamma,G} \times Q_{\Gamma,C}) \times Q_\mathbf{A}), \delta_{\longrightarrow\Gamma})$ *such that:*

For any configuration $C_\mathbf{A}$ *of* $\mathbf{A}$*, the behavior of* $\mathbf{B}_{\longrightarrow\Gamma}$ *on an initial configuration satisfying* $\langle\Gamma(0,0)\rangle_{\mathbf{B}_{\longrightarrow\Gamma}} = ((M_{G,\mathbf{A}_\Gamma}, M_{C,\mathbf{A}_\Gamma}), C_\mathbf{A}(0))$ *and* $\forall x > 0$, $\langle\Gamma(x,0)\rangle_{\mathbf{B}_{\longrightarrow\Gamma}} = (q_{0,\mathbf{A}_\Gamma}, C_\mathbf{A}(x))$ *is such that:*

$$\forall (x,t) \in \mathbf{N} \times \mathbf{N},\ \langle\Gamma(x,t)\rangle_{\mathbf{B}_{\longrightarrow\Gamma}} = (\langle\Gamma(x,t)\rangle_{\mathbf{A}_\Gamma}, \langle(x,t)\rangle_\mathbf{A}).$$

Proof It remains only to define $\delta_{\longrightarrow\Gamma} = (\delta_{\longrightarrow\mathbf{A}_\Gamma}, \delta_{\longrightarrow\mathbf{A}})$. For any states a_1, a and a_3 of $Q_\mathbf{A}$, and states q_1, q_2 and q_3 of $(Q_{\Gamma,G} \times Q_{\Gamma,C})$, (ν, μ) will denote the result of $\delta_{\longrightarrow\mathbf{A}_\Gamma}((q_1,a_1),(q_2,a),(q_3,a_3))$.
If $\nu \in Q_{\Gamma,G,\ell} \cup Q_{\Gamma,G,r}$ then $\delta_{\longrightarrow\mathbf{A}}((q_1,a_1),(q_2,a),(q_3,a_3)) = a$.
And if $\nu \in Q_{\Gamma,G,v}$, then $\delta_{\longrightarrow\mathbf{A}}((q_1,a_1),(q_2,a),(q_3,a_3)) = \delta_\mathbf{A}(a_1,a,a_3)$.
By this way state transitions of $\mathbf{A}$ only occur on vertices of Γ which have their first component in a state of $Q_{\Gamma,G,v}$.
Clearly, such a result may be stated for all the grids of definition 3. □

The theorem 1 allows us to move evolution of any cellular automaton $\mathbf{A}$ in any place ("cellular computable") of the space time diagram. There exist various ways to define grids. The Figure 4 shows how to put such a trellis in the part of the plane between angles $\frac{\pi}{4}$ and $\frac{3\pi}{8}$. Thus, if computations of $\mathbf{A}$ belongs to the part of plane between $\frac{\pi}{4}$ and $\frac{\pi}{2}$, then those of $\mathbf{B}_{\longrightarrow\Gamma}$ are twisted in the part between $\frac{\pi}{4}$ and $\frac{3\pi}{8}$. Another grid will allows us to put them between $\frac{3\pi}{8}$ and $\frac{\pi}{2}$, choosing as right move vector $(1,3)$ and as left move vector $(-1,1)$, for instance. It is possible to show that every uniform grids (in which elementary moves are linear combinations of basic ones) may be set up by cellular automata (see (Mazoyer, 1996)). As it is easy to translate a grid, in some way, any computation may be moved of any rational angle of the plane.

3. Grids and computations

As noticed in section 1, we must choose a notion of input and output. Our choice to consider grids leads us to the following definition:

Definition 4

Let Σ be a finite alphabet, and f a function of $\Sigma^\star$ into $\Sigma^\star$, we say that a cellular automaton $\mathbf{A} = (Q_\mathbf{A}, \delta_\mathbf{A})$ with a distinguished subset $Q_{\mathbf{A},f}$ of $Q_\mathbf{A}$ computes f if there exist two applications $\Phi : \Sigma \longrightarrow Q_\mathbf{A}$ and $\Psi : Q_{\mathbf{A},f} \longrightarrow \Sigma$ such that for any word $a_1 \dots a_n$ of $\Sigma^\star$ we have:

- *for any initial configuration C of $\mathbf{A}$ such that $\langle i,i\rangle_\mathbf{A} = \Phi(a_i)$ (for $i \in \{1,\dots,n\}$) and $\langle i,i\rangle_\mathbf{A} = q_0$, the quiescent state (for $i > n$),*
- *there exists m times $t_1,\dots,t_m$ such that:*
  - *$\langle 0,t\rangle \in Q_{\mathbf{A},f}$ if and only if $t \in \{t_1,\dots,t_m\}$,*
  - *and $(\Psi(\langle 0,t_1\rangle),\dots,\Psi(\langle 0,t_m\rangle)) = f(a_1,\dots,a_n)$.*

In order to emphasize the role of the time (for us the move $\nearrow_\Gamma + \nwarrow_\Gamma$,we define a grid by its left and temporal moves:

Notation 1

The grid Γ is also denoted par $\daleth$. In the sequel, we shall use indifferently Γ or $\daleth$, the second component indicating the number of left moves or the number of temporal moves. We note:
$\Gamma = (\Gamma_x, \Gamma_y)$ ($x \in \mathrm{N}$ and $y \in \mathrm{N}$) and $\daleth = (\daleth_u, \daleth_t)$ ($u \in \mathbb{Z}$ and $t \in \mathrm{N}$). with the relations $u = x - y$ and $t = x + y - u$ (see Figure 5).
In addition, when the grid is uniform, to emphasize the role of time we denote $\Gamma_{a,b;c,d}$ by $\daleth_{a,b;a+c,b+d}$.
We resume our notations by:

Grid	$\Gamma(x,y) : \mathrm{N}^2 \longrightarrow \mathbb{Z}\times\mathrm{N}$	$\daleth(u,t) : \mathbb{Z}\times\mathrm{N} \longrightarrow \mathbb{Z}\times\mathrm{N}$ $u = x - y$ $t = x + y - u$
Nodes	$\Gamma(x,y)$	$\daleth(u,t)$
Moves	$\nearrow_{\Gamma,(x,y)}, \nwarrow_{\Gamma,(x,y)}$	$\nearrow_{\daleth,(x,y)}, \nwarrow_{\daleth,(x,y)}$
Uniform	$\Gamma_{a,b;c,d}$	$\daleth_{a,b;c+a,b+d}$

In the scope of grids of section 2 (see definition 1), we strengthen the previous definition as follows (in this case definition 4 appears as the definition 5 in the special case where Γ is the basic graph of dependencies):

Definition 5

Let Σ be a finite alphabet, and f a function of $\Sigma^\star$ in $\Sigma^\star$, we say that a cellular automaton $\mathbf{A}_\Gamma = (Q_\Gamma \times Q_C, \delta_{\mathbf{A}_\Gamma})$ with a distinguished subset $Q_{C,f}$ of Q_C, setting up the grid Γ, computes f if there exist two applications $\Phi : \Sigma \longrightarrow Q_C$ and $\Psi : Q_{C,f} \longrightarrow \Sigma$ such that for any word $a_1 \ldots a_n$ of $\Sigma^\star$ we have:

- *for any initial configuration C of $\mathbf{A}_\Gamma$ such that $\langle \beth(i,0)\rangle_{\mathbf{A}} = \Phi(a_i)$ (for $i \in \{1, \ldots, n\}$) and $\langle \beth(i,0)\rangle_{\mathbf{A}} = q_0$, the quiescent state, (for $i > n$),*
- *there exists m times $t_1, \ldots, t_m$ such that:*
  - $\langle \beth(0,t)\rangle \in Q_{\mathbf{A},f}$ *if and only if* $t \in \{t_1, \ldots, t_m\}$,
  - *and* $\Psi(\langle \beth(0,t_1)\rangle \ldots \Psi(\langle \beth(0,t_m)\rangle = f(a_1, \ldots, a_n)$.

The Figure 6 illustrates how to obtain whether "$3t$ is a prime number" on the grid $\Gamma_{2,2;-1,1}$, called $\beth_{1,1;0,3}$ by its right and temporal moves, depicted on figure 4 using the algorithm of chapter *Computations on cellular automata.*

By this way, using theorem 1, we are allowed to put a computation on any grid Γ_1 or any area defined by a grid. In order to compose computation, we need to consider that the output of the grid $\beth_1$ are the inputs of another half grid Γ_2. Thus the first question is: "Given $\beth_1(0,t)$ ($t \in \mathbf{N}$), does exist any grid Γ_2 such that $\beth_2(x,0) = \beth_1(0,x)$ and, for all $t \in \mathbf{N}$, $\beth_2(0,t) = (0, \theta(t))$" (for some increasing function θ)? If for any $t \in \mathbf{N}, \beth_1(0,t)$ belongs to the first quarter of the plane the answer is yes as shown by the following proposition:

Proposition 1

For any function $\chi = (\chi_x, \chi_y)$ from $\mathbf{N}$ into $\mathbf{N}^2$ satisfying $\forall x \in \mathbf{N}$, $\chi_x(x+1) > \chi_x(x) \geq 0$ and $\chi_y(x+1) > \chi_y(x) \geq 0$ "simply set up" by a cellular automaton $\mathbf{G}$, there exists a grid Γ_χ constructed by a cellular automaton $\mathbf{A}_\chi$ and an increasing function θ from $\mathbf{N}$ into $\mathbf{N}$ such that for all $x \in \mathbf{N}$:

- $\beth_\chi(x,0) = \chi(x)$
- $\beth_\chi(0,x) = (0, \theta(x))$.

Proof

By "set up" by $\mathbf{G}$, we understand that there exists a cellular automaton $\mathbf{G} = (Q_G \times Q, \delta_G)$ setting up a grid Γ_G such that, for all integers $n \in \mathbf{N}$, $\chi(n) = \beth_G(0,n)$. We shall define later what we understand by "simply".

Let us describe the points of this proof:

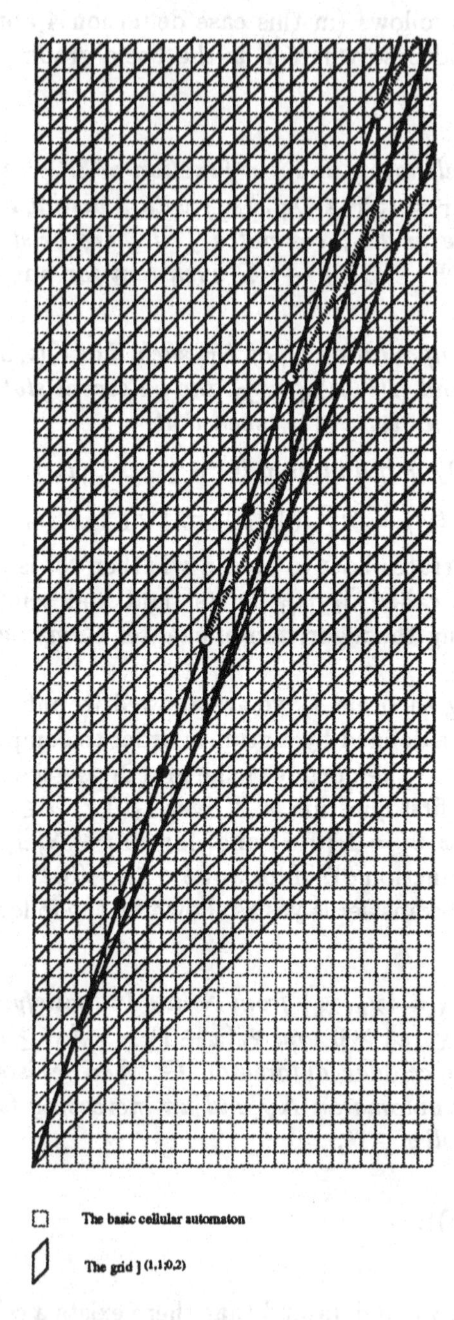

Figure 6. **Recognizing whether $3t$ is prime on the grid $\Gamma_{\nearrow+\nearrow,\nwarrow}$.**

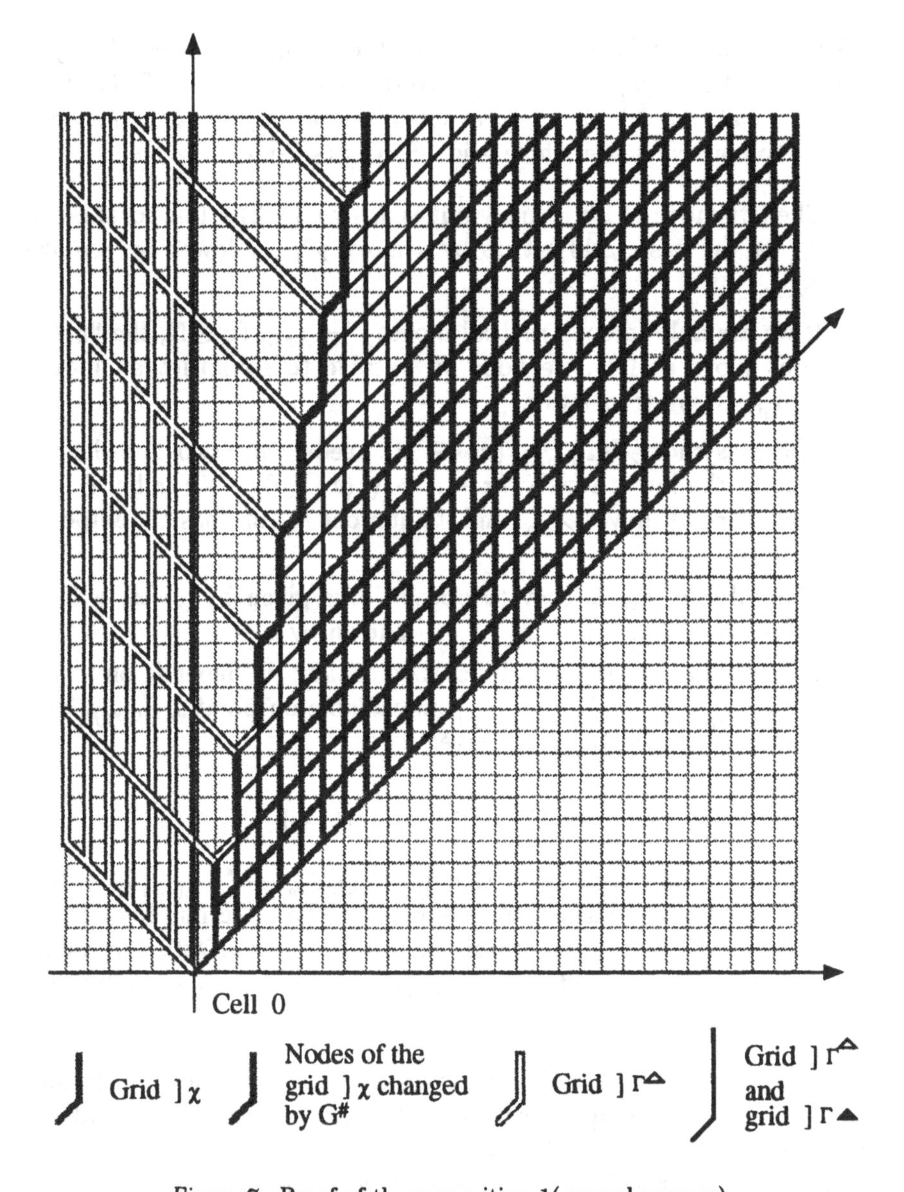

Figure 7. Proof of the proposition 1(general process).

- a - First of all, we observe that the automaton **G** sets up the grid Γ_χ and we need the grid $\rceil_\chi$ in which points $\rceil(0,t)$ ($t \in \mathbf{N}$) are marked (they indicated our given function χ). Thus we need to transform our automaton **G** in another one setting up $\rceil_\chi$ with a special subset of states indicating $\rceil(0,t)$.

- b - Then, we construct a grid $\Gamma^{\vartriangle}$ in which all elementary moves $\nearrow_{\Gamma^{\vartriangle},(x,y)}$ are equal to $\uparrow_{G,(x-y,0)}$ and $\nwarrow_{\Gamma^{\vartriangle},(x,y)}$ to $(0,2)$ and for which $\Gamma^{\vartriangle}(x,0) = \beth_{G,(0,x)}$ (for all $x \in \mathbf{N}$). By the previous observation, points $\beth_{G,(0,x)}$ are known. To construct the whole grid $\Gamma^{\vartriangle}$ results from:

  - i - The left move of $\Gamma^{\vartriangle}$ is known: it is always $(0,2)$ a temporal move $\uparrow_{\Gamma_0}$ of the basic grid set up by a move $\nearrow_{\Gamma_0}$ followed by a move $\nwarrow_{\Gamma_0}$.
  - ii - If we are able to construct all moves $\nearrow_{\Gamma^{\vartriangle},(x,0)}$ ($x \in \mathbf{N}$, the "first" right move of $\Gamma^{\vartriangle}$), we hope to be able to construct all moves $\nearrow_{\Gamma^{\vartriangle},(x,y)}$) by translation.
  - iii - To construct $\nearrow_{\Gamma^{\vartriangle},(x_0,0)}$ ($x_0 \in \mathbf{N}$) needs only two knowledges (due to the hypothesis $\chi_y(x+1) > \chi_y(x) \geq 0$): its number ξ_{x_0} of "basic" right moves $\nearrow_{\Gamma_0}$, and its number η_{x_0} of "basic" temporal moves $\uparrow_{\Gamma_0}$.

    The value of ξ_{x_0} is easy to obtain: it is given by $\chi_x(x_0+1) - \chi_x(x_0) = \beth_{u_{G,(0,(x_0+1))}} - \beth_{u_{G,(0,(x_0))}}$ known by a).

    The value of η_{x_0} is more difficult to obtain: its depends on $\Gamma_{y_{G,((x_0+1),(x_0+1))}} - \Gamma_{y_{G,((x_0),(x_0))}}$ unavailable on point $\beth_{G,(0,x_0)}$. But this value is also $\uparrow_{G,(x_0,x_0)}$, given by

$$\nearrow_{G,(x_0,x_0)} + \nwarrow_{G,(x_0+1,x_0)} .$$

    It is important to observe that in $\nearrow_{G,(x_0,x_0)} + \nwarrow_{G,(x_0+1,x_0)}$, the number of right basic moves is greater than the number of left basic moves by the hypothesis $\chi_y(x+1) > \chi_y(x) \geq 0$. The value η_{x_0} may be obtained if we know $\nearrow_{G,(x_0,x_0)}$, $\nwarrow_{G,(x_0+1,x_0)}$ and, thus, $\beth_G(1,t)$ ($t \in \mathbf{N}$).

- c - Following the remark of b)iii), we transform our automaton $\mathbf{G}$ in another one $\mathbf{G}^{\natural}$ setting up $\beth_\chi$ with special subsets of states indicating $\beth(0,t)$, $\beth(1,t)$, $\nearrow_{G,(t,t)}$ and $\nwarrow_{G,(t+1,t)}$.
- d - Coming back to grid $\Gamma^{\vartriangle}$, we observe that the points $\beth_{\Gamma^{\vartriangle}}$ have no reason to have 0 as x-coordinate. Thus, we transform this grid to another one $\Gamma^{\blacktriangle}$ such that:
  $\forall x > 0, \forall t \geq 0, \beth_{\Gamma^{\blacktriangle},(x,t)} = \beth_{\Gamma^{\vartriangle},(x,t)}$,
  $\forall t \geq 0, \beth_{\Gamma^{\blacktriangle},(0,t)} = (0, \beth_{x_{\Gamma^{\vartriangle},(1,t)}}) + \beth_{\Gamma^{\vartriangle},(1,t)}$,
  $\forall x < 0, \forall t \geq 0, \beth_{\Gamma^{\blacktriangle},(x,t)} = (x,-x) + \beth_{\Gamma^{\blacktriangle},(0,-x)}$.

The various involved grids are shown on Figure 7.

We begin to define $\mathbf{G}^{\natural}$, cellular automaton $\mathbf{A}^{\vartriangle}$ setting up the grid $\Gamma^{\vartriangle}$ and then cellular automaton $\mathbf{A}^{\blacktriangle}$ setting up the grid $\Gamma^{\blacktriangle}$.

1. *Construction of* $\mathbf{G}^\sharp$ Since $\mathbf{G}$ constructs a grid, it is of the form $(Q_G \times Q, \delta_G)$, we consider that states in Q_G are also a product of states and define $Q_G^\sharp$ as $Q_G \times \{\mathbf{0}, \mathbf{1}, \mathbf{0} \to \mathbf{1}, >\}$ and $Q^\sharp = Q$, thus
$$\mathbf{G}^\sharp = ((Q_G \times \{\mathbf{0}, \mathbf{1}, \mathbf{0} \to \mathbf{1}, >\}) \times Q, \delta_{\mathbf{G}^\sharp}).$$
Intuitively, our choice is:
a state with its second component in $\mathbf{0}$ has its first component in $Q_{G,v}$ and indicates a point of $\rceil(0,t)$,
a state with its second component in $\mathbf{1}$ has its first component in $Q_{G,v}$ and indicates a point of $\rceil(1,t)$,
a state with its second component in $\mathbf{0} \to \mathbf{1}$ has its first component in $Q_{G,r}$ or $Q_{G,\ell}$ and indicates a move $\nearrow_{G,(t,t)}$ or $\nwarrow_{G,(t+1,t)}$,
a state with its second component in $>$ has its first component in $Q_{G,v}$ and indicates a point of $\rceil(x,t)$ with $i > 1$.
The state (M_G, M_C) of $\mathbf{G}$ appearing in the definition 1, is replaced by $((M_G, \mathbf{0}), M_C)$.
If $\delta_G((q_{G,1}, q_1), (q_{G,2}, q_2), (q_{G,3}, q_3)) = (q_G^\star, q^\star)$, we define $\delta_{\mathbf{G}^\sharp}((q_{G,1}, \xi_1, q_1), (q_{G,2}, \xi_2, q_2), (q_{G,3}, \xi_3, q_3))$ as $(q_G^\star, \xi^\star, q^\star)$ with:
If $\xi_1 = \mathbf{0}$ and $q_G^\star \in Q_{G,r}$, then $\xi^\star = \mathbf{0} \to \mathbf{1}$,
if $\xi_2 = \mathbf{1}$ and $q_G^\star \in Q_{G,\ell}$, then $\xi^\star = \mathbf{0} \to \mathbf{1}$,
if $\xi_1 = \mathbf{0} \to \mathbf{1}$, $q_{G,1} \in Q_{G,r}$ and $q_G^\star \in Q_{G,v}$, then $\xi^\star = \mathbf{1}$,
if $\xi_3 = \mathbf{0} \to \mathbf{1}$, $q_{G,3} \in Q_{G,\ell}$ and $q_G^\star \in Q_{G,v}$, then $\xi^\star = \mathbf{0}$,
in other cases, $\xi^\star = >$.
It is easy to see that $\mathbf{G}^\sharp$ has the expected behavior.
2. *Construction of* $\mathbf{A}^\Delta$ We begin by indicating how from $\rceil_{\Gamma^\Delta,(x,t)}$, $\rceil_{\Gamma^\Delta,(x+1,t)}$, $\nearrow_{\Gamma^\Delta,(x,t)}$ and $\nwarrow_{\Gamma^\Delta,(x-1,t)}$ we deduce the same information for $x-1$ and $t+1$ (this corresponds to the point b) ii)). We choose that a temporal move is made of move $\nearrow_{\Gamma_0}$ followed by $\nwarrow_{\Gamma_0}$ (and not the reverse, assuming that the underlying rid is the basic trellis). The Figure 8 illustrates this point.
First let us observe what is known:
points of $\nearrow_{\Gamma^\Delta,(x,t)}$ with their first component in a state of $Q_{\Gamma^\Delta,G,r}$,
points of $\nwarrow_{\Gamma^\Delta,(x-1,t)}$ with their first component in a state of $Q_{\Gamma^\Delta,G,\ell}$,
points of $\rceil_{\Gamma^\Delta,(x,t)}$ and $\rceil_{\Gamma^\Delta,(x-1,t+1)}$ with their first component in a state of $Q_{\Gamma^\Delta,G,v}$.
The process involves the following signals:
   - Point $\rceil_{\Gamma^\Delta,(x,t)}$ sends a signal $\beta_{\blacklozenge,r}$ (if the first move (on the basic grid)) of $\nearrow_{\Gamma^\Delta,(x,t)}$ is $\nearrow_{\Gamma_0}$ or $\beta_{\blacklozenge,\ell}$ if we are in case $\nwarrow_{\Gamma_0}$. This signal achieves a basic temporal move $\uparrow_{\Gamma_0}$.
   - After its move, $\beta_{\blacklozenge,r}$ (or $\beta_{\blacklozenge,\ell}$) disappears and creates a signal α. This signal α achieves move $\nearrow_{\Gamma_0}$ (case of $\beta_{\blacklozenge,r}$) or $\nwarrow_{\Gamma_0}$ (case of $\beta_{\blacklozenge,\ell}$).

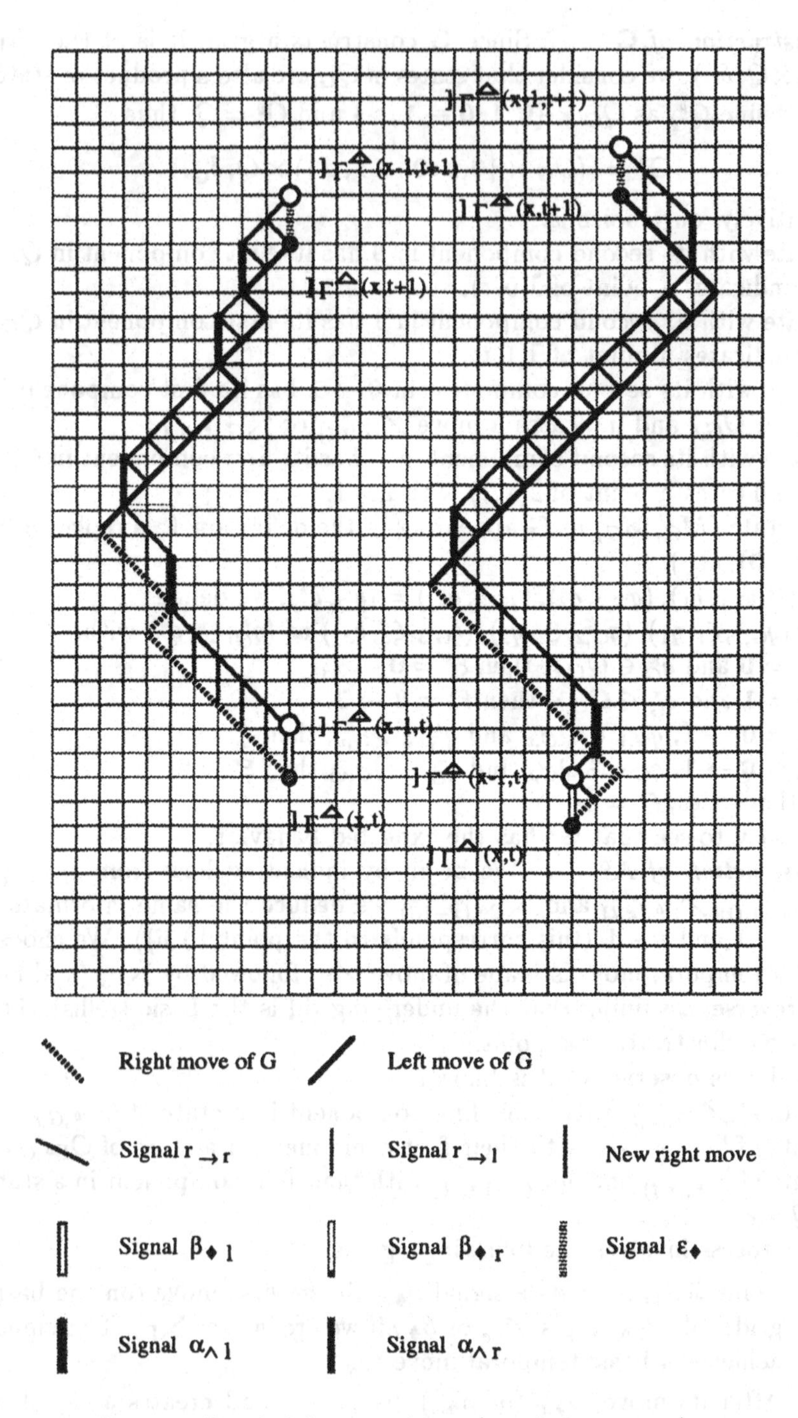

Figure 8. Proof of the proposition 1(translation process).

- Every time that, from some site (ξ, η) in $\nearrow_{\Gamma^\Delta,(x,t)}$ or $\nwarrow_{\Gamma^\Delta,(x+1,t)}$, the first component of state is:
in $Q_{\Gamma^\Delta,G,r}$ (corresponding to $\nearrow_{\Gamma_0}$) a signal $\beta_{r\to r}$ is created, excepted if the first component of site $(\xi+1, \eta-1)$ was in $Q_{\Gamma^\Delta,G,\ell}$.
In a more dynamical point of view, every time that signal α (at the previous step) makes a move $\nearrow_{\Gamma_0}$ not preceeded by a move $\nwarrow_{\Gamma_0}$, it emits signal $\beta_{r\to r}$. Signal $\beta_{r\to r}$ makes a basic move to the left, $\nwarrow_{\Gamma_0}$.
in $Q_{\Gamma^\Delta,G,\ell}$ (corresponding to $\nwarrow_{\Gamma_0}$) a signal $\beta_{r\to\ell}$ is created, excepted if the first component of site $(\xi-1, \eta-1)$ was in $Q_{\Gamma^\Delta,G,r}$.
In a more dynamical point of view, every time that signal α (at the previous step) makes a move $\nwarrow_{\Gamma_0}$ not preceeded by a move $\nearrow_{\Gamma_0}$, it emits signal $\beta_{r\to\ell}$. Signal $\beta_{r\to r}$ makes a basic temporal move, $\uparrow_{\Gamma_0}$.
- Moves of signal α are:
    - When α meets $\beta_{r\to r}$, it goes to the right by a basic move to the right, $\nearrow_{\Gamma_0}$.
    - When α meets $\beta_{r\to\ell}$, it goes to the left by a basic move to the left, $\nwarrow_{\Gamma_0}$.
    - When α meets neither $\beta_{r\to r}$ nor $\beta_{r\to\ell}$, it always meets (from the opposite direction) the trace of $\nearrow_{\Gamma^\Delta,(x,t-x)} \cup \nwarrow_{\Gamma^\Delta,(x+1,t)}$ achieving either $\nearrow_{\Gamma_0}$ or $\nwarrow_{\Gamma_0}$. In this case, it becomes a signal $\alpha_{\wedge,r}$ or $\alpha_{\wedge,\ell}$. Such a signal after a temporal move ($\uparrow_{\Gamma_0}$), becomes α and makes a right ($\nearrow_{\Gamma_0}$) or left ($\nwarrow_{\Gamma_0}$) move.
- Finally, point $\daleth_{\Gamma^\Delta,(x-1,t+1)}$ emits a signal $\epsilon_\blacklozenge$, killing α (only if you wish to achieve the process here).

By this way, signal α is always two units under $\nearrow_{\Gamma^\Delta}$ $(x,t) \cup \nwarrow_{\Gamma^\Delta}$ $(x-1,t)$ in the space time diagram.
To set up all points $\daleth_{\Gamma^\Delta,(x,1)}$ from points $\daleth_{\Gamma^\Delta,(\xi,0)}$, it is suficient to identify signal $\epsilon_\blacklozenge$ with one of the two signals $\beta_{\blacklozenge,r}$ and $\beta_{\blacklozenge,\ell}$ according to the next move of $\nearrow_{\Gamma^\Delta,(x+1,0)}$ and, thus, suppressing $\epsilon_\blacklozenge$, signal α is never killed.
Signals α, $\alpha_{\wedge,r}$ and $\alpha_{\wedge,\ell}$ are put in $Q_{\Gamma^\Delta,G,r}$. Signals $\beta_{\blacklozenge,r}$ and $\beta_{\blacklozenge,\ell}$ are in $Q_{\Gamma^\Delta,G,\ell}$. All previously defined signals are in $Q_{\Gamma^\Delta,G} \setminus (Q_{\Gamma^\Delta,G,r} \cup Q_{\Gamma^\Delta,G,\ell} \cup Q_{\Gamma^\Delta,G,v})$.
In conclusion, knowledge of $\daleth_{\Gamma^\Delta,(x,0)}$, $\nearrow_{\Gamma^\Delta,(x,0)}$ and $\nwarrow_{\Gamma^\Delta,(x,0)}$ is sufficient to set up the whole grid Γ^Δ. This knowledge is given by states of $\mathbf{G}^\sharp$ (point 1)). Formally we define $\Gamma^\Delta = (Q_{G,\Gamma^\Delta} \times Q_{C,\Gamma^\Delta}, \delta_{\Gamma^\Delta})$ by:

- $Q_{G,\Gamma^\Delta,r} = Q_{G,\mathbf{G}^\sharp,r} \cup \{\alpha, \alpha_{\wedge,r}, \alpha_{\wedge,\ell}\}$.
- $Q_{G,\Gamma^\Delta,\ell} = Q_{G,\mathbf{G}^\sharp,\ell} \cup \{\beta_{\blacklozenge,r}, \beta_{\blacklozenge,\ell}\}$.

- $Q_{G,\Gamma^\blacktriangle,v} = Q_{G,\mathbf{G}^!,v}$.
- $Q_{G,\Gamma^\blacktriangle} = Q_{G,\mathbf{G}^!} \cup \{\beta_{r\to r}, \beta_{r\to\ell}\}$.
- $Q_{C,\Gamma^\blacktriangle} = Q_{C,\mathbf{G}^!} \cup Q_{C,\Gamma^{\blacktriangle\star}}$. In $Q_{C,\Gamma^{\blacktriangle\star}}$, we may put any new compuations occuring on grid Γ^Δ since states of $Q_{C,\mathbf{G}^!}$ may be considered as states of Γ^Δ when the first component is in a state of $Q_{G,\mathbf{G}^!,v}$, distinguishing points of $\daleth_{\mathbf{G}^!,(1,t)}$ by point 1).

But to generalize the following process to the construction of all points $\daleth_{\Gamma^\blacktriangle}(x,t)$ ($t \in \mathbf{N}\setminus\{0\}$) is not direct. Roughly speaking, "simply" in the proposition 1 indicates this possibility. Let us discuss this possibility. Due to the finite nature of the automata located on cells, we need to have a finite number of states. This implies that the signals α_t, corresponding to the construction of $\daleth_{\Gamma^\blacktriangle}(x,t+1)$ from $\daleth_{\Gamma^\blacktriangle}(x,t)$ and $\daleth_{\Gamma^\blacktriangle}(x-1,t)$, must have a construction area bounded on the whole space-time diagram. For example consider the following case where $\uparrow_{G,(x,0)}= (4,4)$, $\nearrow_{G,(x,0)}= (2-x, 2-x)$ and $\nwarrow_{G,(x,0)}= (2+x, 2+x)$ in which we do not see, for signal α_t, how choose between signals $r_{\to r}$ and $r_{\to\ell}$ if the two occur. This difficulty may be overcome by setting signal α_{t+1} directly from signal α_t. In the simple case where signal $\nearrow_{G,(x,0)} + \nwarrow_{G,v}$ is made of signals $\nearrow_{\Gamma_0}$, followed by signals $\nwarrow_{\Gamma_0}$, followed by $\nearrow_{\Gamma_0}$, finally followed by signals $\nwarrow_{\Gamma_0}$ as showed by the Figure 9.

But a new difficulty arises. The area $\mathbf{H}_{(x,t)}$ defined by $\nearrow_{\Gamma^\blacktriangle,(x,t-x)}$, $\nwarrow_{\Gamma^\blacktriangle,(x-1,t-x+1)}$ and $\nearrow_{\Gamma^\blacktriangle,(x-1,t-x+1)}$, $\nwarrow_{\Gamma^\blacktriangle,(x-2,t-x+2)}$ may be very large and overlap as many areas $\mathbf{H}_{(x^\star,t^\star)}$ as we want. This means that the growth ratio of areas $\mathbf{H}_{(x,t)}$ must not be too fast increasing and decreasing.

Nevertheless the previous remarks show that it is possible to set up $\mathbf{A}^\Delta$ in numerous cases. For instance, when,

- When $\nearrow_{G,(x,0)}$ and $\nwarrow_{G,(x,0)}$ are constant.
- When $\nearrow_{G,(x,0)}$ is linear in x and $\nwarrow_{G,(x,0)}$ is constant.
- When $\nearrow_{G,(x,0)}$ and $\nwarrow_{G,(x,0)}$ are both linear in x.
- When $\nearrow_{G,(x,0)}$ is increasing in x and $\nwarrow_{G,(x,0)}$ is constant.

The general case remains open.

3. *Construction of* $\mathbf{A}^\blacktriangle$ As in point 1), we may transform cellular automaton Γ^Δ into a cellular automaton $\Gamma^{\Delta'}$ in which points $\daleth_{\Gamma^\blacktriangle}(0,t)$ and $\daleth_{\Gamma^\blacktriangle}(1,t)$ are distinguished. The new cellular automaton $\mathbf{A}^\blacktriangle = (Q_{G,\mathbf{A}^\blacktriangle} \times Q_{C,\mathbf{A}^\blacktriangle}, \delta_{\mathbf{A}^\blacktriangle})$ is such that:
$Q_{G,\mathbf{A}^\blacktriangle,v} = Q_{G,\mathbf{A}^\Delta,v} \cup \{\blacktriangle_v\}$,
$Q_{G,\mathbf{A}^\blacktriangle,r} = Q_{G,\mathbf{A}^\Delta,r} \cup \{\blacktriangle_{r,1}\}$,
$Q_{G,\mathbf{A}^\blacktriangle,\ell} = Q_{G,\mathbf{A}^\Delta,\ell} \cup \{\blacktriangle_{\ell,1}, \blacktriangle_{\ell,2}\}$,

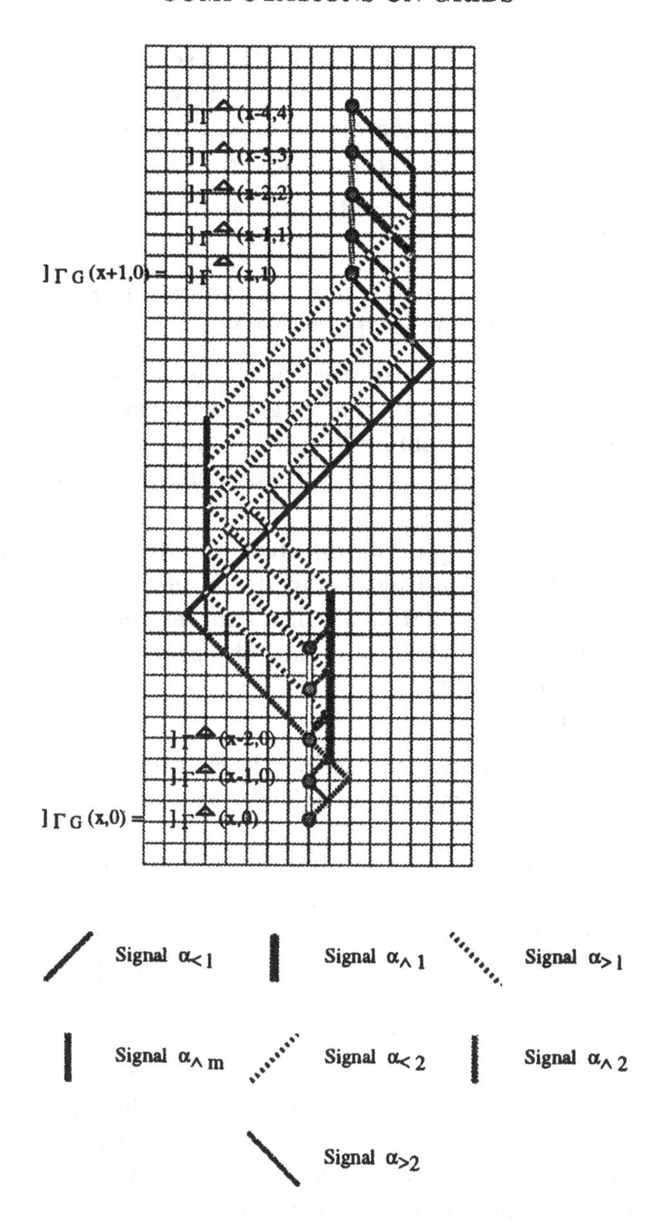

Figure 9. Proof of the proposition 1(stack overflow in holes).

$Q_{C,\mathbf{A}^{\blacktriangle}} = Q_{C,\mathbf{A}^{\vartriangle}} \cup Q_{C,\mathbf{A}^{\blacktriangle\star}}$.
The indices 1 are used to set up $\rceil_{\mathbf{A}^{\blacktriangle}}(0,t)$ and indices 2 to $\rceil_{\mathbf{A}^{\blacktriangle}}(x,t)$ $(x < 0)$.
We begin with the construction of $\rceil_{\mathbf{A}^{\blacktriangle}}(0,t)$. Point $(0,0)$, which is $\rceil_{\mathbf{A}^{\vartriangle}}(0,0)$ and $\rceil_{\mathbf{G}}(0,0)$, sends signal $\blacktriangle_{r,1}$ which remains on cell 0 achiev-

ing only moves $\uparrow$. Points of the form $\daleth_{\Gamma^\vartriangle}(1,t)$ send signal $\blacktriangle_{\ell,1}$ which runs to the left by $\nwarrow$ moves. At their meeting, they create points $\daleth_{\Gamma^\blacktriangle}(0,t)$ in state $\blacktriangle_v$ the first coordinate of which is 0.
To construct other points ($\daleth_{\Gamma^\blacktriangle}(x,t)$ with $x < 0$) is easy:

- A site in state $\blacktriangle_v$ sends to the left signal $\blacktriangle_{\ell,2}$ (with only $\nwarrow_{\Gamma_0}$ moves).
- A point receiving $\blacktriangle_{\ell,2}$ enters state $\blacktriangle_v$ and sends $\blacktriangle_{\ell,2}$ (with only $\nwarrow_{\Gamma_0}$ moves) and $\blacktriangle_{r,1}$ (with $\uparrow_{\Gamma_0}$ moves).

The grid set up by $\mathbf{A}^\blacktriangle$ has the expected behavior and we take it as $\mathbf{A}_\chi$. Let us close this proof by a remark. Looking to vertices of a grid set up both by $\mathbf{G}^\sharp$ and $\mathbf{A}^\blacktriangle$, we do not obtain a grid $\Gamma_\cup$ such that $\daleth_{x,\Gamma_\cup}(0,t) = 0$. But, we may, using a trick similar to point 1), consider only one vertices out of two, defining a new grid.

The use of $Q_{C,\Gamma^{\vartriangle}\star}$ and $Q_{C,\mathbf{A}^{\blacktriangle}\star}$ will appear in the following example. □

As an example of the proposition 1 and its proof (use of $Q_{C,\Gamma^{\vartriangle}\star}$ and $Q_{C,\mathbf{A}^{\blacktriangle}\star}$), we construct an automaton which computes the function which returns value 1 on entries 1^n where n is the p-th prime number and p is also prime and value 0 else.
The idea is very simple:

- a - We use the Fischer's algorithm described in chapter *Computations on cellular automata* and we modify it slightly, considering that the input is 1^n.
- b - We put it on the grid $\Gamma_{2,2;-1,1}$, obtaining an evolution similar to the one depicted on the figure 6 (grouping cells 3 by 3).
- c - On this new grid, the nodes $\daleth_{\Gamma_{2,2;-1,1}}(0,t)$ have state indicating if integer t is or not prime (point $\daleth(0,t)$ corresponds to the input string 1^t). We consider points $\daleth(0,t(x))$ with $t(x)$ is the x-th prime number as points $\chi(x)$ of the proposition 1. The grid $\Gamma_{2,2;-1,1}$ does not play exactly the role of grid Γ_G of the proof of proposition 1. But, looking to this proof, we see that we do not need all the grid Γ_G but only to points $\daleth_G(0,t)$ and $\daleth_G(1,t)$. Using the same feature as in point *Construction of* $\mathbf{G}^\sharp$, repeating moves $\nearrow_{\mathbf{G}}$ and $\nwarrow_{\mathbf{G}}$ until to meet a site $\daleth_{\Gamma_{2,2;-1,1}}(0,\theta)$ with θ prime, we set up the points $\daleth_G(0,t)$ and $\daleth_G(1,t)$ and moves $\nearrow_{G,(x,x)}$ and $\nwarrow_{G,(x,x)}$ between them. This allows us to use the end of the proof of proposition 1, observing that moves $\nearrow_{G,(x,x)}$ are increasing and that moves $\nwarrow_{G,(x,x)}$ are constant. By this way, we obtain grids $\Gamma^\vartriangle$ and, then, $\Gamma^\blacktriangle$.
- d - Finally, we put once again Fischer's algorithm on grid $\Gamma^\blacktriangle$, previously obtained. The new computations occur on points $\daleth_{\Gamma^\blacktriangle}(x,t)$ with $x \geq 0$. By this way, on points $\daleth_{\Gamma^\blacktriangle}(0,t^\star)$ (of the form $(0,\theta(t^\star))$) is indicated if $t^\star$ is prime.

Thus a point $(0, \tau)$ is marked, if and only if it corresponds to a point $\mathfrak{I}(0, t(x))$ with x prime and, thus to an input string 1^n for which n is the x-th prime number.

- e - The previous process does not give a real time recognizer. To obtain such a real time recognizer, we observe that all the involved grids are regular and thus by a grouping operation it is possible to obtain a real time recognizer.

4. Asynchronous computations

4.1. POSITION OF THE PROBLEM

Due to their definition, cellular automata appear as a model of massive synchronous parallelism. This hypothesis of synchronism may be viewed as not realistic. If we consider a complex cellular automaton the transitions of which are given not by a look-up table but by some program to access transition results needs variable amounts of time. Clearly all the system may become synchronous if any cell waits for a delay (previously computed if possible!) long enough to insure that all cells have got their new state. By this way, we decrease the performance of the system: the worst case becomes the common case. On the other hand, the same remark is available for the communication time between cells depending on the length of the encoding of the state. Thus we give a definition of cellular automata taking into account these remarks and keeping true the discrete and regular nature of the system. In order to distinguish delays due to computations from those due to communications, we introduce a new set of communicating letters.

Definition 6

1. *An asynchroneous cellular automaton on* $\mathbb{Z}$ *is* $CA_a = (Q, A, \delta, \gamma, \theta)$ *where:*
   - Q *is a finite set (*states*),* q_0 *denoting the so-called quiescent state.*
   - A *is a finite set (*letters*),* a_0 *denoting the so-called blank.*
   - $\delta : A \times Q \times A \longrightarrow Q \times A$ *is the* transition function, *denoted by* (δ_Q, δ_A).
   - $\gamma : A \longrightarrow \mathbf{N}$ *is the* communication cost function.
   - $\theta : A \times Q \times A \longrightarrow \mathbf{N}$ *is the* computation cost function.
2. *A* temporal configuration *is an application from* $\mathbb{Z}$ *into* $(A \times \mathbf{N})^{<\mathbf{N}} \times Q \times (A \times \mathbf{N})^{<\mathbf{N}}$ *(where* $(A \times \mathbf{N})^{<\mathbf{N}}$ *denotes the finite sequences of couple of elements of* A *and integers), denoted by*
   $C_a = \{(\lhd\, (a_\ell(z, \zeta), \mu_\ell(z, \zeta)) \mid \zeta \in \{1, \ldots, \zeta_{z,\ell}\} \rhd, q(z),$

$\triangleleft (a_r(z,\zeta),\mu_r(z,\zeta)) \mid \zeta \in \{1,\dots,\zeta_{z,r}\} \triangleright, \nu(z)) \mid z \in \mathbb{Z}\}$

We understand:

- $q(z)$ *as the state of cell* z.
- $\triangleleft (a_\ell(z,\zeta),\mu_\ell(z,\zeta)) \mid \zeta \in \{1,\dots,\zeta_{z,\ell}\} \triangleright$ *as the ordonned sequence of letters* $(a_\ell(z,\zeta))$ *received (and not used) by cell* z *from cell* $z-1$. *The letter* $a_\ell(z,1)$ *is the first (last in time) received. The letter* $a_\ell(z,\zeta)$ *was received since* $\mu_\ell(z,\zeta)$ *units of time.*
- $\triangleleft (a_r(z,\zeta),\mu_r(z,\zeta)) \mid \zeta \in \{1,\dots,\zeta_{z,\ell}\} \triangleright$ *as the ordonned sequence of letters* $(a_r(z,\zeta))$ *received (and not used) by cell* z *from cell* $z+1$. *The letter* $a_r(z,1)$ *is the first (last in time) received. The letter* $a_r(z,\zeta)$ *was received since* $\mu_r(z,\zeta)$ *units of time.*
- $\nu(z)$ *is the elapsed time since the last transition of states of cell* z.

From a temporal configuration C_a, *an asynchronous cellular automaton evolves to another temporal configuration* $G(C_a)$ *defined by*
$G(C_a) = \Big\{(\triangleleft (a^\star_\ell(z,\zeta),\mu^\star_\ell(z,\zeta)) \mid \zeta \in \{1,\dots,\zeta^\star_{z,\ell}\} \triangleright, q^\star(z),$
$\triangleleft (a^\star_r(z,\zeta)\mu^\star_r(z,\zeta)) \mid \zeta \in \{1,\dots,\zeta^\star_{z,r}\} \triangleright, \nu^\star(z)) \mid z \in \mathbb{Z}\Big\}$
defined by:
First for all $z \in \mathbb{Z}$ *we achieve first the following process, updating lists of a cell ignoring neighbor cells:*

- *i* - *We call* $\mathbf{CT}(z)$ *the conjunction of the following conditions:*
  - $\mu_\ell(z,2) \geq \gamma(a_\ell(z,1))$,
  - $\mu_r(z,2) \geq \gamma(a_r(z,1))$,
  - $\nu(z) \geq \theta(a_\ell(z,1), q_z, a_r(z,1))$,

  meaning that the next information to proceed are get (the letters sent by neighbors have reached the cell and the cell has finished its current computation using the previous received couple of letters).
- *ii* - *If* $\mathbf{CT}(z)$ *is false then:*
  - $\nu^\star(z) = \nu(z) + 1$.
  - $q^\star(z) = q(z)$.
  - *for* $\zeta \in \{1,\dots,\zeta_{z,\ell}\}$, $\mu^\star_\ell(z,\zeta) = \mu_\ell(z,\zeta) + 1$.
  - *for* $\zeta \in \{1,\dots,\zeta_{z,\ell}\}$, $a^\star_\ell(z,\zeta) = a_\ell(z,\zeta)$
  - *for* $\zeta \in \{1,\dots,\zeta_{z,r}\}$, $\mu^\star_r(z,\zeta) = \mu_r(z,\zeta) + 1$.
  - *for* $\zeta \in \{1,\dots,\zeta_{z,r}\}$, $a^\star_r(z,\zeta) = a_r(z,\zeta)$.
  - $\zeta_{z,\ell} = \zeta_{z,\ell}$.
  - $\zeta_{z,r} = \zeta_{z,r}$.

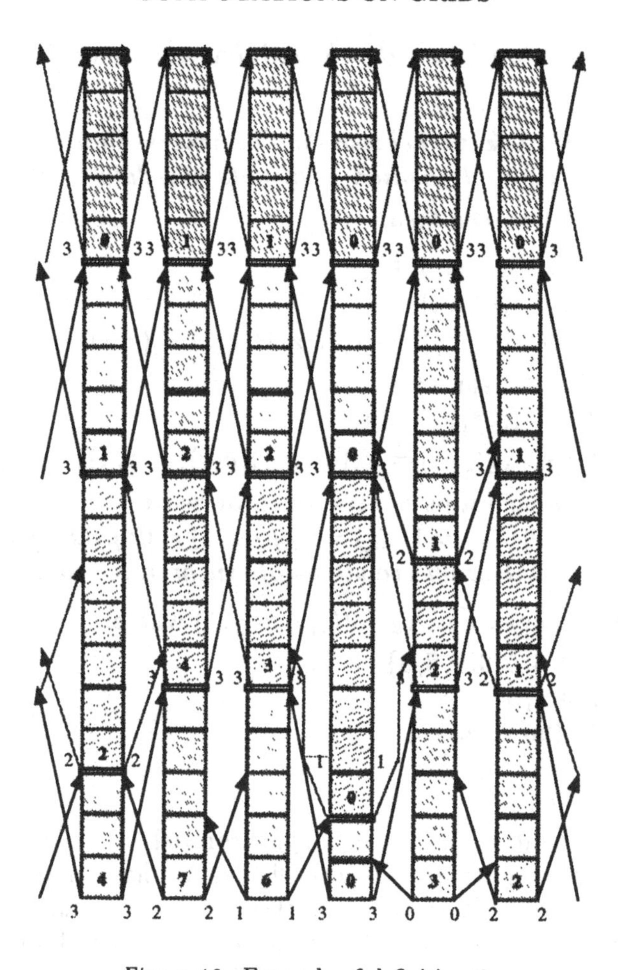

Figure 10. Example of definition 6.

We have nothing to do except to take into account the elapsed time.

- *iii* - *If* $\mathbf{CT}(z)$ *is true then:*

  - $\nu^\star(z) = 0$
  - $q^\star(z) = \delta(a_\ell(z,1), q(z), a_r(z,1))$
  - *for* $\zeta \in \{1, \ldots, \zeta_{z,\ell} - 1\}$, $\mu_\ell^\star(z,\zeta) = \mu_\ell(z, \zeta+1) + 1$
  - *for* $\zeta \in \{1, \ldots, \zeta_{z,\ell} - 1\}$, $a_\ell^\star(z,\zeta) = a_\ell(z, \zeta+1)$
  - *for* $\zeta \in \{1, \ldots, \zeta_{z,r} - 1\}$, $\mu_r^\star(z,\zeta) = \mu_r(z, \zeta+1) + 1$
  - *for* $\zeta \in \{1, \ldots, \zeta_{z,r} - 1\}$, $a_r^\star(z,\zeta) = a_r(z, \zeta+1)$
  - $\zeta_{z,\ell}^\star = \zeta_{z,\ell} - 1$
  - $\zeta_{z,r}^\star = \zeta_{z,r} - 1$

We take into account the elapsed time and we achieve a cell transition.

Second for all $z \in \mathbb{Z}$ we achieve the new process in which a cell which has achieved a state transition, transmits new data to its neighbors :

- iv - If $\mathbf{CT}(z)$ *is true then:*

- $\zeta^\star_{z-1,\ell} = \zeta_{z-1,\ell} + 1$
- $\mu^\star_\ell(z-1, \zeta^\star_{z-1,\ell}) = 0$
- $a^\star_\ell(z-1, \zeta^\star_{z-1,\ell}) = \delta_A(a_\ell(z,1), q(z), a_r(z,1))$
- $\zeta^\star_{z+1,r} = \zeta_{z+1,r} + 1$
- $\mu^\star_r(z+1, \zeta^\star_{z+1,r}) = 0$
- $a^\star_r(z+1, \zeta^\star_{z+1,r}) = \delta_A(a_\ell(z,1), q(z), a_r(z,1))$

3. A temporal configuration is said synchronized *if*
$\forall z \in \mathbb{Z},\ \nu(z) = 0;\ \zeta_{z,\ell} = 1;\ \zeta_{z,r} = 1;\ \mu_\ell(z,1) = 0;\ \mu_r(z,1) = 0$
We restrict ourselves on evolutions of synchronized configurations.

We give an example. Let $\mathbf{A}_a = (Q, A, \delta, \gamma, \theta)$ where:

- $Q = \{0,1,2,3,4,5,6,7,8,9\}$
- $A = \{0,1,2,3,4,5\}$
- $\delta(a_\ell, q, a_r) = (\max\ (a_\ell, a_r), \lceil \frac{q}{2} \rceil)$
- $\gamma(a) = a$
- $\theta(q) = \lceil \log_2 (q) \rceil$

An example of evolution of $\mathbf{A}_a$ on a synchronized periodic configuration is depicted on the Figure 10. On this figure, we have marked parity of waiting delays by colors, a thin line mark when the computational time is consumed and a thick line when all data have been received allowing a new computation. We observe that the value of $\zeta_{z,\ell}$ and $\zeta_{z,r}$ is always 1 except from the cells 3 and 5. For example we describe data of cell 3 at time 5: $\zeta_{3,\ell} = 1$; $a_\ell(3,1) = 2$; $q_3 = 6$; $\zeta_{3,2} = 2$; $a_r(3,1) = 3$; $a_r(3,2) = 1$.

We observe that, in some way, every cell evolves "as soon as possible". On Figure 10, we have represented by dotted lines times on which all cells have achieved (for the first time) their t-th transition.

4.2. FREEZING AN AREA

In a first time, we forget the definition 6 and we ask ourselves what may happen if a cell of a one dimensional cellular automaton knows that it will be unable to achieve any computation during $\theta(q)$ units of time. The simpler way as depicted on the figure 11, is to freeze some area of computation in order to send through the network this impossibility. This is done by sending in both directions a freezing signal ($\mathbf{S}_F$) with slope 2 and, then, an unfreezing signal ($\mathbf{S}_{U_f}$) at speed 1.

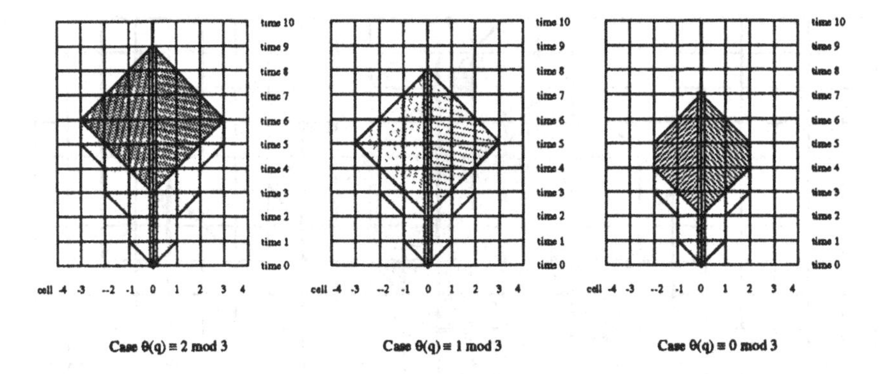

Figure 11. Freezing left and right areas due to computation time.

It is easy to see that such a process depends on the remaining of $\theta(q)$ by 3. More precisely, we have:

- a - If $\theta(q) \equiv 2 \mod 3$, $\mathbf{S}_{U_f}(2)$ is sent $\theta(q)$ div 3 units of time after $\mathbf{S}_F$. When $\mathbf{S}_{U_f}(2)$ reaches $\mathbf{S}_F$, it kills it and sends immediately to the opposite direction an unfreezing signal at maximal speed $\mathbf{S}_U$.
- b - If $\theta(q) \equiv 1 \mod 3$, $\mathbf{S}_{U_f}(1)$ is sent $(\theta(q)$ div $3) - 1$ units of time after $\mathbf{S}_F$. When $\mathbf{S}_{U_f}(1)$ reaches $\mathbf{S}_F$, it follows it one time and, then, it kills it and sends immediately to the opposite direction an unfreezing signal at maximal speed $\mathbf{S}_U$.
- c - If $\theta(q) \equiv 0 \mod 3$, $\mathbf{S}_{U_f}(0)$ is sent $(\theta(q)$ div $3) - 1$ units of time after $\mathbf{S}_F$. When $\mathbf{S}_{U_f}(0)$ reaches $\mathbf{S}_F$, it kills it and sends, after one time, to the opposite direction an unfreezing signal at maximal speed $\mathbf{S}_U$.

What happens, when we consider a delay of the form $\gamma(a)$? The situation is very close to the previous one, considering that, now, the forbidden time for the two neighbors is $\gamma(a) - 1$. Thus the process is identical replacing $\theta(q)$ by $\gamma(a) - 1$ on the two neighbors: this is done by sending a signal to the two neighbors $\mathbf{S}_{C_f}(\gamma(a) - 1)$, indicating this fact.

By this way to set an asynchronous cellular automaton on a synchronous one is to manage all the freezing areas. We also observe that, due to the finite nature of any machine, we have the ability to label signals $\mathbf{S}_F$, $\mathbf{S}_{U_f}$ and $\mathbf{S}_U$ by the label of the freezing origin cell and by the number of cell it remains to be visited ($\{\theta(q) \mid q \in Q\}$ and $\{\gamma(a) \mid a \in A\}$ are bounded).

Let M be the maximum of the values of $\{\theta(q) \mid q \in Q\}$ and $\{\gamma(a) \mid a \in A\}$, To solve problems we use the same trick as in (Choffrut and Čulik II, 1984): to each cell is associated the remaining of its number (in the line) by $M + 1$.

The figure 12 illustrates the freezing process of several areas.

We summarize the ideas of this section, remarking that the number of possible delays is bounded:

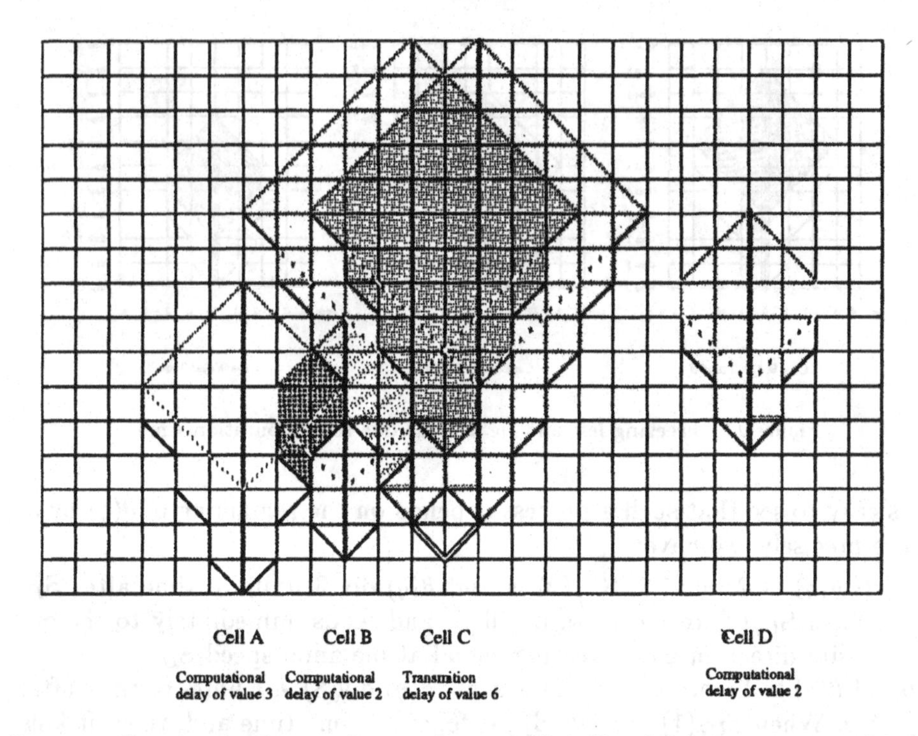

Figure 12. Freezing simultaneously several areas: cell A, B and D freeze a computational area corresponding to delay 3,2 and 2; cell C must send a letter with delay 6 and send a freezing signal of value 5.

1. Computational delay are taken into account via a freezing process involving 3 signals $\mathbf{S}_F$, $\mathbf{S}_U$, $\mathbf{S}_{U_f}$. The value of the delay is put in states.
2. Delays due to transmission are considered as computational delays of the neighbors: they involved a finite number of letters $\mathbf{S}_{C_f}(\gamma(a)-1)$.
3. To avoid collapses between signal sent by different cells, all the previous signals are also labeled by the remainder of the number of the cell in the line by $M+1$.

By this way, we multiply the number of states by $M(M+1)$ and the number of letters by $2 \times 4M(M+1)$.

Finally, we observe that this freezing process do not solve our problem: for example, on figure 12, cell A (or cell B) must keep in mind many values (possibly in an unbounded number) of letters received from its left (right) neighbors. This problem is (partially) solved using grids in the next section.

4.3. DYNAMIC GRIDS

The main idea is to define a grid the size of holes of which depends on the different delays. The Figure 13 illustrates this process in the case of the

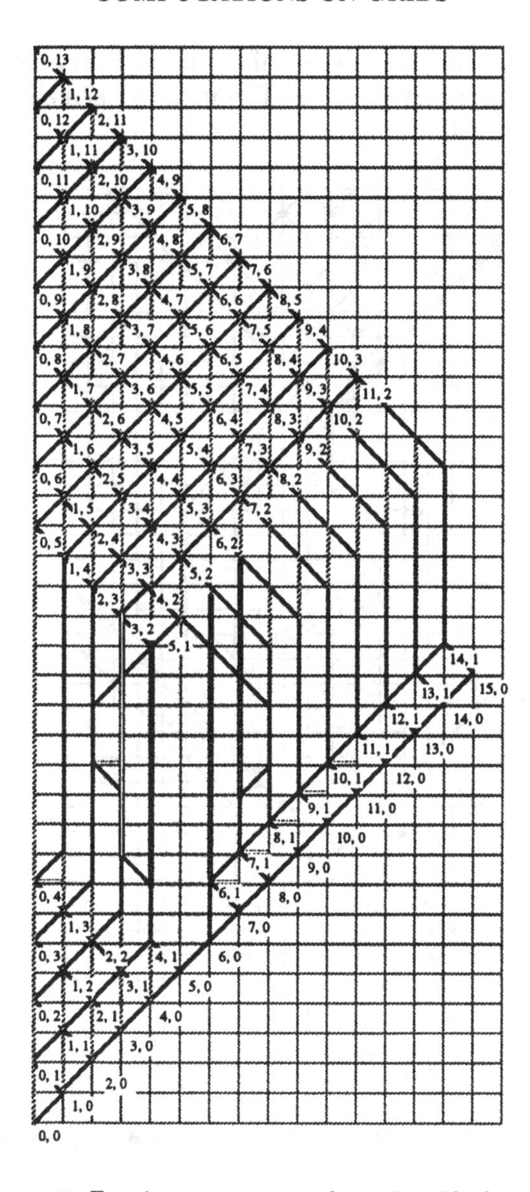

Figure 13. Freezing an area on a dynamic grid: the grid.

only delay is the one of cell A of Figure 12. We observe that the delay has now the ability to propagate itself both to the right and to the left. We describe the involved process.

The usual grid of a trellis automaton on which computations occur has a delay corresponding to the frozen area. Thus in both directions of influence, the holes become greater in the time. As shown on the Figure 12, the "height" of hole (in fact spatial move to the right or the left) is increased by the value of the delay minus one unit of time. Every cell keeps in mind its own time and the received letter. The involved signals are besides $\mathbf{S}_F$, $\mathbf{S}_U$, $\mathbf{S}_{U_f}$ already indicated:

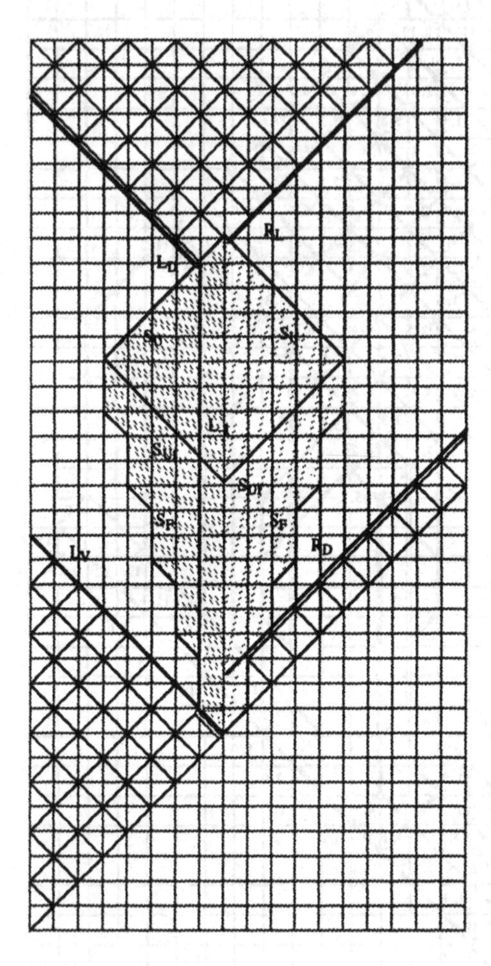

Figure 14. Signals setting up a frozen area due to a computational delay.

- a - L_V is sent to the left at maximal speed. It indicates to construct "big" holes on the left: right moves becomes a succession of underlying right and left moves (finally in -b- ending by a left one). L_V runs to the left until it meets one of the (left or right) signals $\mathbf{S}_F$.
- b - L_{-1} is used to indicate the end of the influence of the delay on the left. Its movement is: it achieves one left move then is make a succession of right and left moves (in the three cases of the figure 11) until it reaches the (right) signal $\mathbf{S}_U$; then it goes at maximal speed to the left indicating to" frozen right moves" to go one time to the left.

- c - R_D is used to construct "big" holes on the right. It remains 2 unit of times (right and left moves) on the "freezing" cell, and then goes to the right at maximal speed: left moves becomes a succession of underlying left and right moves (finally in -d- ending by a right one).
- d - R_L is used to indicate the end of the influence of the delay on the right. It is sent by the 'freezing" cell at the end of the delay. It runs at maximal speed indicating to" frozen right moves" to go one time to the right.

On the Figure 13, the number indicate the site on which occurs the i, j computation (the j-th computation on cell i). The so-constructed grid is called a *dynamic grid* because it depends of the computation and, thus, of the inputs.

It remains to see how various delay may combine themselves: to indicate how some frozen areas kill other ones or modify them. All situations are indicated on the Figure 15 which shows the dynamic grid corresponding to Figure 12.

- i - Every signal L_V or R_D meeting a freezing area (a signal $\mathbf{S}_F$) is killed.
- ii - A signal L_D meeting freezing area (a signal $\mathbf{S}_F$) creates a signal R_D.
- iii - A signal R_L meeting freezing area (a signal $\mathbf{S}_F$) creates a signal L_V.

We observe on the Figure 15 that a great delay influences all the line in both directions, but that another great delays do not add their disadvantage but only combine them. For example, on Figure 15, the total delay induced on the line by 40 units of time is reduced to 16 units. In any case, this process is greatly more performing that the naive one: to make a grid with holes of size 10.

5. Conclusion

We have seen that the strongly constrained nature of the dependencies in one dimensional cellular automata allows us to twist the dependencies obtaining by this way the notion of grids. The new point is to study how to twist grids. In fact four types of studies arise:

1. To construct "rational" grids in which the size of the holes are constant (as in the section 3) or defined by a family of finite automata (as in (Mazoyer, 1996)).
2. To study the impossibility of some grids. For example all "recursive" grids (the function $\mathbf{N} \times \mathbf{N}$ into $\mathbf{N} \times \mathbf{N}$, giving the sizes of the (i, j)-th hole, is recursive) may be not all set up (think to an exponential tower size).
3. To construct more complex grids, constrained by another one dimensional cellular automaton (see (Heen, 1996)).

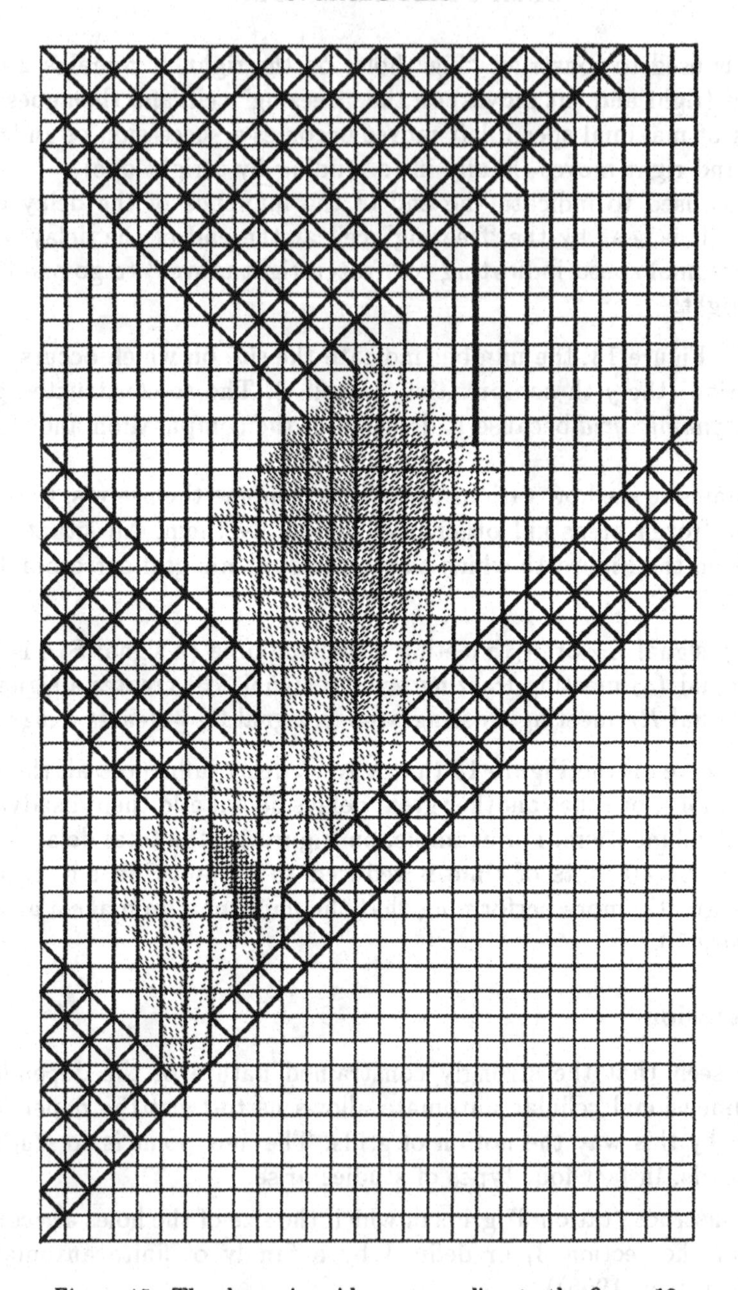

Figure 15. The dynamic grid corresponding to the figure 12.

4. To set up grids depending on the computations themselves (and, thus, of the inputs) as the dynamic grids of the section 4.

In addition, it is possible to show that all recursive functions are computable using grids (Mazoyer, 1996), this situation may be worthy to study in the case of asynchronous computations; An important point is, now, to develop some software (programming cellular language ?) using these ideas.

References

Atrubin A. An iterative one-dimensional real-time multiplier. *Term paper for App. Math. 298*, Stanford University, 1962.

Choffrut C. and Čulik II K. On real time cellular automata and treillis automata. *Actae Informaticae*: 393–407, 1884.

Ibarra O. Kim S. and Moran S. Sequential machine characterization of trellis automata and applications. *SIAM J. on Computing*Vol. no. 14: 349–365, 1985.

Čulik II K., Gruska J. and Salomaa A. Systolic treillis automaton (for VLSI). *Research Report CS-81-34*. Department of Computer Science, University of Waterloo, 1981.

Heen O. Economie de ressources sur Automates Cellulaires. *Diplôme de Doctorat*, Université Paris 7 (in french), 1996.

Mazoyer J. Entrées et sorties sur lignes d'automates. in *Algorithmique parallèle*. Cosnard M., Nivat M. and Robert Y. Eds, Masson, 47–65, 1992.

Mazoyer J. Computations on one-dimensional cellular automaton. *Annals of Mathematics and Artificial Intelligence* Vol. no. 16: 285–309, 1996.

Mazoyer J. and Terrier V. Signals in one dimensional cellular automata. To appear in *Theoretical Computer Science*, 1998.

Minsky M. *Finite and infinite machines*, Prentice-Hall, 1962.

Part 3
Computational power

Traditionally the computational power of a computation model is expressed by means of complexity classes, which are language classes. Thus studying it gives rise to language recognition and decidability problems. Cellular automata are the best formalized models of massive parallelism and a lot of papers have been devoted to evaluated their efficiency and to compare their performance to those of other models, especially Turing machines.

Their infinite space and parallel behavior suggest different notions of recognition. They nevertheless lead to few different complexity classes (in particular linear and real time complexity classes), which, astonishingly, leave open a few robust problems.

The following contributions draw up the state-of-the-art (of the issue) for one-dimensional cellular automata, present some developments for dimension 2 and explain some applications to recurrence equations. Moreover, a powerful – and now classical - technique in order to differentiate classes is set out.

Computational power

CELLULAR AUTOMATA AS LANGUAGES RECOGNIZERS

M. DELORME
LIP ENS Lyon
46, Allée d'Italie, 69364 Lyon Cedex 07, France

AND

J. MAZOYER
LIP ENS Lyon
46, Allée d'Italie, 69364 Lyon Cedex 07, France

1. Introduction

The computational power of a computation model may be roughly defined by "what it is able to compute". At this level, cellular automata have the same computational power as Turing machines, PRAM or boolean circuits for example. In order to get more subtle understanding and results, one has to compare their performances, on the computational functions or problems, according to criteria, which depend, more or less, on their own features or on features of some variants. Time and space are the natural and basic resources of all models, and they give rise to complexity classes, that may be refined according to some characteristics of the model. In the cellular automata case, the input-output location issue is worthy to take into account.

Actually, cellular automata are infinite objects and inputs are encoded as finite words, that means finite sequences of letters. So a first question is: where to start to give the input word to the system? An origin cell has to be chosen. But another question arises: how to go on with this - necessary sequential - process of entering the following input letters? It can be done time after time on the same cell (sequential input mode, which corresponds to the time "sequentiality") or, at the same initial time, each input letter a_i connected to a distinct cell at site s_i in such a way that $i \neq j$ implies $s_i \neq s_j$ and that the letters order should be expressed by links between the corresponding sites (parallel input mode, which corresponds to some "space sequentiality"). Let us note that this distinction, which leads to major differences for cellular automata, does not make sense for

Turing machines. This comes from the fact that such machines have only a finite number of heads (so a number necessarily independent of the input length) while a cellular automaton has as many cells as the input length, whatever the length. Another classic problem is the output one: where or how to locate the result of a computation. To limit the difficulty, it is usual to restrict the complexity measure to decision problems which only require canonical "yes/no" outputs. Finally to evaluate the performances of a computation model on a decision problem comes down to evaluate the perfomances of the model as language recognizer.

A lot of papers have been dedicated to language recognition. We will sum up the "state of art" in stressing connections with complexity, which is developed in the following O. Ibarra's contribution.

It is not difficult to verify that any one-dimensional cellular automaton $\mathcal{A}$ can be simulated by a one-tape-Turing machine $\mathcal{M}_{\mathcal{A}}$, in such a way that if $s_{\mathcal{A}}$ (resp. $s_{\mathcal{M}\mathcal{A}}$) and $t_{\mathcal{A}}$ (resp. $t_{\mathcal{M}\mathcal{A}}$) denote the space and the time used by $\mathcal{A}$ for some computation on a finite configuration which terminates (resp. space and time for $\mathcal{M}_{\mathcal{A}}$ to simulate it),

- $s_{\mathcal{M}\mathcal{A}} = s_{\mathcal{A}}$,
- $t_{\mathcal{A}} \leq t_{\mathcal{M}\mathcal{A}} \leq k_0 t_{\mathcal{A}} s_{\mathcal{A}}$, where k_0 is some constant integer only depending on the neighborhood size (Smith, 1960).

These results have some interesting and deciding consequences:

- For cellular automata the "space" resource will not produce proper and new space complexity classes. So only time resource remains in the lists.
- Considering space unbounded computations will give rise to time complexity classes high enough in the Turing hierarchy, and cellular automata have no chance of accelerating computations. Thus, to hope for cellular automata specific classes, requires to bound the computation space.

  Nevertheless that does not make the unbounded case futile. Because it allows to specify the links between the Turing and cellular automata models in the computability frame, where it is the computation potential that is studied, and to translate some classical issues in the Turing machines field to the cellular automata one.
- It immediately follows from the above relations that computation needing exponential time on cellular automata are also devoid of interest.
- Suppose that the cellular automaton space is constant, equal to the input length n, and that $t_{\mathcal{A}} = n^k$, $k \in \mathbb{N}$. Then, $n^k \leq t_{\mathcal{M}\mathcal{A}} \leq n^{k+1}$. This shows that to get a significant speed up with cellular automata, considering linear time, which moreover gives the maximal acceleration, is enough.

  We will see that taking constant space equal to the input space suffices to get the interesting classes and open questions, the most exciting of which being: is linear time distinct from real time? is there any relevant time distinct from real time?

We know that the first neighbors neighborhood allows to simulate any other neighborhood. But this neighborhood is not minimal. Paying attention to minimal ones leads defining and studying one-way-cellular automata and comparing them to the classical ones, especially through their complexity classes. There, still remain tickling open problems.

We will now give definitions, examples and results, then recall some proofs methods already used in other chapters of this volume, and develop another one, which does not appear elsewhere in this book. We will put to light the links between the obtained classes and those of the Chomsky's hierarchy. All this will be done in the one dimensional case, which has been well studied, although interesting issues remain open. Finally, we will make an incursion into the two dimensional case which starts again to be explored (see (Smith, 1971c), the next chapter by O. Ibarra and (Terrier, 1998)).

2. Language recognition on CA and OCA

Let us recall that a one-dimensional cellular automaton $\mathcal{A}$ is defined by a couple (S, δ) where S is the set of states and δ the local transition function from S^i into S, with $i = 3$ in the case of cellular automata (CA for short) and $i = 2$ for one-way-cellular automata (OCA for short). Communications in both cases are represented on Figure 1 below:

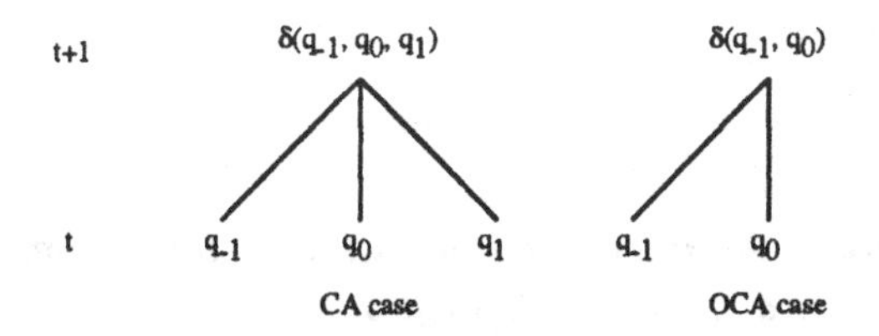

Figure 1. Communications in CA and OCA cases

In order to specify languages by CA or OCA, it is necessary to make clear how such a machine behaves on a finite word in such a way that it can decide whether the checked word belongs to the considered language or not. So, initial and accepting or rejecting configurations have to be defined. That set up two problems. The first one is: are space unbounded computation allowed? the second one: how are the words given to the automaton?

2.1. BOUNDED AND UNBOUNDED COMPUTATIONS

One has already remarked that, in order to get significant time complexity results by means of cellular automata, one has to limit computations to bounded ones. What does that precisely means? Intuitively one wants to recursively master the number of "active" cells during a computation on a word, and this, according to the length word. Actually we will speak of bounded computation when the number of active cells is the input size. It will be sufficient to capture the significant classes, that is *real* and *linear time classes.*

Let us notice that in the OCA case the active cells will to be found on the right of the initial cell, as well as in the CA case for bounded computations. But, let also note that to simulate a CA on which computations (on finite initial configurations) are not bounded by a CA on which the active cells are always on the right of the first initial active cell is possible in folding the space, as for Turing machines. So, finally, in case of language recognition, it is sufficient to restrict configurations to applications from $\mathbf{N}^i$ into S, what will be done in the sequel.

2.2. PARALLEL AND SEQUENTIAL INPUTS

Let Σ be a finite alphabet, L a language on Σ and $\mathcal{A} = (S, \delta)$ a cellular automaton (CA or OCA). The $\mathcal{A}$-configuration at time t, $t \geq 0$, will be denoted c_t.

First of all, as we have to cope with finite data, we suppose that $\mathcal{A}$ has a quiescent state q_e. Secondly, the Σ letters have to be encoded as states of $\mathcal{A}$. It is done by means of some given function ϕ together with two other special encoding functions ϕ_{begin} and ϕ_{end} which allow to distinguish the beginning and the end of the input word.

Definition 1 *Parallel and sequential modes*

- *In parallel mode, the initial configuration of the automaton corresponding to a word $a_1 a_2 \ldots a_n$ is defined by:*
  $c_0(1) = \phi_{begin}(a_1)$, $c_0(n) = \phi_{end}(a_n)$,
  $c_0(i) = \phi(a_i)$, $1 < i < n$ *and* $c_0(i) = q_e$ *for* i, $i > n$.
- *In sequential mode, we have to specify the state of the first cell for the n first configurations, and we have to distinguish OCA and CA.*
  - *In the OCA case,*
    $c_0(1) = \phi_{begin}(a_1)$, $c_{n-1}(1) = \phi_{end}(a_n)$,

    $c_i(1) = \phi(a_{i+1})$, $0 < i < n-1$ *and* $c_0(i) = q_e$ *for* i, $1 < i$.

- *In the CA case, we have to take into account the result of the t to $t+1$ computation step on the first cell, which is the entry cell. So,*

  $$c_0(1) = \phi_{begin}(a_1),\ c_{n-1}(1) = \phi'_{end}(a_n),$$

  $c_i(1) = \phi'(a_{i+1})$, $0 < i < n-1$ *and* $c_0(i) = q_e$ *for* i, $1 < i$, *where* $\phi'(a_i)$ *and* $\phi'_{end}(a_n)$ *depend on* $\phi(a_i)$ *and on the computation of the automaton on the letters already entered.*

Intuitively, a word will be said recognized or accepted by some automaton when the automaton, starting at time 0 from a convenient initial configuration will enter, after a finite time, a configuration carrying some special information. Where does this information can be found? When is it reasonable to expect it? In order to better understand conventions adopted in the definitions to come, it can be worthy to well conceive how information goes according to the model.

2.3. DEPENDENCIES IN CA AND OCA

Look at Figure 2. The patterns on cell x do not represent the state of x but only the numbers (coded as patterns for a better visibility) of the cells information in x is depending on. Let us set up some notations. From now on we will refer to CA working in bounded parallel (sequential) mode by PCA (SCA), to OCA working in bounded parallel (sequential) mode by POCA (SOCA), and to corresponding devices working in unbounded mode as uPCA, uSCA, uPOCA and uSOCA.

It can help to note that:

- On PCA, the whole initial information, contained in the n active cells of the initial configuration, induces information on every active cell as soon as the $(n-1)$-th computation step.
- On SCA, the whole information induced by the input is on the initial cell as soon as the last letter is entered.
- On POCA, cell i only gets information from the cells situated at its left, while on SOCA each cell i holds the whole entered information. These facts make some simulations intricate enough.

We are now able to formalize what will mean for a word to be accepted or recognized by some of the above types of cellular automata.

2.4. ACCEPTANCE OR RECOGNITION

Cellular automata implied in recognition procedures have some distinguished states: a quiescent state and some other states called the accepting states. Figure 3 illustrates the following definitions, which do not depend

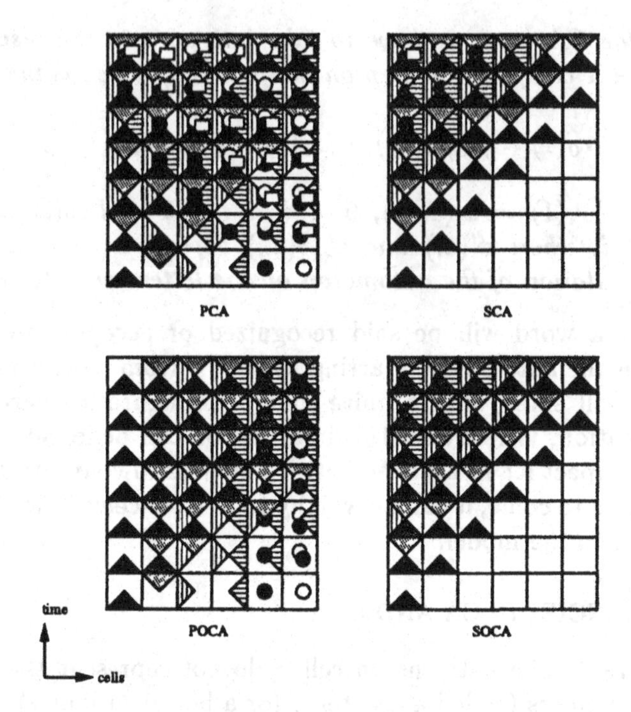

Figure 2. Dependencies on CA and OCA

on the input mode, but on the CA and OCA cases as well as on the computation type.

Definition 2 *Word acceptance*

- *A word u is accepted or recognized by a CA, $\mathcal{A}$, if, $\mathcal{A}$ starting at time 0 from the initial configuration defined by u, there exists a time t such that the initial cell enters an accepting state at t. This definition is valid in both bounded and unbounded cases.*
- *A word u is accepted or recognized by an OCA, $\mathcal{A}$, if, $\mathcal{A}$ starting at time 0 from the initial configuration defined by u, there exist a time t and a positive integer m such that the m-th cell enters an accepting state at t. In case of bounded computation, $m = |u|$, while in the unbounded case, $m \geq |u|$*

Incidentally, note that we implicitly only consider non empty words.

What happens when a word is not recognized? In the bounded case, as the states number is finite, there is a time $t_{\mathcal{A}}$ such that cell 1 (in case of CA or cell m in case of OCA) will never enter any accepting state if that has not happend before $t_{\mathcal{A}}$ computation steps, and then if a word is not recognized, it is known to be outside the checked language. In the unbounded case, if

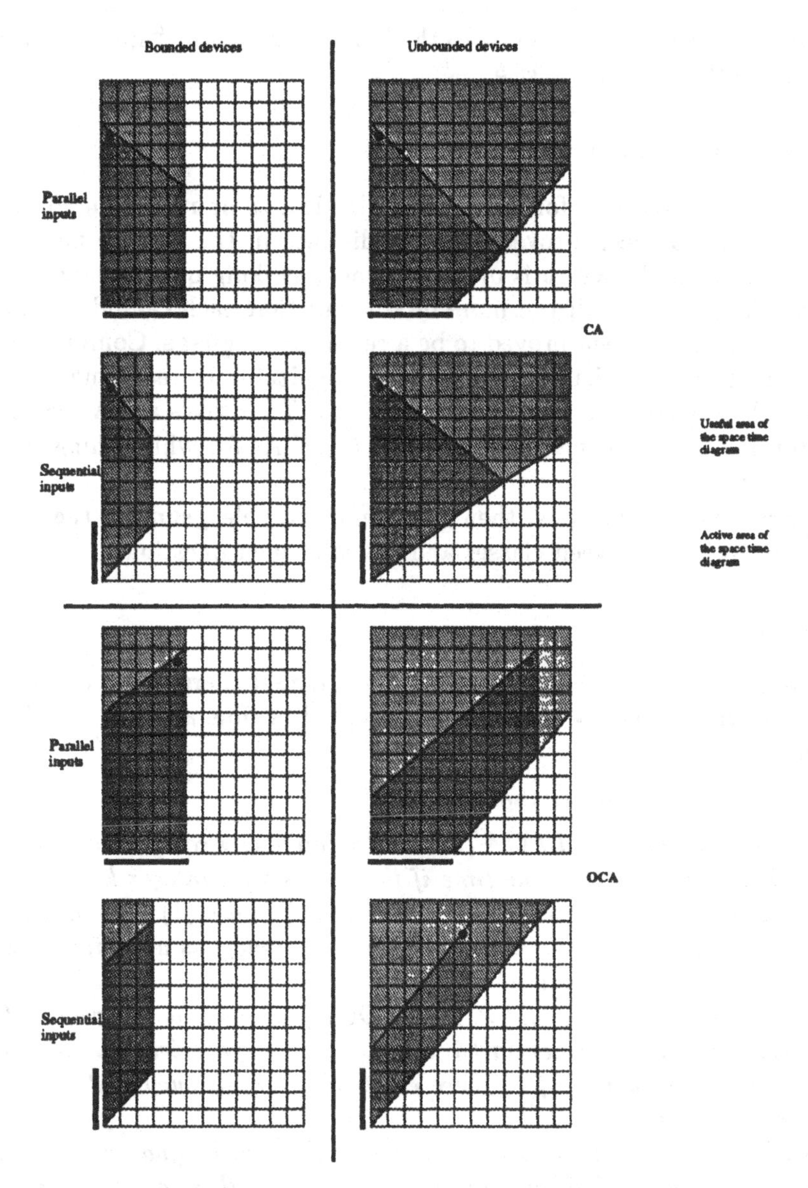

Figure 3. Useful (darker colored) and active areas of space-time diagrams and location of acceptance (indicated by a black point), in case of CA and OCA according to the input mode. The black straight lines represent the inputs lengths.

no rejecting state is specified, only accepting states are significant, and so only accepted words have a status.

What about recognition times? We already have said that two classes are of major interest: the classes of languages recognized in linear or real time, that is in time proportional to the input length or in "minimal" time. Let us make the notions precise.

3. Linear and real time

These classes are distinct for most of the classical computation models. Up to now, it has not been proved they are distinct in the cellular automata case in parallel mode, which is one of the most striking question remaining open in this field. It is all the more interesting that each candidate, up to date, has ultimately been proved to be a real time language. Computability theory asserts there exist languages not recognizable in real time on CA, but no such language has never been exhibited. That gives a new reason to pay attention to the sequential mode and to OCAs for which things are a little clearer.

We start with definitions, then give some examples, sum up the established connections between classes and the still open questions.

3.1. LINEAR TIME

A language L is said to be a linear time language if there exists an integer k, $k \geq 1$, such that each word u of L is accepted in time at most $k|u|$. More precisely:

Definition 3 *Linear time languages*

- *L belongs to the class $LPCA$ ($LSCA$) of languages recognized in bounded parallel (sequential) linear time if there exists an integer k such that, for each u of L, there exists t, $t \leq k|u|$, such that $c_t(1)$ is an accepting state. Definitions in the unbounded case is analogous and gives rise to the classes $uLPCA$ and $uLSCA$.*
- *L belongs to the class $LPOCA$ ($LSOCA$) of languages recognized in bounded parallel (sequential) linear time if there exists an integer k such that, for each u of L, there exists t, $t \leq k|u|$, such that $c_t(|u|)$ is an accepting state.*
- *L belongs to the class $uLPOCA$ ($uLSOCA$) of languages recognized in unbounded parallel (sequential) linear time if there exists an integer k such that, for each u of L, there exists t, $t \leq k|u|$, such that $c_{t+|u|}(t)$ is an accepting state.*

3.2. REAL TIME

A language L is said to be a real time language when each of its words is accepted "as soon as possible" or in "minimum" time. Thus, in this case, unbounded computations are no more relevant.

Definition 4 *Real time languages*

- *L belongs to the class RPCA (RSCA) of languages recognized in parallel (sequential) real time if, for each u of L,* $c_{|u|-1}(1)$ *is an accepting state.*
- *L belongs to the class RPOCA of languages recognized in parallel real time if, for each u of L,* $c_{|u|-1}(|u|)$ *is an accepting state.*
- *L belongs to the class RSOCA of languages recognized in sequential real time if, for each u of L,* $c_{2|u|-1}(|u|)$ *is an accepting state.*

These definitions are illustrated on Figure 4. Some remarks could complete them. First, in the one-way case, information going only to the right, it is natural to take a decision when the whole initial information reaches the last active cell. But in the two-way case, we could have chosen any active cell, or rather, in view of uniformity, a "central" one. That would have set up the problem of knowing the central cell(s), which is not trivial for OCA and changes the problematics. On the other hand, if the input is not encoded by means of ϕ_{end}, we have to add a unit time to give the automaton the time to know that the input is completed. So, in the above definitions we have to consider successively $c_{|u|}(1)$, $c_{|u|}(|u|)$ and $c_{2|u|}(|u|)$.

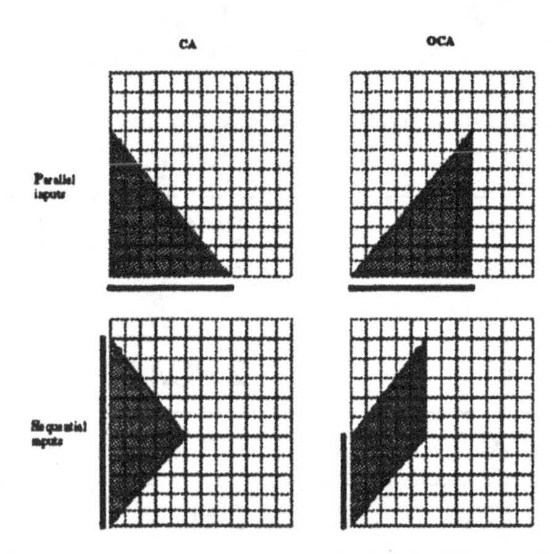

Figure 4. Useful active areas of space-time diagrams and location of acceptance, in case of CA and OCA according to the input mode, in real-time recognition.

Notions we are mainly interested in are now sum up in Figure 5, with their abbreviations[1].

The notations we have chosen are not the historical ones nor the ones used in the other chapters of this volume. We are nevertheless keeping them up in the sequel because they are very expressive and simple, and we give in

[1]In the Figure 5, CA and OCA obviously mean 1-CA and 1-OCA: we have chosen not to mention the dimension when it is 1.

Automaton type	CA OCA 2-CA 2-OCA
Input mode	P (parallel) S (sequential)
Time	L (linear) R (real) (any)
Space	u (unbounded) (bounded)

Figure 5. Usual abbreviations.

the following array the correspondance with those used elsewhere through the volume.

Input mode	*Automaton type*	*Present notations*	*Other notations in this book*
		1D-CA	
Parallel	CA	PCA	LCA
Parallel	OCA	POCA	OLCA
Sequential	CA	SCA	LIA
Sequential	OCA	SOCA	OLIA
		2D-CA	
Parallel	CA	2-PCA	MCA
Parallel	OCA	2-POCA	OMCA
Sequential	CA	2-SCA	MIA
Sequential	OCA	2-SOCA	OMIA

4. Results

Let us start with some examples in order to situate well known languages in the different introduced classes or to put to light some other ones used to separate some of these classes. The reader is refered to the source papers for proofs or to (Delorme and Mazoyer, 1994), where the main results are proved, and the usual proof methods are explained[2].

4.1. EXAMPLES

They are gathered together in the two following arrays. The letters which can be found inside are to be understood the following way: [*a*] signifies that the proof is built by means of signals, [*b*] that the result is inferred from (Čulik, 1989) and [*c*] that it is a consequence of (Čulik *et al.*, 1981).

Languages	*RPOCA*	*RPCA*	*RSCA*
$\{a^n / n \ prime\}$	− (Čulik *et al.*, 1981)	+	+ (Fisher, 1965)
$\{a^{2^n} / n \geq 0\}$	− (Čulik *et al.*, 1981)	+	+ (Choffrut and Čulik, 1984)
$\{a^n b^n c^n / n \geq 1\}$	+ (Dyer, 1980)	+	+ [a]
$\{a^n b^{n+m} a^m / m, n \geq 1\}$	+ (Čulik, 1989)	+	+ [a]
$\{uu / u \in \Sigma^+\}$	?	+	+ (Cole, 1969)
$\Sigma^*\{uu / u \in \Sigma^+\}$	?	+	− (Cole, 1969)
$\{uu^R / u \in \Sigma^+\}$	+ (Dyer, 1980)	+	+ (Cole, 1969)
$\Sigma^*\{uu^R / u \in \Sigma^+\}$ [3]	+ [b]	+	− (Cole, 1969)
$\{u^{\lvert u \rvert} / u \in \Sigma^+\}$	− [c]	+	+ (Terrier, 1995)
$\Sigma^*\{u^{\lvert u \rvert} / u \in \Sigma^+\}$	?	+ (Terrier, 1995)	− (Terrier, 1995)

[2]But, there, the proofs, though largely inspired by the original ideas are, generally, not the initial ones.

If $|u|_1$ denotes the number of 1 occurrences in the word u, let us denote L the set $\{u/u = 1^n0^n,\ n > 0\} \cup \{u/u = 1^n0y10^n, y \in \{0,1\}^*,\ n > 0\}$, and K the language $\{u/\ u \in \{0,1\}^+,\ |u|_1 = |Bin(u)|_1\}$. Then we have:

Languages	$RPOCA$	$RPCA$	$RSCA$
K	− [c]	+ (Hemmerling, 1986) (Terrier, 1995)	?
L	+ [a]	+	+ [a]
LL	− (Terrier, 1996)	+	+ [a]

The main results in language recognition domain have focused on complexity classes comparisons. The known inclusion or no-inclusion relations as well as some still open problems are summarized in Figure 6, where the numbers labeling some of the arrows refer to papers or give easy justifications, as follows.

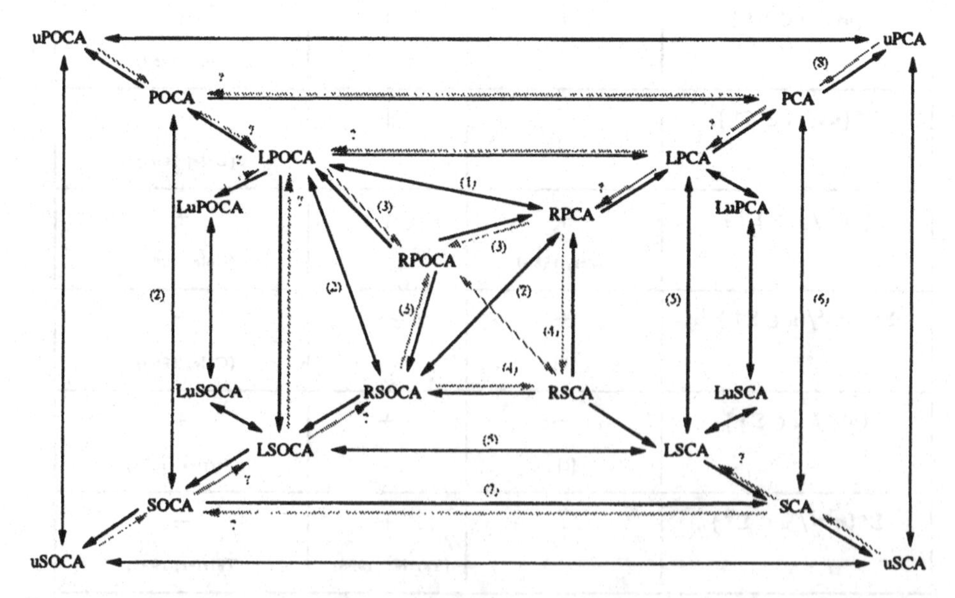

Figure 6. Results graph, for one-dimensional cellular automata. Dark arrows mean inclusion, grey ones no-inclusion and grey ones with question mark, obviously, open question.

- (1) results from (Choffrut and Čulik, 1984) and (Ibarra and Jiang, 1987),
- (2) Proofs can be found in (Ibarra and Jiang, 1987) or (Delorme and Mazoyer, 1996),
- (3) $\{a^n / \, n \, prime\}$ does not belong to *RPOCA* (Čulik *et al.*, 1981) but belongs to *RSCA* (Fisher, 1965) which is included in *RPCA*,
- (4) $\Sigma^*\{uu^R/ \; u \in \Sigma^+\}$[4] does not belong to *RSCA* but belongs to *RPOCA* (Cole, 1969),
- (5) Proofs can be found in (Choffrut and Čulik, 1984), (Ibarra and Jiang, 1987) or (Delorme and Mazoyer, 1996),
- (6) is founded on the following results, (Mazoyer, 1994): Every language recognized on a CA (uCA) (P as S) in time t(n) is recognized on a CA (uCA) (S as P) in time t(n)+n.
- (7) is founded on the following propositions which use both folding of the working area and cells grouping (Choffrut and Čulik, 1984), (Mazoyer, 1994),
    - Every language recognized on a SOCA (POCA) in time t(n) is recognized on a SCA (PCA) in time t(n).
    - Every language recognized on a SCA (PCA) in time t(n) is recognized on a uSOCA (uPOCA) in time 2t(n).

  We have here to put to light an important remark about the working area and the fact we go from bounded to unbounded computations through some simulations, which may lead to ambiguities or errors. Actually, in the last proposition proof, ***the simulations need the working area to be enlarged.***
- (8) is inferred from classical theorems of Computability.

The graph of figure 6 shows the crux of difficulty the node "*RPCA* = *LPOCA* = *RSOCA*" remains. Note also that, up to now, each candidate for differentiating *RPCA* and *LPCA* has been proved belonging to *RPCA*! Figure 7 is an other way to present the former results, better focusing on the open remaining problems, and including the comparisons with the Chomsky's classes.

4.2. COMPARISONS WITH THE CHOMSKY'S HIERARCHY CLASSES

The known results are summarized in the following table, where *Rat* denotes the class of rational (or regular) languages, *Alg* (*AlgL*) the one of algebraic (or context free) (linear algebraic (or linear context free)) languages and *CS* the one of context-sensitive languages (see (Hopcroft and Ullman, 1979)). Moreover, [d] means that the result is a consequence of two facts: *PCA* =

[4] where $|uu^R| \geq 3$ and $|\Sigma| \geq 2$

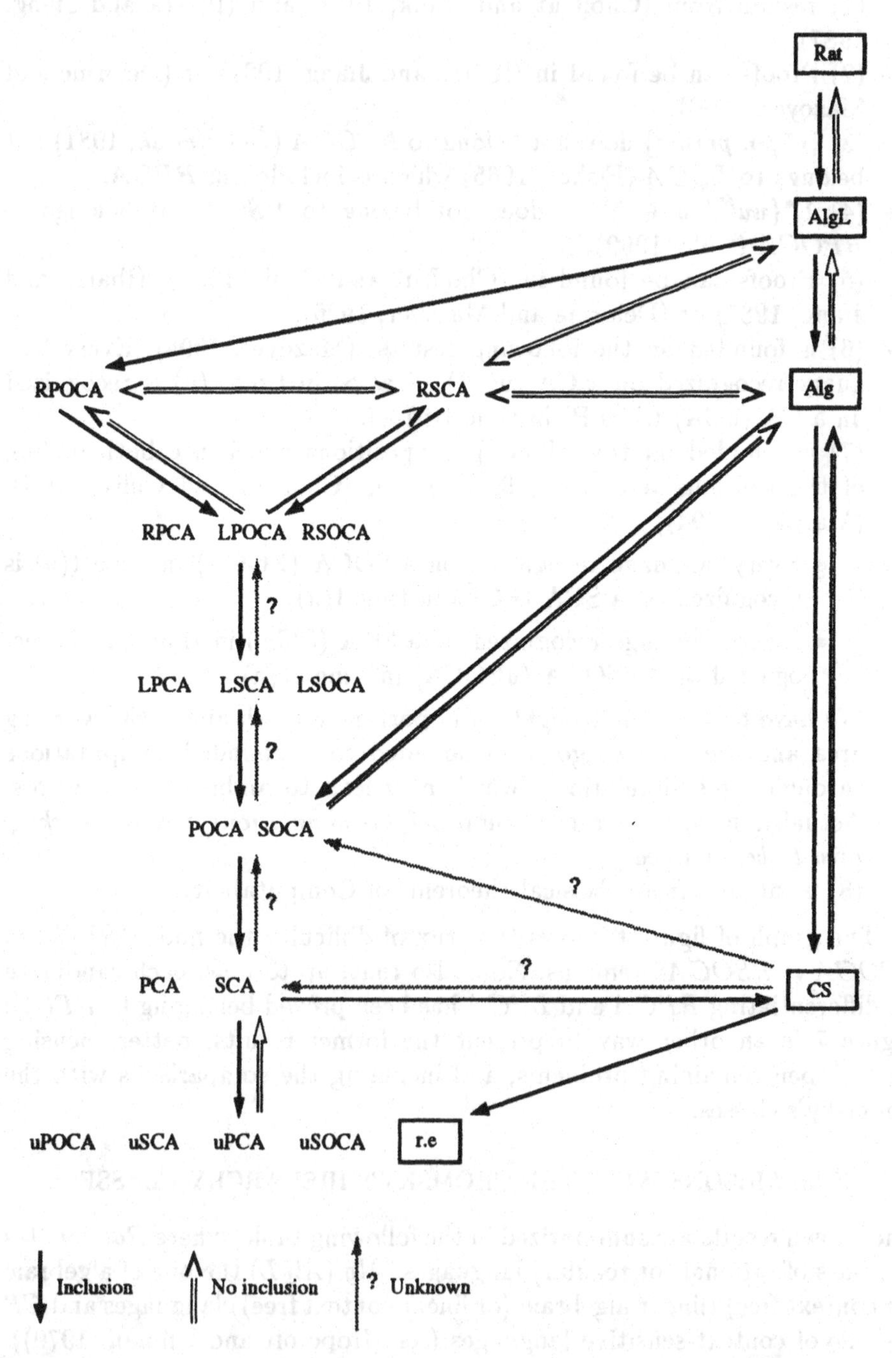

Figure 7. **Main complexity classes and their relations, for one-dimensional cellular automata.**

$DSPACE(n)$ (that can be deduce from simulations in (Smith, 1960) and $CS = NSPACE(n)$ (Wagner and Wechsung, 1986) for example. We can add that the unary languages in $RPOCA$ are the unary rational languages (Smith, 1971a).

$Rat \subseteq RPOCA$	$Rat \subseteq RSCA$
$AlgL \subseteq RPOCA$ (Smith, 1971a)	$Alg \subset SOCA$ (Ibarra and Jiang, 1987)
$Alg \not\subset RSCA$ (Cole, 1969)	$RSCA \not\subset Alg$ (Cole, 1969)
$RPOCA \not\subset Alg$ (Cole, 1969)	$Alg \not\subset RPOCA$ (Terrier, 1996)
$PCA = SCA \subseteq CS$	

So, connections between the Chomsky's hierarchy and the interesting classes via cellular automata are not quite easy to interpret. According to Figure 7, algebraic languages could to be low in complexity, and it could be a gap between Alg and CS, but how really deep? It is an exciting question.

4.3. CLOSURE PROPERTIES

In formal language theory, closure properties are naturally studied, and they, actually, can lead to results which resist to other investigation methods. The following table shows some results known, up to now, for the complexity classes we priviledged. Actually more is known, and the reader is refered to the next chapter by O. Ibarra, but we have chosen to emphasize these ones in order to put light on the result just below.

Classes	reversal	Concatenation	Boolean Op. (Smith, 1971a)
PCA	+ (Ibarra and Jiang, 1987)	+ (Ibarra and Jiang, 1987)	+
POCA	+	+	+
LSOCA	+	?	+
RPCA	?	?	+
RPOCA	+ (Choffrut and Čulik, 1984)	− (Terrier, 1996)	+
RSCA	− (Cole, 1969)	− (Cole, 1969)	+

Among the results proved in (Ibarra and Jiang, 1988), we have to stress the following one which is examplary:

RPCA is closed under reversal if and only if $LPCA = RPCA$.

An immediate consequence of the *RPCA* closure under reversal would be that *RPCA* is closed under concatenation. Both questions are open since 1971 (Smith, 1971a).

5. About proofs

Most proofs are founded on the main following methods: *"folding" the working area, grouping cells, using signals and synchronization.*

The only *algebraic* method is due to S. Cole. It consists in estimating the number of equivalence classes of relations E_k defined as follows: let Σ be an alphabet, L a language on Σ and k an integer, then two words u and v are equivalent modulo E_k if and only if for every word w , $|w| \leq k$, on Σ, $uw \in L \leftrightarrow vw \in L$. We refer to the chapter in this volume by V. Terrier, who develops and uses analogous processes. As for us, we will illustrate the first cited method in proving that $uLPCA = LPCA$.

5.1. FOLDING THE WORKING AREA

Actually, to prove $uLPCA = LPCA$ comes down to prove $uLPCA \subseteq LPCA$. So, suppose that L is in $uLPCA$. That means that there exists a cellular automaton $\mathcal{A}$ and a positive integer k_L such that each word u in

L, $|u| = n$, is recognized in t_u steps with $t_u \leq k_L n$. We can see the useful computation areas on Figure 8, with $k_L = 2$ and $k_L = 3$.

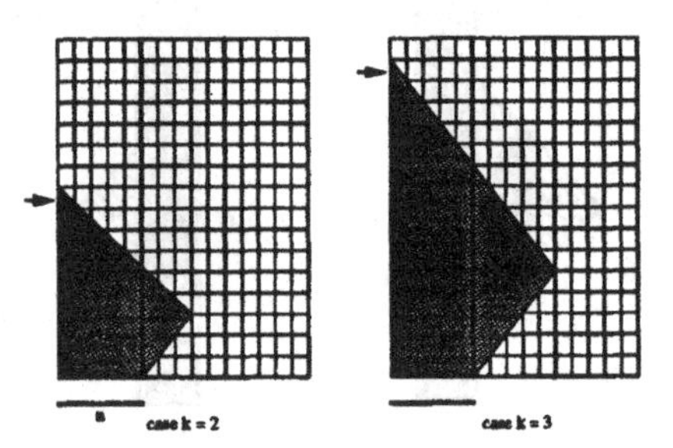

Figure 8. Useful computation areas during the recognition of a word in parallel unbounded mode and linear time kn.

We are looking for an automaton, built from $\mathcal{A}$ and such that the useful computation area would be the first stripe of length equal to the input length. The case $k_L = 3$ in Figure 8 invites to simply fold back the second n-width stripe on the first one and to build an automaton the states of which are 2-uples such that the 2-th component of $\langle x, t\rangle$ is the state $\langle 2n - x + 1, t\rangle$ of $\mathcal{A}$.

More generally, if k_L is odd and $k_L = 2k' + 1$, it is not difficult to verify that the useful computation area is inscribed inside $k' + 1$ stripes of width n, up to the right edge of the $(k' + 1)$-th stripe. The idea to define the new automaton is to generalize the above remark in folding over the stripes of width n as shown on Figure 9, and to consider as states $(k' + 1)$-uples such that, if x is any cell of the first stripe of width n, the j-th component of $\langle x, t\rangle$ is the state of the cell of the j-th stripe which is superimposed on x in the folding (see Figure 10).

More generally the following lemma holds, from which $uLPCA \subseteq LPCA$ can be easily deduced, as soon as it is remarked that if L is recognized in time $k_L n$ then it is recognized in time cn, $c \in \mathbb{N}$, $c \geq k_L$, which implies that assuming k_L odd is not restrictive.

Fact 1

Let k be any positive integer and $\mathcal{A} = (Q_\mathcal{A}, \delta_\mathcal{A})$ a cellular automaton working in parallel mode, with a distinguished state q_e such that $q_e = \delta(q_1, q_2, q_3)$ implies $q_1 = q_2 = q_3 = q_e$.
Then there exists a cellular automaton $\mathcal{A}^{(k)} = ({Q_\mathcal{A}}^{k+1}, \delta_{\mathcal{A}^{(k)}})$ such that: for each integer n, $n \in \mathbb{N}$ and sequence $(x_1, x_2, \ldots, x_n)$ of $Q_\mathcal{A}$ states, if the configurations $c_\mathcal{A}(\bar{x})$ and $c_{\mathcal{A}^{(k)}}(\bar{x})$ are defined by

- *for each z, $z \in \mathbb{Z}$, $z < 0$ and $z > n$, $c_\mathcal{A}(\bar{x})(z) = q_e$ and $c_{\mathcal{A}^{(k)}}(\bar{x})(z) = (q_e, q_e, \ldots, q_e)$,*
- *for $z \in \{1, \ldots, n\}$, $c_\mathcal{A}(\bar{x})(z) = x_z$ and $c_{\mathcal{A}^{(k)}}(\bar{x})(z) = (x_z, q_e, \ldots, q_e)$,*

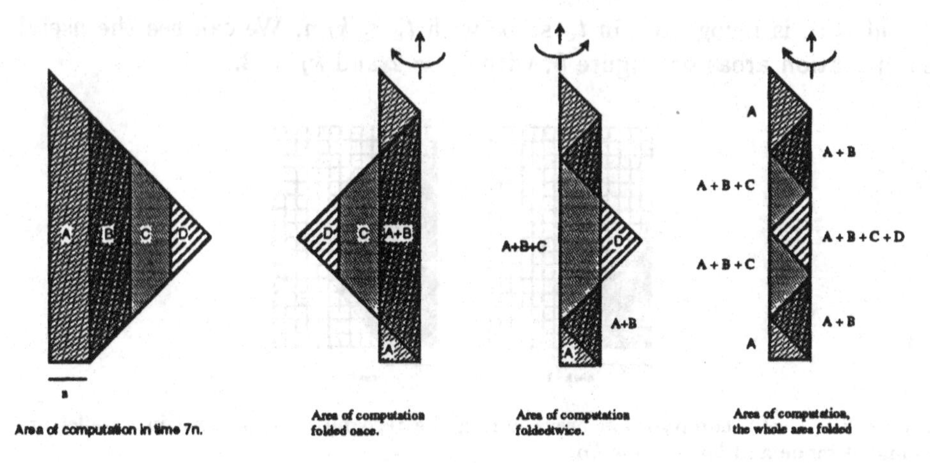

Figure 9. Folding over the computation area.

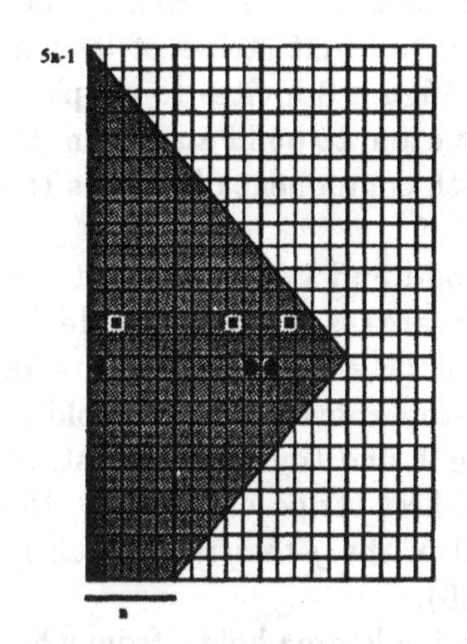

Figure 10. Examples of cells that will be superimposed in the folding.

then,

$$\langle 1, (2k+1)n-1 \rangle_{\mathcal{A}^{(k)}} = (\langle 1, (2k+1)n-1 \rangle_{\mathcal{A}}, q_2, \ldots, q_{k+1})$$

where $q_i, 2 \leq i \leq k+1$ can be any state in Q.

In order to define $\delta_{\mathcal{A}^{(k)}}$, let us set
$\delta_{\mathcal{A}^{(k)}}\left((q_1^\ell, \ldots, q_{k+1}^\ell), (q_1^c, \ldots, q_{k+1}^c), (q_1^r, \ldots, q_{k+1}^r)\right) = (q_1^\star, \ldots, q_{k+1}^\star)$. We will distinguish two cases according whether i, $1 \leq i \leq k+1$ is odd or even (which respectively corresponds to the fact that the i-th stripe of the useful computation area is folded from the right of the first one or from the left). Figure 11 illustrates the idea under the formal definition. Actually, it holds:

– $(\langle x, t \rangle_{\mathcal{A}^{(k)}})_i = \langle x + (i-1)n, t \rangle_{\mathcal{A}}$ when i is odd,

– $(\langle x,t\rangle_{\mathcal{A}^{(k)}})_i = \langle in - x + 1, t\rangle_A$ when i is even.

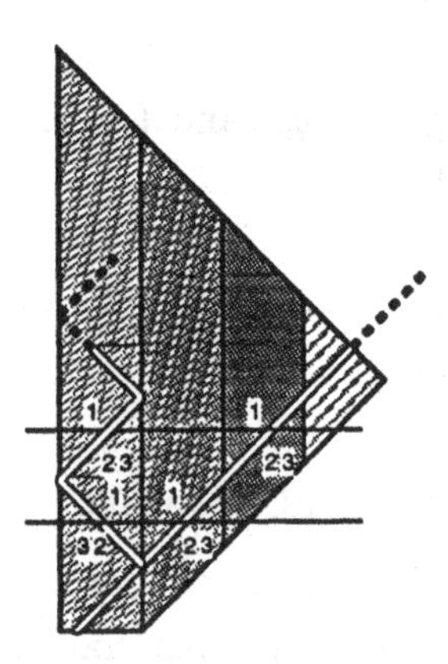

Figure 11. Basic features in order to conceive $\delta_{\mathcal{A}^{(k)}}$.

Let us now precisely define $\delta_{\mathcal{A}^{(k)}}$.

1. **i odd** *In this case, the useful computation area has been folded over an even number of times and the computation is done in the initial ($\mathcal{A}$) order.*
   - i - If $(q_1^\ell, \ldots, q_k^\ell) \neq (q_e, \ldots, q_e)$ and $(q_1^r, \ldots, q_k^r) \neq (q_e, \ldots, q_e)$, then, $q_i^\star = \delta_A(q_i^\ell, q_i^c, q_i^r)$.
     Concerned cells are not edge ones.
   - ii - If $(q_1^\ell, \ldots, q_k^\ell) = (q_e, \ldots, q_e)$ and $(q_1^r, \ldots, q_k^r) \neq (q_e, \ldots, q_e)$, then $q_i^\star = \delta_A(q_{i-1}^c, q_i^c, q_i^r)$ with $q_0^c = q_e$.
     The considered cell is on the left edge of the active area, $q_1^\star = (q_e, q_1^c, q_1^r)$, and the hypothesis made on q_e ensures that (q_e, q_e, q_e) marks the limits of the computation area.
   - iii - If $(q_1^\ell, \ldots, q_k^\ell) \neq (q_e, \ldots, q_e)$ and $(q_1^r, \ldots, q_k^r) = (q_e, \ldots, q_e)$, then $q_i^\star = \delta_A(q_i^\ell, q_i^c, q_{i+1}^c)$.
     Concerned cells are on the right edge of the computaion area.
   - iv - If $(q_1^\ell, \ldots, q_k^\ell) = (q_e, \ldots, q_e)$ and $(q_1^r, \ldots, q_k^r) = (q_e, \ldots, q_e)$, then $q_i^\star = \delta_A(q_{i-1}^c, q_i^c, q_{i+1}^c)$.
     This corresponds to the case $n = 1$.

2. **i even** *In this case the useful computation area has been folded over an odd number of times, the computation is done in the order opposite to the initial one. The different considered possibilities are those of the previous case.*
   - i - If $(q_1^\ell, \ldots, q_k^\ell) \neq (q_e, \ldots, q_e)$ and $(q_1^r, \ldots, q_k^r) \neq (q_e, \ldots, q_e)$, alors $q_i^\star = \delta_A(q_i^r, q_i^c, q_i^\ell)$.

- ii - If $(q_1^\ell, \ldots, q_k^\ell) = (q_e, \ldots, q_e)$ and $(q_1^r, \ldots, q_k^r) \neq (q_e, \ldots, q_e)$, then $q_i^\star = \delta_A(q_i^r, q_i^c, q_{i+1}^c)$.

- iii - If $(q_1^\ell, \ldots, q_k^\ell) \neq (q_e, \ldots, q_e)$ and $(q_1^r, \ldots, q_k^r) = (q_e, \ldots, q_e)$, then $q_i^\star = \delta_A(q_{i-1}^c, q_i^c, q_i^\ell)$.

- iv - If $(q_1^\ell, \ldots, q_k^\ell) = (q_e, \ldots, q_e)$ and $(q_1^r, \ldots, q_k^r) = (q_e, \ldots, q_e)$, then $q_i^\star = \delta_A(q_{i-1}^c, q_i^c, q_{i+1}^c)$.

5.2. USING SIGNALS AND SYNCHRONIZATION

This method is used in proving $LPCA \subseteq LSCA$ for example. We will only give a hint of the proof by means of figures, namely Figure 12.

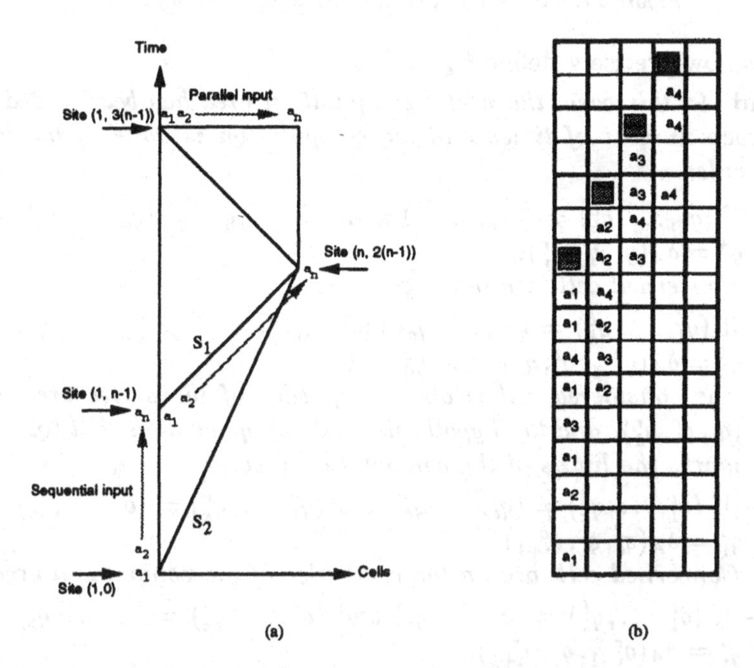

Figure 12. Data moves in order to convert a sequential input into a parallel one, in linear time.

Suppose that $u = a_1 \ldots a_n$ is a word recognized on some PCA, $\mathcal{A}$, in linear time. If we want it to be accepted on some SCA $\mathcal{B}$, we have to set up some process which converts the sequential input corresponding to u into some configuration realizing the parallel input corresponding to u on $\mathcal{B}$. It is shown on Figure 12 a.

When the first input letter arrives on the cell 1 at time 0, cell 1 sends a signal S_2, at speed 1/2. When the last input letter a_n enters cell 1 at time $n-1$, then cell 1 sends a signal S_1 at maximal speed, which meets signal S_2 on cell n at time $2(n-1)$. And then a synchronization is launched, via a Firing Squad optimal time solution, which ends up, at time $3(n-1)$, with the wanted parallel input on the first n cells. The recognition process on $\mathcal{A}$ can start, and the word u will obviously be recognized on $\mathcal{B}$ in linear time. Figure 12 b shows how the input letters successively appear on the diagonal stemming from site $(1,0)$.

5.3. GROUPING CELLS

This method is used to prove $uLSOCA = LSOCA$ for example. It was extensively practised in this book (see in this volume, *An Introduction to Cellular Automata* and *Computations on Cellular Automata*). So we will only recall its principle on Figure 13.

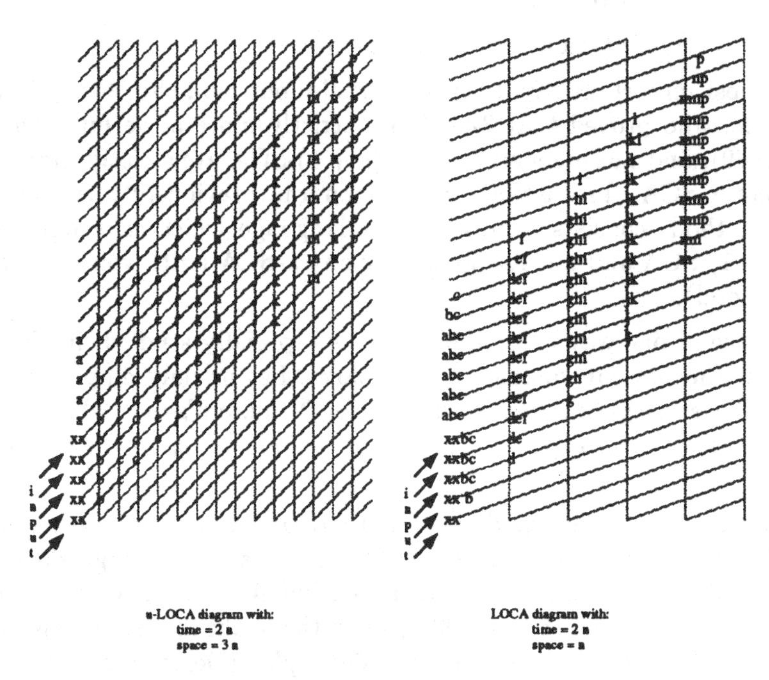

Figure 13. A hint for proving $uLSOCA = LSOCA$.

6. Languages on 2-dimensional cellular automata

Languages on 2-dimensional cellular automata can be understood either as sets of 2-dimensional patterns (also often called images) or as standard languages of finite words. Actually, although some work has been done on images ((Smith, 1971c), (Inoue and Nakamura, 1977) and (Terrier, 1998)), we will here restrict the matter to the latter ones.

In the frame of 2-dimensional cellular automata, inputs and outputs problems naturally become a little more involved, but, moreover, a new one may arise because the standard first neighbors neighborhood of dimension one splits into von Neumann's and Moore's neighborhoods.

6.1. INPUT AND OUTPUT MODES

First, all the authors choose some distinguished cell, which is used to communicate with the outside, that means to enter letters one after the other (sequential mode), to enter the first input letter (parallel mode) or, possibly, to decide the acceptation. But, depending on the way words are encoded on the plane, a second cell has to be chosen, where the recognition will be decided if there is a reason.

Secondly, computations are bounded ones.

The most important result for sequential input mode has been obtained by S. Cole, who proved that 2-SCA are strictly more powerful than SCA (Cole, 1969), and more generally that $(k+1)$-SCA are strictly more powerful than k-SCA. The proof (founded on the previously mentioned algebraic method) consists in producing a language which is recognized by some $(k+1)$-SCA and not recognized by any k-SCA (with von Neumann's neighborhood).

The case of one-way communications is especially studied, in the next chapter of the present volume, by O. Ibarra. Let us only emphasize here that the chosen neighborhood is still the von Neumann's one, which means, in this frame, the vector $((x_c-1, y_c), (x_c, y_c-1))$ if (x_c, y_c) is the considered cell.

Innovation in parallel mode is that there are many ways to enter the input. Some of them are represented on Figure 14, as the distinguished cells. In his following contribution, O. Ibarra uses linelike (llo) or brokenwordlike (bwlo) order, while in (Terrier, 1998), it is the snakelike (snlo) one which is used. The last one is the spirallike order (splo) (Delorme and Mazoyer, 1994), but one can imagine other sort of "wires".

The choice of the accepting cell depends on the way the word is displayed on the plane and also on the communication way. Examples are given on Figure 15.

Let us remark that broking the input word brings a new parameter in, which can be seen as distinguishing two cells: the second one, in fact the accepting cell, telling how the input is entered.

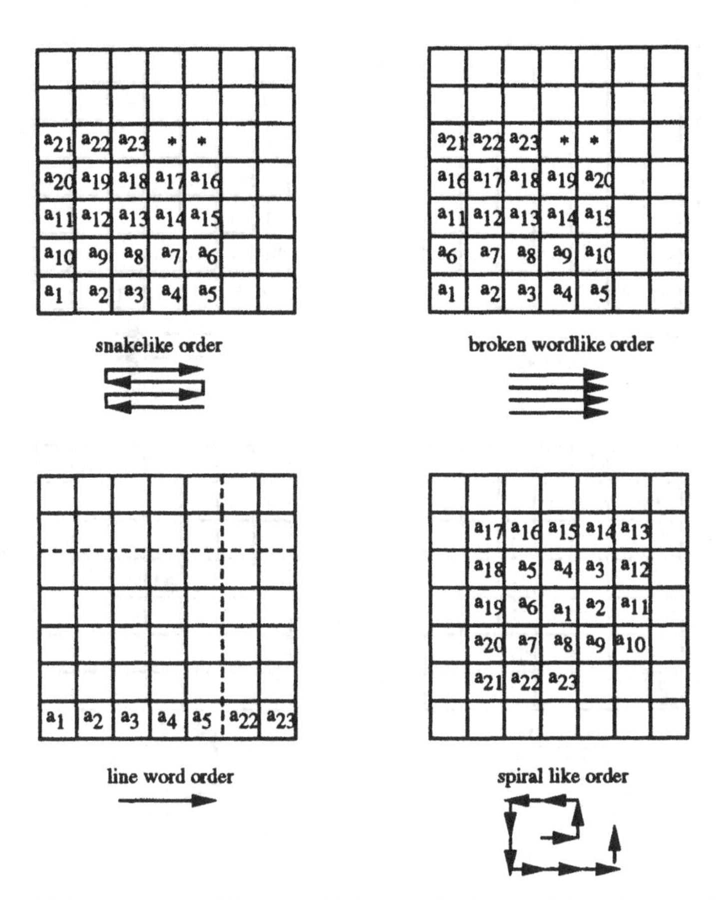

Figure 14. Different standard inputs of finite words, in parallel mode on 2-dimensional cellular automata.

6.2. REAL TIME

What does mean real time in this frame? As usual, that means the minimal time necessary to the accepting cell to know the whole input. We will not give formal definitions, but only some examples illustrating how real time depends upon different computation features.

Figure 16 shows real time in case of cellular automata, checking words of length n in parallel "linelike order" input mode, the accepting cell being the n-th one, in case of Moore's and von Neumann's neighborhoods.

In case of length n words displayed in snakelike order inside a square of size $(\lceil\sqrt{n}\rceil, \lceil\sqrt{n}\rceil)$ as on Figure 15, with the first cell as accepting one, the real time is $\lceil\sqrt{n}\rceil - 1$ in case of Moore's neighborhood and $2(\lceil\sqrt{n}\rceil - 1)$ in case of von Neumann's one.

When words of length n are displayed in spiralike order real time is

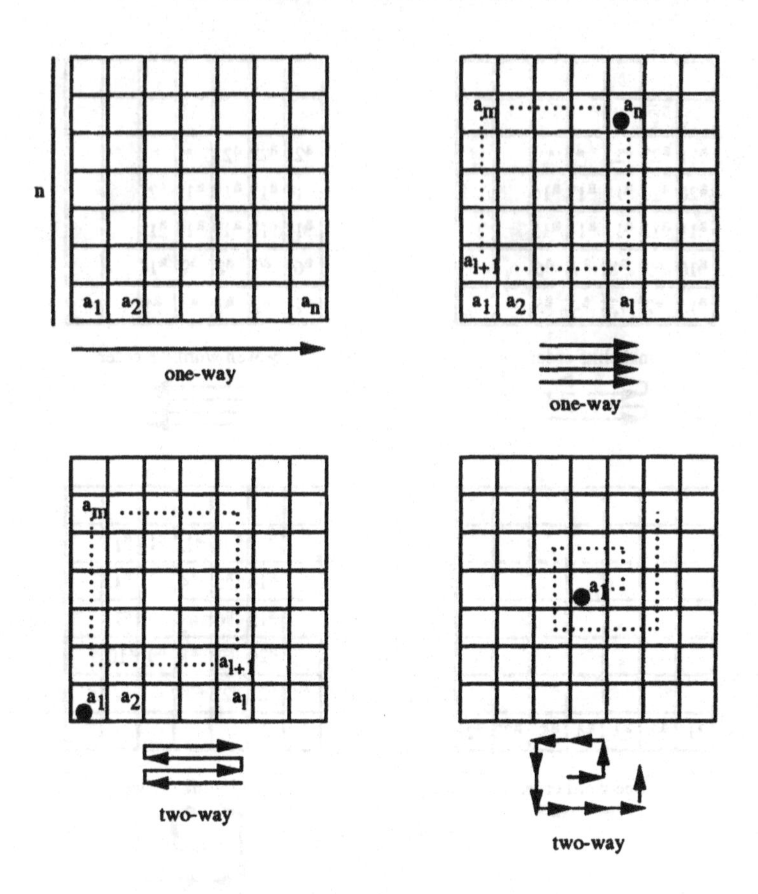

Figure 15. Localization of the accepting cell.

$O(\lceil\sqrt{n}\rceil)$, and it has been shown that rational languages are recognized in real time (Mazoyer, 1998).

6.3. SOME RESULTS AND OPEN QUESTIONS

Let us sum up some interesting comparison results on standard computational complexity classes. Incidentally, some of them show the gap which exists between the dimensions 1 and 2.

1. According to O. Ibarra (see next chapter), the 2-PCA with von Neumann neighborhood, on standard inputs, are strictly more powerful than analogous 2-POCA.
   That comes from the fact that $2POCA_{\rightarrow,VN} = DSPACE(n^{3/2})$ and $2SOCA_{VN} = DSPACE(n^{3/2})$, while $2PCA_{\rightarrow,VN} = DSPACE(n^2)$

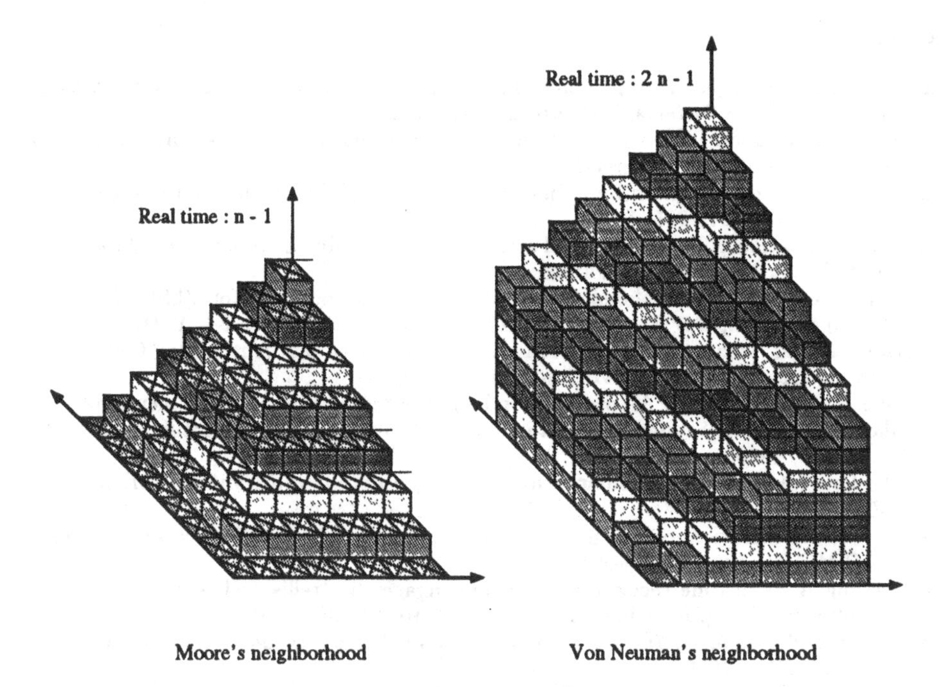

Figure 16. Real time computation examples.

and $2SCA_{VN} = DSPACE(n^2)$. One observes here another efficient proof method, which consists in separately comparing classes determined via cellular automata and classes determined via another computation model (Turing machines par excellence).

2. It is not known whether 2-POCA are more powerful than POCA, nor whether there exists some relation between 2POCA and PCA, apart the fact that $2PCA_{blo,VN} = 2PCA_{snlo,VN}$ which can be deduced from the fact that $DSPACE(n) = 2PCA_{blo,VN}$ and $DSPACE(n) = 2PCA_{snlo,VN}$.
3. Obviously, the famous problem $RPCA =^? PCA$ can be transposed in dimension 2 to the different relevant explicit cases according inputs forms and neighborhoods.

6.4. IMAGES LANGUAGES

To finish, and though we decided to focuse on languages of finite words, it is worthy to stress the following result in (Terrier, 1998):
There exists a two-dimensional language recognized in real time in case of von Neumann's neighborhood, which is not real-time in case of Moore's neighborhood.

References

Albert J. and Čulik II K. A simple universal cellular automaton and its one-way totalistic version. *Complex Systems.* **Vol. no. 1**: 1-16, 1987.

Choffrut C. and Čulik II K. On real time cellular automata and trellis automata. *Acta Informatica* **Vol. no. 21**: 393–407, 1984.

Cole S. Real time computation by n-dimensional arrays. *IEEE Transactions on Computers.* **Vol. no. 4**: 349-365, 1969.

Čulik II K. Variation of the firing squad synchronization problem. *Information Processing Letter.* **Vol. no. 30**: 152-157, 1989.

Čulik II K., Gruska J. and Salomaa A. Systolic Treillis Automaton (for VLSI). *Res. Rep CS. Dept of Computer Science, Univ. of Waterloo.* **Vol. no. 81-34** ,1981.

Delorme M. and Mazoyer J. Reconnaissance de Langages sur Automates Cellulaires. *Research report LIP-IMAG,***Vol. no. 94-46**, 1994.

Delorme M. and Mazoyer J. An overview on language recognition on one dimensional cellular automata. in *Semigroups, Automata and Langages.* World Scientific, J.Almeida Ed: 85-100, 1996.

Dyer C. One-way bounded Cellular Automata. *Information and Control.* **Vol. no. 44**: 54-69, 1980.

Fischer P.C. Generation on primes by a one dimensional real time iterative array. *J.A.C.M.* **Vol. no. 12**: 388-394, 1965.

Hemmerling A. Real time recognition of some langages by trellis and cellular automata and full scan Turing machines. *EATCS.* **Vol. no. 29**: 35-39, 1986.

Hopcroft J. and Ullman J. *Introduction of automata theory, languages and computation.* Addison Wesley, 1979.

Ibarra O., Jiang T. and Vergis A. On the power of one-way communication. *J.A.C.M* **Vol.35 no. 3**: 697-726, 1988.

Ibarra O. and Jiang T. On one-way cellular arrays. *S.I.A.M J. Comput.* **Vol.16 no. 6**: 1135-1154 ,1987.

Ibarra O. and Jiang T. Relating the power of cellular arrays to their closure properties. *Theoretical Computer Science.* **Vol. no. 57**: 225-238,1988.

Ibarra O. Kim S. and Moran S. Sequential machine characterization of treillis and cellular automata and applications. *S.I.A.M J. Comput.* **Vol. no.14**: 426–447, 1985.

Inoue K. and Nakamura A. Some properties of two-dimensional on-line tessalation acceptors. *Information Sciences.* **Vol. no. 13**: 95-121, 1977.

Mazoyer J. Parallel language recognition on a plane of automata. Unplublished paper, 1998.

Mazoyer J. Entrées et sorties sur lignes d'automates. in *Algorithmique parallèle.* Cosnard M., Nivat M. and Robert Y. Eds, Masson, 47–65, 1992.

Mazoyer J. Cellular automata, a computational device. in *CIMPA School on Parallel Computation.* Temuco (Chile), 1994.

Mazoyer J. and Reimen N. A linear speed-up for cellular automata. *Theoretical Computer Science.* **Vol. no. 101**: 58-98, 1992.

Mazoyer J. and Terrier V. Signals on one dimensional cellular automata. To appear in *Theoretical Computer Science*, 1998.

Róka Zs. Simulations between Cellular Automata on Cayley Graphs. To appear in *Theoretical Computer Science*, 1998.

Smith III A. Cellular automata theory. *Technical Report 2.* Stanford University, 1960.

Smith III A. Real time language recognition by one-dimensional cellular automata. *Journal of Computer and System Science.* **Vol. no. 4**: 299-318, 1971.

Smith III A. Simple computation-universal spaces. *Journal of ACM.* **Vol. no. 18**: 339-353, 1971.

Smith III A. Two-dimensional formal languages and pattern recognition by cellular automata. *Proceedings of the 12-eme Annual IEEE Symposium on Switching and Automata Theory.*: 144–152, East Lansing, Michigan, 1971.

Terrier V. Language recognizable in real time by cellular automata. *Complex systems* **Vol. no. 8**: 325–336, 1994.
Terrier V. On real time one-way cellular automata. *Theoretical Computer Science* **Vol. no. 141**: 331–335, 1995.
Terrier V. Language not recognizable in real time by one-way cellular automata. *Theoretical Computer Science*. **Vol. no. 156**: 281–287, 1996.
Terrier V. Two dimensional cellular automata recognizers. To appear in *Theoretical Computer Science*.
Wagner K. and Wechsung G. *Computational Complexity*. Reidel, 1986.

COMPUTATIONAL COMPLEXITY OF CELLULAR AUTOMATA: AN OVERVIEW

O. IBARRA
Department of Computer Science
University of California
Santa Barbara, CA 93106, USA

Abstract.
We give an overview of the sequential as well as the parallel complexity of cellular automata (also called cellular arrays).

Key words: Cellular automaton (array), computational complexity, one-way communication, P-complete, parallel complexity, recurrence equations.

1. Introduction

One of the earliest and simplest models of parallel computation is the cellular automaton, also called cellular array. (We use the former name in this paper.) They have been studied extensively in the literature. Early papers have studied these devices in the context of formal language recognition - their recognition power, closure and decision properties, and their relationships to other models of computation, such as Turing machines, linear bounded automata, pushdown automata, and finite automata. In later papers, the study of these arrays has focused on their abilities to perform numeric and nonnumeric computations in various areas such as computational linear algebra and signal and image processing. Such arrays, whose

[0]Parts of this paper have appeared in "On Some Open Problems Concerning the Complexity of Cellular Arrays" by O. Ibarra and T. Jiang, LNCS 812 Springer Verlag and "On the Parallel Complexity of Solving Recurrence Equations" by O. Ibarra and N. Tran, LNCS 834 Springer Verlag. This work was supported in part by NSF Grant IRI97-00370.

processors need no longer be "finite-state", have also been called systolic arrays.

Here we give an overview of the computational complexity of cellular arrays with respect to well known models of sequential and parallel computation. We discuss results concerning one-way communication versus two-way communication, linear-time versus real-time, serial input versus parallel input, space-efficient simulation of one-way arrays, parallel complexity of cellular arrays, etc. We also point out some important and fundamental open problems that remain unresolved.

The paper has six sections, including this section. In the rest of this section, we recall the definition of a linear cellular array and a related model called linear iterative array. Section 2 looks at the one-way versus two-way question, while Section 3 investigates the complexity of linear-time linear arrays. Section 4 concerns mesh-connected arrays. Section 5 examines the parallel complexity (in terms of the PRAM theory) of one-way cellular arrays. Finally, in Section 6, we give an application to recurrence equations.

1.1. LINEAR CELLULAR ARRAYS

A *linear cellular array* (LCA) (Bucher and Čulik II, 1984), (Kosaraju, 1974), (Smith, 1970), (Smith, 1971), (Smith, 1972) is a one-dimensional array of n identical finite-state machines (called nodes) that operate synchronously at discrete time steps by means of a common clock (see Figure 1). The input $a_1a_2\cdots a_n$, where a_i is in the finite alphabet Σ is applied to the array in parallel at time 0 by setting the states of the nodes to $a_1, a_2, \cdots, a_n$. The state of a node at time t is a function of its state and the states of its left and right neighbors at time $t-1$. We assume that the leftmost (rightmost) node has an "imaginary" left (right) neighbor whose state is \$ at all times. We say that $a_1a_2\cdots a_n$ is accepted by the LCA if, when given the input $a_1a_2\cdots a_n$, the leftmost cell eventually enters an accepting state. The LCA has time complexity T(n) if it accepts inputs of length n within T(n) steps. Clearly, for a nontrivial computation, $T(n) \geq n$. If $T(n) = cn$ for some real constant $c \geq 1$, then the LCA is called a *linear-time* LCA. When $T(n) = n$, it is called a *real-time* LCA. Note that an LCA without time restriction is equivalent to a linear-space bounded deterministic TM.

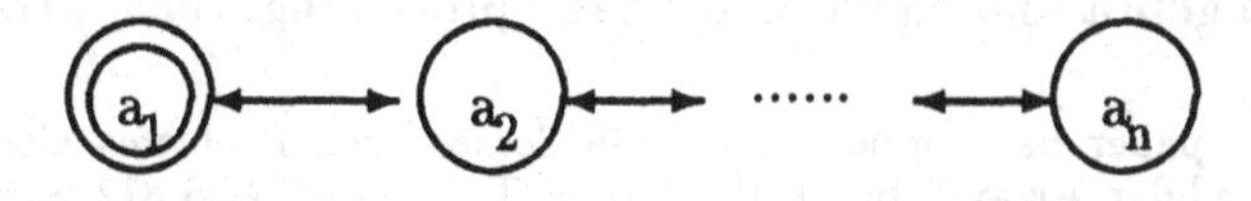

Figure 1. An LCA.

A restricted version of an LCA is the *one-way* linear cellular array (OLCA) (Dyer, 1980), where the communication between nodes is one-way, from left to right. The next state of a node depends on its present state and that of its left neighbor (see Figure 2). An input is accepted by the OLCA if the rightmost node of the array eventually enters an accepting state. The time complexity of an OLCA is defined as in the case of an LCA.

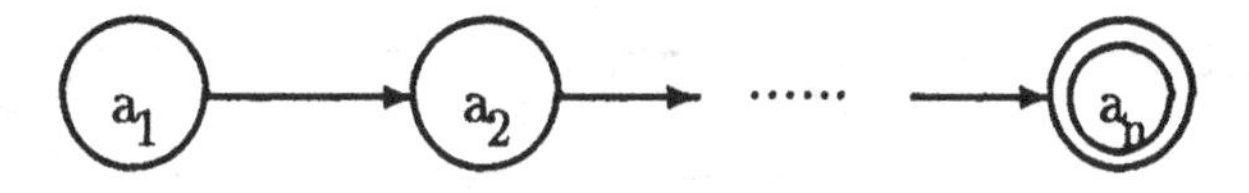

Figure 2. An OLCA.

Although OLCA's have been studied extensively in the past (see, *e.g.*, (Bucher and Čulik II, 1984), (Choffrut and Čulik II, 1984), (Dyer, 1980), (Ibarra *et al.*, 1985b), (Ibarra *et al.*, 1986), (Umeo *et al.*, 1982)) a precise characterization of their computational complexity with respect to space- and/or time-bounded TM's is not known. For example, it is not known whether linear space-bounded deterministic TM's are more powerful than OLCA's, although a positive answer seems likely.

1.2. A RELATED MODEL: LINEAR ITERATIVE ARRAYS

Another simple model that is closely related to linear cellular arrays is the *linear iterative array* (Cole, 1969), (Hennie, 1961), (Ibarra *et al.*, 1985b), (Ibarra *et al.*, 1986). The structure of an LIA is similar to an LCA, as shown in Figure 3. (Note that here we assume that the size of the array is bounded by the length of the input. In some papers, the array is assumed to be infinite.) The only difference between an LIA and an LCA is that in an LIA the input $a_1 a_2 \cdots a_n\$$ is fed serially to the leftmost node. Symbol a_i, $1 \leq i \leq n$, is received by the leftmost node at time $i-1$; after time $n-1$, it receives the endmarker \$. That is, \$ is not consumed and always available for reading. At time 0, each cell is in a distingushed quiescent state q_0. As in an LCA, the state of a node at time t is a function of its state and the states of its left and right neighbors at time $t-1$. For the leftmost node, the next state depends on its present state, input symbol and its right neighbour. An OLIA (the one-way version of the LIA) is defined in a straightforward way.

For a nontrivial computation, the time complexity of an LIA is at least n, and the time complexity of an OLIA is at least $2n$. An LIA operating n steps is called a real-time LIA and an OLIA operating in $2n$ steps is called a *pseudo-real-time* OLIA.

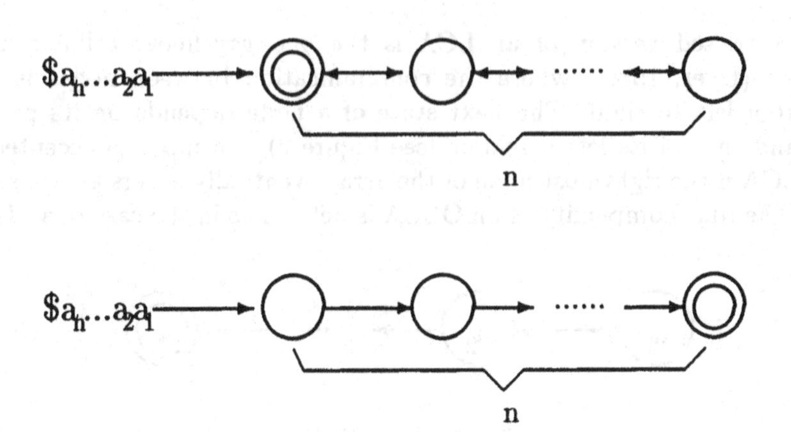

Figure 3. An LIA and an OLIA.

It is relatively easy to show that an LIA and an LCA can efficiently simulate each other. We will focus on fast simulations between LCA, OLCA, and OLIA in this paper.

1.3. MESH-CONNECTED ARRAYS

Mesh-connected cellular arrays (MCA's) and *mesh-connected iterative arrays* (MIA's) are the two-dimensional analogs of LCA's and LIA's. Here we are mostly interested in the arrays with one-way communication. A one-way mesh-connected cellular array (OMCA) and a one-way mesh-connected iterative array (OMIA) are shown in Figure 4.

To simplify the presentation, we introduce the following notations.

1. For any class C of machines and function $T(n)$, $C(T(n))$ denotes the machines in C operating in time $T(n)$.
2. Let C_1 and C_2 be two classes of machines. $C_1 \subseteq C_2$ means that every machine M_1 in C_1 can be simulated by some machine M_2 in C_2. $C_1 \subset C_2$ means that $C_1 \subseteq C_2$ and there is a machine in C_2 that cannot be simulated by any machine in C_1.

2. One-way versus two-way for linear arrays

The question of whether one-way communication reduces the power of a linear array has remained open for over a decade. In particular, we do not know if OLCA = LCA and if OLIA = LIA. The following result shows that an LIA and an LCA can simulate each other with a delay of at most n steps (Ibarra *et al.*, 1985b).

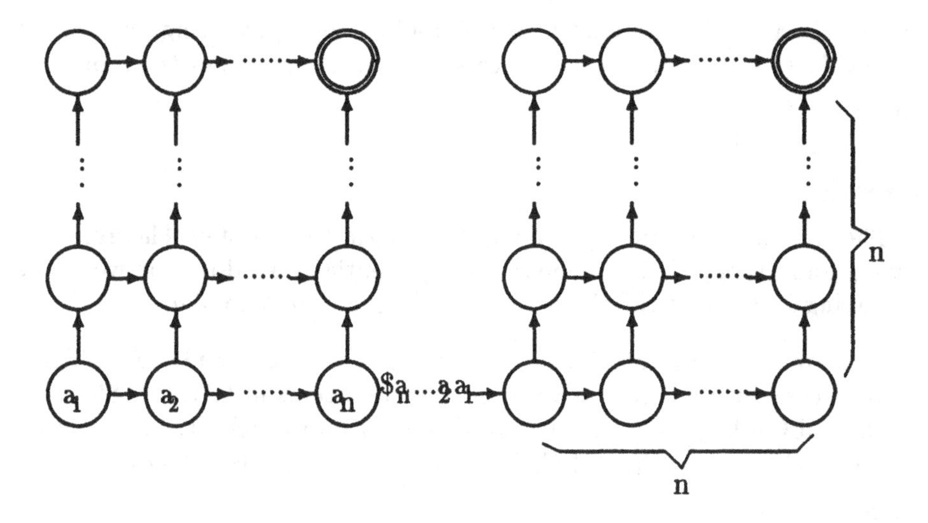

Figure 4. An OMCA and an OMIA.

Theorem 1

For any $T(n) \geq n$,

1. $LCA(T(n)) \subseteq LIA(T(n) + n)$;
2. $LIA(T(n) + n) \subseteq LCA(T(n) + n)$.

Hence LIA = LCA. It seems difficult to precisely characterize the computational complexity of an OLCA or an OLIA. Nevertheless, it has been shown in [CHAN88b] that OLIA's are actually very powerful since they can simulate linear time-bounded alternating TM's.

Theorem 2

1. *Every linear time-bounded alternating TM can be simulated by an OLIA.*
2. *$NSPACE(n^{1/2}) \subseteq OLIA$.*
3. *Every language accepted by a multihead two-way PDA operating in $c^{n/\log n}$ time (for some constant c) can be accepted by an OLIA. Thus, every context-free language is accepted by an OLIA.*
4. *The class of languages accepted by OLIA's is an AFL closed under intersection, complementation, and reversal. (An AFL is a family of languages containing at least one nonempty language which is closed under the operations of union, concatenation, Kleene +, ϵ-free homomorphism, inverse homomorphism, and intersection with regular sets.)*

Clearly, OLCA $\subseteq$ OLIA. On the other hand, it has also been shown (quite surprisingly) that every OLIA can be simulated by an OLCA (Ibarra and Jiang, 1987). The difficulty arises from the fact that in an OLIA, *every* node of the array has access to *each* symbol of the input string, whereas in an OLCA, the i-th cell can only access the first i symbols of the input.

Theorem 3

OLCA = OLIA.

Remark 1

The simulation of an OLIA by an OLCA (in the proof of Theorem 3 in (Ibarra and Jiang, 1987)) involves an exponential slow-down. It would be interesting to know if the slow-down can be made polynomial.

It follows that the results in Theorem 2 also hold for OLCA's. These results answer in the affirmative some open questions in (Dyer, 1980), e.g., whether OLCA languages are closed under operations such as concatenation and reversal, and whether OLCA's accept context-free languages.

Corollary 1

1. *Every linear time-bounded alternating TM can be simulated by an OLCA.*
2. *NSPACE($n^{1/2}$) $\subseteq$ OLCA.*
3. *Every language accepted by a multihead two-way PDA operating in $c^{n/\log n}$ time (for some constant c) can be accepted by an OLCA. Thus, every context-free language is accepted by an OLCA.*
4. *The class of languages accepted by OLCA's is an AFL closed under intersection, complementation, and reversal.*

It seems unlikely that OLCA = LCA. On the other hand, by the above corollary and the fact that LCA = DSPACE(n), proving OLCA $\subset$ LCA would imply NSPACE($n^{1/2}$) $\subset$ DSPACE(n), which would be an improvement of Savitch's well-known result (Savitch, 1970). This should explain why the one-way communication versus two-way communication problem for linear arrays is hard.

3. Complexity of linear-time linear arrays

The following important issues concerning LCA's operating in $O(n)$ time still remain unresolved as of today.

- Is LCA(n) = LCA(O(n))?
- Is every context-free language accepted by a real-time LCA?
- Is the class of real-time LCA languages closed under reversal? under concatenation?
- Is LCA(O(n)) = LCA?

  The first question was asked in (Bucher and Čulik II, 1984). The last 3 questions were raised in (Smith, 1972). In fact, the following seemingly easier question is also open:

- Is LCA(n) = LCA?

Below we review the partial results that have been obtained in this area. In particular, we will show that

- LCA(n) = OLCA(O(n)) $\supset$ OLCA(n).
- LCA(n) = LCA(O(n)) if and only if the class of real-time LCA languages is closed under reversal.
  This result is rather interesting since it provides a method of proving/disproving the equivalence of two classes of LCA's in terms of a closure operation.
- If the class of real-time LCA languages is closed under reversal, then it is also closed under concatenation.
- OLCA $\subset$ LCA implies LCA(O(n)) $\subset$ LCA.

3.1. LINEAR-TIME VERSUS REAL-TIME FOR THE ARRAYS

The following result (Choffrut and Čulik II, 1984) holds because, when the alphabet is unary, a real-time OLCA can only accept a regular language and a linear-time OLCA can accept the language $\{0^{2^n} | n > 0\}$.

Theorem 4
OLCA(n) $\subset$ OLCA(O(n)).

We now consider relations between linear/real -time LCA, OLCA, and OLIA. The proofs make essential use of the following speed-up lemma, which is interesting in its own right (Ibarra *et al.*, 1985a) (see also (Bucher and Čulik II, 1984), (Ibarra *et al.*, 1986), (Smith, 1972)).

Lemma 1
Let $T(n) \geq 0$ and M be an OLIA (LCA, OLCA) operating in time $2n + T(n)$ $(n + T(n), n + T(n))$. We can effectively construct an OLIA (LCA, OLCA) equivalent to M which operates in time $2n + T(n)/k$ $(n + T(n)/k, n + T(n)/k)$ for any positive integer k.

It follows from the above lemma that any linear-time OLIA (LCA, OLCA) can be converted to one that operates in time $(2+\epsilon)n$ $((1+\epsilon)n, (1+\epsilon)n)$ for any positive real constant ϵ.

The relations between linear time-bounded LCA, OLCA, and OLIA are summarized in the following theorem (Ibarra and Jiang, 1988b).

Theorem 5

1. *LCA(n) = OLCA(O(n)) = OLIA(2n).*
2. *LCA(O(n)) = OLIA(O(n)).*

Hence, $LCA(O(n)) \supset OLCA(O(n))$ if and only if
$OLIA(O(n)) \supset OLCA(O(n))$.
This contrasts with the equivalence of OLIA and OLCA when there is no restriction on the time.

Next we establish the relations between the open questions listed above. The next theorem is from (Ibarra and Jiang, 1988b).

Theorem 6

$OLIA(O(n)) = OLIA(2n)$ if and only if the class of pseudo-real-time OLIA languages is closed under reversal.

From Theorems 5 and 6, we have the following corollary.

Corollary 2

$LCA(O(n)) = LCA(n)$ if and only if the class of real-time LCA languages is closed under reversal.

Interestingly, the class of real-time OLCA languages is closed under reversal (Choffrut and Čulik II, 1984). Whether it is closed under concatenation is still open. The following partial result is shown in (Ibarra and Jiang, 1987).

Theorem 7

The concatenation of two real-time OCA languages is a linear-time OLCA language.

We do not know if the same result holds for LCA languages, *i.e.*, whether the concatenation of two real-time CA languages is a linear-time LCA language. However, it has been shown that if the class of real-time CA languages is closed under reversal, then it is also closed under concatenation (Ibarra and Jiang, 1988b).

Theorem 8

If the class of pseudo-real-time OLIA (or real-time LCA) languages is closed under reversal, then it is also closed under concatenation.

When the alphabet is unary, the closure is known (Ibarra and Jiang, 1988b).

Theorem 9

The class of unary real-time LCA languages is closed under concatenation.

It would be interesting to know whether the class of unary real-time LCA languages is identical to the class of unary linear-time LCA languages. We conjecture the answer to be negative. Since real-time LCA's can accept fairly difficult unary languages (*e.g.*, the set of primes, the set of perfect squares, etc), it is even hard to find a promising candidate language.

3.2. LINEAR-TIME LCA VERSUS NONLINEAR-TIME LCA

Many conjecture that LCA(O(n)) $\subset$ LCA. Note that Paterson's result (*i.e.*, $NTIME_1(n^2) \subseteq DSPACE(n)$, where $NTIME_1(n^2)$ is the class of languages accepted by one-tape nondeterministic Turing machines in time $O(n^2)$) (Paterson, 1972) does not help unless we can show that the class of linear-time LCA languages is properly contained in $NTIME_1(n^2)$, which is probably even harder. The next result relates this conjecture to the one-way versus two-way problem. The implication holds because LCA(O(n)) = OLIA(O(n)) $\subseteq$ OLCA.

Corollary 3
If OLCA $\subset$ LCA, then LCA(O(n)) $\subset$ LCA.

The question of whether LCA(n) $\subset$ LCA seems to be easier to answer. Of course, a negative answer to questions concerning the closure properties under reversal or concatenation would imply a positive answer to this problem, because we know that the class of LCA languages is closed under both reversal and concatenation.

3.3. A SUMMARY OF OPEN QUESTIONS AND RESULTS

The following diagram summarizes what are currently known about the closure properties of various LCA, OLCA and OLIA language classes and their inclusion relations. Here, a single box means the closure under reversal (but the closure under concatenation is open), and a double box means the closure under both reversal and concatenation. A vertical bold line means inclusion and a line with marker / means that the inclusion is proper.

4. The complexity of one-way mesh-connected arrays

We now consider mesh-connected arrays, especially the ones with one-way communication. Theorem 3 can be easily extended to OMCA and OMIA.

Theorem 10
OMCA = OMIA.

The one-way communication versus two-way communication question can be answered for mesh-connected arrays, because of the following space-efficient simulation result for OMCA's and OMIA's (Chang *et al.*, 1989).

Theorem 11
OMCA = OMIA $\subseteq$ *DSPACE*$(n^{3/2})$.

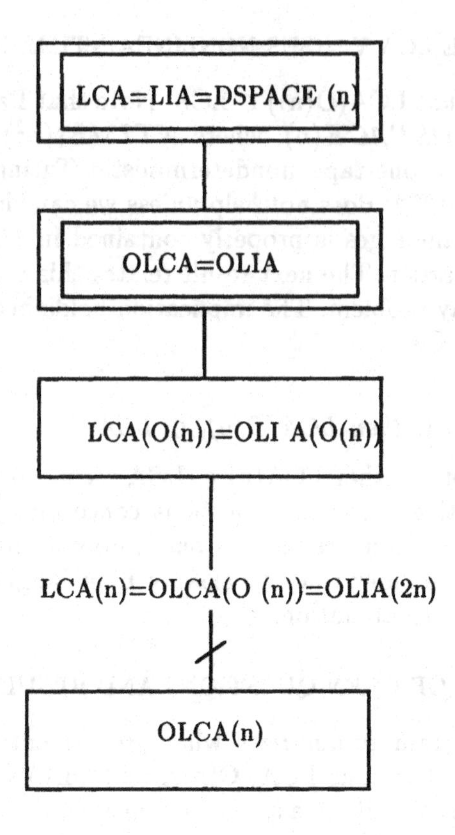

Figure 5. The status diagram.

Remark 2

It is unclear if the the above space bound is the best possible. In fact, we do not know if OMCA's are more powerful than OLCA's. We also do not know the relationship between OMCA's and LCA's.

Since MCA = MIA = DSPACE(n^2), we obtain that one-way mesh-connected arrays are weaker than their two-way counterparts. In general, for time-bounded one-way arrays, we have the following simulation.

Theorem 12

For every $T(n)$, $OMCA(T(n)) \subseteq DSPACE(n \log T(n))$ and and $OMIA(T(n)) \subseteq DSPACE(n \log T(n))$.

The next theorem gives an upper bound on the time complexity of a one-way array.

Theorem 13

Each OMCA (or OMIA) operates in $O(c^n)$ time for some constant $c > 0$.

Note that Theorems 12 and 13 together do not imply Theorem 11. Both Theorems 12 and 13 actually work for a larger class of arrays called *uniform conglomerates* (Chang *et al.*, 1989). Theorem 11 was obtained by exploring the regularity of a mesh.

OMCA's and OMIA's are quite powerful. They can accept fairly complex languages efficiently. For example, the following result can be shown (Ibarra *et al.*, 1986):

Theorem 14

OMIA's (OMCA's) can accept context-free languages in $2n - 1$ time ($3n - 1$ time), which is optimal with respect to the model of computation.

Theorem 14 is an improvement of a result in (Kosaraju, 1974) which showed that context-free languages can be accepted by two-way mesh-connected iterative arrays in linear time.

In our definition of an MCA (or an MIA, or their one-way versions), the number of nodes is the square of the length of the input. It is also interesting to consider mesh-connected arrays where the number of nodes is equal to the length of the input. Denote these models as MCA_1 and MIA_1. The one-way version of these arrays are shown in Figure 6.

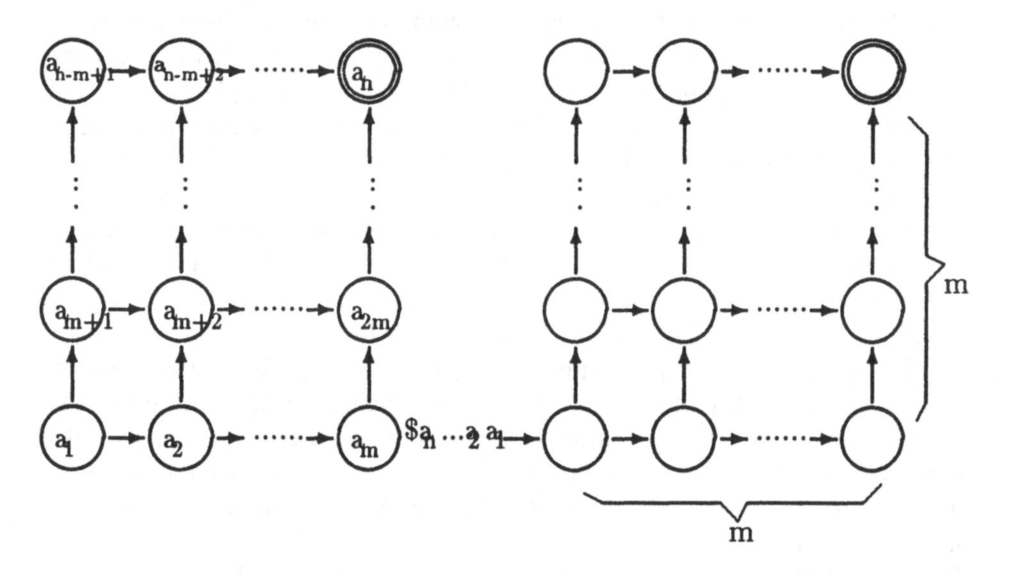

Figure 6. An $OMCA_1$ and an $OMIA_1$. (Here, $m = n^{1/2}$.)

We do not know if $MCA_1 = OMIA_1$. Clearly each $OMCA_1$ can be simulated by an $OMIA_1$. It seems difficult to prove the converse. The technique in the proof of Theorem 3 does not work anymore since an $OMCA_1$ can only count up to $O(c^{n^{1/2}})$. We also do not know if $OMCA_1 = OLCA$, if

$\text{OMIA}_1 = OLIA$, and if OMCA_1's and OMIA_1's can simulate linear time-bounded alternating TM's. The best we have is the following, which can be shown using the ideas in Theorem 2.

Theorem 15
Every $O(n^{1/2})$ time-bounded alternating TM can be simulated by an OMIA_1.

It is not obvious how to extend the above result to OMCA_1.

5. The parallel complexity of cellular arrays

In this section, we look at the parallel complexity of real-time OLCA's and pseudo-real-time OLIA's. We show that it is unlikely that the classes of languages accepted by these arrays are contained in the class NC, which is defined as follows: NC^i is the class of languages accepted by uniform boolean circuits of polynomial size and depth $O(\log^i n)$, and $NC = \bigcup_{i\geq 1} NC^i$ (Ruzzo, 1981), (Cook, 1985). Thus, it is unlikely that a general technique can be found that maps any real-time OLCA (or pseudo-real-time OLIA) algorithm into a parallel random-access machine (PRAM) algorithm that runs in polylogarithmic time using a polynomial number of processors.

Formally, we show that there is a language accepted by a real-time OLCA (and by a pseudo-real-time OLIA) that is P-complete. Hence, if such a language is in NC, then P (= the class of languages accepted by deterministic Turing machines in polynomial time) equals NC, which is widely believed to be unlikely.

It is convenient to introduce a machine model called *resetting* deterministic linear-bounded automaton (DLBA). Recall that a DLBA is equivalent to a deterministic linear space-bounded Turing machine (TM). A *resetting* DLBA is a restricted DLBA which operates as follows: the machine starts on the left end of the input tape $a_1 \ldots a_n$ in a distinguished reset state, r. Then it scans the tape from left-to-right (advancing one tape square to the right in each step), changing states and reading/rewriting the tape just like a DLBA. The machine either halts in an accepting state after processing a_n, or it resets to the left end of the tape in the reset state, r, to begin a new left-to-right *sweep*.

It is important to note that the reset state is always r. If this were not the case, i.e., if we allow the machine to reset to a different state (e.g., allowing the machine to reset to the state it enters after processing a_n), then one can easily show that such a machine is equivalent to a DLBA. It is easy to see that DLBA's are equivalent to LCA's and resetting DLBA's are equivalent to OLIA's (and, hence, also to OLCA's). Thus, it is also an open problem whether resetting DLBA's are equivalent to DLBA's.

A resetting DLBA that makes $S(n)$ sweeps on the input before accepting is called an $S(n)$-sweep resetting DLBA. Note that $S(n) \geq n$ (not n+1, since the input to the resetting DLBA does not have an endmarker).

Theorem 16

There is an n-sweep resetting DLBA that accepts a P-com plete language L.

We sketch the idea behind the proof of the above theorem. Let $\mathbf{Z}$ be a single-tape deterministic TM that runs in polynomial time, p(n). We construct a resetting DLBA M which accepts an input of the form $y = ID_1\#ID_2\#\cdots\#ID_k d^s$, where ID_1 is an initial instantaneous description (ID) of $\mathbf{Z}$ on some input x, ID_k is an accepting ID, ID_{i+1} is the direct successor of ID_i, and d^s is a string of dummy symbols. The dummy symbols are used to pad the accepting sequence of ID's so that M can carry out the simulation of $\mathbf{Z}$ within n sweeps, where n is the length of y. M uses two "markers". On the first sweep, it places the first marker at the beginning of ID_1 and the second marker at the first #. M makes left-to-right sweeps moving the markers appropriately to check that ID_{i+1} is the direct successor of ID_i. We omit the details.

It is easy to show that any language accepted by an n-sweep resetting DLBA can be accepted by a linear-time OLIA. Hence, the language L of Theorem 16 can be accepted by a linear-time OLIA. By padding the strings in L with (new) dummy symbols, we can convert L to a new language L' that can be accepted by a pseudo-real-time OLIA and L' is also P-complete. By Theorem 5, L' is accepted by a linear-time OLCA. Again, by padding L' with dummy symbols, we can convert L' to another language L" that can be accepted by a real-time OLCA and L" is also P-complete. Thus, we have the following corollary.

Corollary 4

Real-time OLCA's and pseudo-real-time OLIA's can accept P-complete languages.

6. Application to recurrence equations

Many computational problems can often be expressed in terms of recurrence equations. Examples are problems in computational linear algebra, signal processing, and dynamic programming. Thus, efficient sequential and parallel algorithms for solving recurrence equations are of great practical interest.

In this section, we use Theorem 16 to show that recurrence equations, even the simple ones, are not likely to admit fast parallel algorithms, i.e., they are not likely to be in NC. Consider the following recurrence equation:

$$\begin{aligned} R(0,0) &= c \\ R(i,j) &= f(a_i, R(i-1,j), b_j, R(i,j-1)) \\ &\qquad \text{for } 0 \le i \le n, 0 \le j \le m \text{ such that } i+j \ge 1, \end{aligned} \tag{1}$$

where c, a_r, b_s ($1 \le r \le n, 1 \le s \le m$) and the values of the $R(i,j)$'s are symbols from some fixed finite alphabet, and f is a finite function of four arguments. We assume without loss of generality that $m \le n$. For notational convenience, let $a_0 = b_0 = \epsilon$, and the boundary conditions $R(i,-1) = R(-1,i) = \epsilon$, where ϵ is a dummy symbol. The objective is to compute $R(n,m)$. Note that we can have f depend also on $R(i-1,j-1)$; however, this dependence can be removed by a simple coding technique.

Clearly, this recurrence equation can be solved by a parallel algorithm in linear time using a linear number of processors by computing along the diagonals of the recurrence table $R(i,j)$. We show that it is unlikely that it can be solved by a parallel algorithm in polylogarithmic time using a polynomial number of processors, i.e., it is unlikely that it belongs in the class NC.

Theorem 17

There is a recurrence equation of the form (1) that accepts a P-complete language. Thus, it is unlikely that such a recurrence can be solved by a parallel algorithm in polylogarithmic time using a polynomial number of processors.

Proof. By Theorem 16, it suffices to show that the computation of an n-sweep resetting DLBA can be reduced to solving a recurrence.

Let M be an n-sweep resetting DLBA with state set $Q = \{1, 2, \ldots, s\}$, input alphabet Σ, worktape alphabet Γ, and transition function δ. Note that $\Sigma \subsetneq \Gamma$, and we assume a distinguished symbol $\epsilon \in \Gamma$. Assume that the resetting state is 1 and s is the only accepting state. Since M moves right after each atomic move, we can write the transition function in the form $\delta(q,a) = [q', a']$. This means that M in state q reading a enters state q' after rewriting a by a'.

Given an input $a_1 \ldots a_n$, we define recurrence R in the following manner: informally, $R(i,j)$ represents the pair of state and symbol of M after processing the ith tape cell in sweep $j+1$. We denote by $st(R(i,j))$ the first component of $R(i,j)$, and $sym(R(i,j))$ the second component. The sweeps are numbered $0, \ldots, n-1$ (we assume the $0th$ position of the tape always contains the symbol $a_0 = \epsilon$).

$$R(0,0) = [1, \epsilon] \tag{2}$$

$$\begin{aligned}
R(i,0) &= f(a_i, R(i-1,0), R(i,-1)) \\
&= \delta(st(R(i-1,0)), a_i) \text{ for } 1 \leq i \leq n \\
R(0,j) &= f(a_0, R(-1,j), R(0,j-1)) \\
&= [1,\epsilon] \text{ for } 1 \leq j \leq n-1 \\
R(i,j) &= f(a_i, R(i-1,j), R(i,j-1)) \\
&= \delta(st(R(i-1,j)), sym(R(i,j-1))) \\
& \quad \text{for } 1 \leq i \leq n, 1 \leq j \leq n-1
\end{aligned}$$

Note that recurrence above is of the form (1) with the b_j's set to ϵ.

Remark 3

We can show the converse, i.e., given a recurrence equation, we can construct a resetting DLBA to simulate the evaluation of $R(n,m)$. We sketch the construction of the resetting DLBA M. The input to M is the string $b_1 b_2 \ldots b_m \# a_1 a_2 \ldots a_n$ (of length $n+m+1$). M creates a new track on the tape to record $R(i,j)$ in cell $i+m+1$ in sweep j. For M to compute $R(i,j)$, it needs $a_i, R(i-1,j), b_j, R(i,j-1)$. Now a_i and $R(i,j-1)$ are in cell $i+m+1$. $R(i-1,j)$ was just computed and written in the previous cell, and M can remember this information in the state. Thus, if b_j is also available, M can compute $R(i,j)$, record it in cell $i+m+1$, and remembers it in the state (replacing $R(i-1,j)$). To access b_j, M can use a marker to mark a tape cell position. The marker is initially placed in position 1 at the beginning of the first sweep (i.e., sweep 0), and M remembers b_1 in the state. In the second sweep (sweep 1), M moves the marker to position 2, thus marking b_2, and now remembers b_2 in the state, etc. Thus, M can compute $R(n,m)$.

In the recurrence equation (1), $R(i,j)$ depends on both $R(i-1,j)$ and $R(i,j-1)$, i.e., in general, in the function f, $R(i-1,j)$ and $R(i,j-1)$ interact. We now look at the special case when the equation is of a restricted form:

$$\begin{aligned}
R(0,0) &= R_0 \\
R(i,j) &= g_1(a_i, R(i-1,j)) \text{ } \mathbf{op} \text{ } g_2(b_j, R(i,j-1))
\end{aligned} \tag{3}$$

where

1. R_0 is a finite set.
2. $R(i,j)$ is a finite set for all i and j, and the set of all such sets is finite, independent of n and m.
3. **op** is an operation on sets.

4. g_1 and g_2 are distributive, i.e.,

$$g_1(a_i, R(i-1,j)) = \{g_1(a_i,x) | x \in R(i-1,j)\}$$

and

$$g_2(b_j, R(i,j-1)) = \{g_2(b_j,x) | x \in R(i,j-1)\}.$$

We will show that even for this case, there is such a recurrence equation that accepts a P-complete language. We give an example where **op** is *set intersection*. We can modify the proof of Theorem 17 as follows:

Given an input $a_1 \ldots a_n$, we define recurrence R in the following manner: informally, $R(i,j)$ represents the quadruple of states and symbols of M before *and* after processing the *ith* tape cell in sweep j. We denote by $st1(R(i,j))$, $sym1(R(i,j))$, $st2(R(i,j))$, and $sym2(R(i,j))$ the first, second, third, and fourth component of $R(i,j)$, respectively.

$$\begin{aligned}
R(0,0) &= \{[1,\epsilon,1,\epsilon]\} \\
R(i,0) &= g_1(a_i, R(i-1,0)) \cap g_2(\epsilon, R(i,-1)) = \\
&\quad \{[1, a_i, \delta(st2(R(i-1,0)), a_i)]\} \cap Q \times \Sigma \times Q \times \Sigma \\
&\quad \text{for } 1 \le i \le n \\
R(0,j) &= g_1(a_0, R(-1,j)) \cap g_2(\epsilon, R(0,j-1)) = \\
&\quad \{[1,\epsilon,1,\epsilon]\} \cap Q \times \Gamma \times Q \times \Gamma \text{ for } 1 \le j \le n-1 \\
R(i,j) &= g_1(a_i, R(i-1,j)) \cap g_2(\epsilon, R(i,j-1)) = \\
&\quad \{[st2(R(i-1,j)), \gamma, \delta(st2(R(i-1,j)), \gamma)] | \gamma \in \Gamma\} \cap \\
&\quad \{[q, sym2(R(i,j-1)), \delta(q, sym2(R(i,j-1)))] | q \in Q\} \\
&\quad \text{for } 1 \le i \le n, 1 \le j \le n-1
\end{aligned} \tag{4}$$

Clearly the recurrence above is of the restricted form (3), where the b_j's are set to ϵ. Note that because the sweeping machine is deterministic, the result of any intersection is a singleton set. Thus, we have

Corollary 5

There is a recurrence equation of the restricted form that accepts a P-complete language.

Thus, it is unlikely that such an equation can be solved by a parallel algorithm in polylogarithmic time using a polynomial number of processors.

References

Bucher B. and Čulik II K. On real time and linear time cellular automata. *R.A.I.R.O. Informatique théorique/Theoretical Informatics* **Vol.18 no.4**:307–325, 1984.

Chandra C., Kozen D. and Stockmeyer L. Alternation. *J. ACM* **Vol.28 no.1**:114–133, 1981.

Chang C., Ibarra O. and Palis M. Efficient simulations of simple models of parallel computation by space-bounded TMs and time-bounded alternating TMs. *Theoretical Computer Science* **Vol. no.68**:19–36, 1989.

Chang C., Ibarra O. and Vergis A. On the power of one-way communication. *J. ACM* **Vol. no.35**: 697–726, 1988.

Choffrut C. and Čulik II K. On real-time cellular automata and trellis automata. *Acta Inform.* **Vol. no.21**: 393–409, 1984

Cole S. Real-time computation by n-dimensional iterative arrays of finite-state machines. *IEEE Trans. Comput.* **Vol. 18 no.4**: 346–365, 1969.

Cook S. A taxonomy of problems with fast parallel algorithms. *Information and Control* **Vol. no.64**: 2–22, 1985.

Čulik II K., Ibarra O. and Yu S. Iterative Tree Arrays with Logarithmic Depth. *Intern. J. Computer Math.* **Vol. no.20**: 187–204, 1986.

Dyer C. One-way bounded cellular automata. *Information and Control* **Vol. no.44**: 54–69, 1980.

Hennie F. *Iterative arrays of logical circuits* MIT Press, Cambridge, Mass., 1961.

Hopcroft J. and Ullman J. *Introduction to automata theory, languages, and computation.* Addison-Wesley, 1979.

Ibarra O., Kim S. and Moran S. Sequential machine characterizations of trellis and cellular automata and applications. *SIAM J. Computing* **Vol. no.14**: 426–447, 1985.

Ibarra O., Palis M. and Kim S. Some results concerning linear iterative (systolic) arrays. *J. of Parallel and Distributed Computing* **Vol. no. 2**: 182–218, 1985.

Ibarra O., Kim S. and Palis M. Designing Systolic Algorithms using sequential machines. *IEEE Trans. on Computers* **Vol.C35 no. 6**: 31–42, 1986.

Ibarra O. and Jiang T. On one-way cellular arrays. *SIAM J. on Computing* **Vol. no. 16**: 1135–1154, 1987.

Ibarra O. and Palis M. Two-dimensional systolic arrays: characterizations and applications. *Theoretical Computer Science* **Vol. no. 57**: 47–86, 1988.

Ibarra O. and Jiang T. Relating the power of cellular arrays to their closure properties. *Theoretical Computer Science* **Vol. no. 57**: 225–238, 1988.

Kosaraju S. On some open problems in the theory of cellular automata. *IEEE Trans. on Computers* **Vol. C no. 23**: 561–565, 1974.

Paterson N. Tape bounds for time-bounded Turing machines. *Journal of Computer and System Sciences* **Vol. no. 6**: 116–124, 1972.

Ruzzo W. On uniform circuit complexity. *Journal of Computer and System Sciences* **Vol. no. 22**: 365–383, 1981.

Savitch W. Relationships between nondeterministic and deterministic complexities. *Journal of Computer and System Sciences* **Vol. no. 4**: 177–192, 1970.

Smith III A. Cellular automata and formal languages. *Proc. 11th IEEE Ann. Symp. on Switching and Automata Theory* 216–224, 1970.

Smith III A. Cellular automata complexity trade-offs. *Information and Control* **Vol. no. 18**: 466–482, 1971.

Smith III A. Real-time language recognition by one-dimensional cellular automata. *Journal of Computer and System Sciences* **Vol. no. 6**: 233–253, 1972.

Umeo H., Morita K. and Sugata K. Deterministic one-way simulation of two-way real-time cellular automata and its related problems. *Inform. Process. Lett.* **Vol. no. 14**: 159–161, 1982.

A COUNTING EQUIVALENCE CLASSES METHOD TO PROVE NEGATIVE RESULTS

V. TERRIER
GREYC Université de Caen
Esplanade de la Paix, 14032 Caen, France

Abstract. In this paper we review the techniques developed to prove limitation on the real time recognition ability of Iterative Array and One-way Cellular Automata. The different proofs are based on the same principle which uses equivalence relations and counting arguments.

1. Introduction

Cellular Automata (CA) appear to be one of the most relevant models for massively parallel computation. Various models have been defined depending on the input mode and how the cells are interconnected. In particular from one dimensional CA where the input mode is parallel and the communication is two-way, two restricted versions have been introduced : Iterative Array (IA) with sequential input mode and One-way Cellular Automata (OCA) with one-way communication.

In order to investigate their power and their limitations cellular automata are studied especially as language recognizers. A lot of interest focuses on lower complexity classes : real time and linear time. It must be stressed that simple questions as whether unrestricted time CAs bounded in space are more powerful than real time CAs or whether unrestricted time OCAs bounded in space are as powerful as unrestricted time CAs bounded in space remain open. Behind these questions lies the relationship between time and space. Nevertheless Ibarra and Jiang (Ibarra and Jiang, 1988) have set a connection between complexity classes equivalence and closure properties : they have shown that real time CAs are closed under reversal if and only if real time CAs are as powerful as linear time CAs and it would imply that real time CAs are closed under concatenation. For the restricted versions of CAs it has been shown that from linear time IAs and CAs are equivalent and that linear time OCAs are equivalent to real time CAs (Choffrut and Čulik, 1984) while real time IAs and real time OCAs are incomparable and less powerful than real time CAs (Cole, 1969) (Čulik *et al.*, 1981). In particular for unary language real time OCAs are not more powerful than Finite Automata. It has also been proved that real time OCAs are reversible (Choffrut and Čulik, 1984) and contain all linear context free languages (Smith, 1970).

Actually the negative results on the recognition ability of real time IAs and real time OCAs are based on the same arguments. The aim of this paper is to review this arguments. They are based on equivalence relations which characterize the behavior of real time IAs and real time OCAs. More precisely the principle to prove limitation on the real time recognition ability of IAs and OCAs is the following. Let define both an equivalence relation R_A on Σ^* induced by any cellular array A of some particular type (real time IA or real time OCA) with input alphabet Σ, and an equivalence relation R_L on Σ^* induced by any language $L \subset \Sigma^*$. These two equivalence relations must be defined in such a way that if L is recognized by A then R_A refines R_L (which means that if $(u,v) \in R_A$ then $(u,v) \in R_L$). Thus the number of equivalence classes of R_A sets an upper bound on the number of equivalence classes of R_L which can be distinguished. In this way a language whose number of equivalence classes is greater than the number of equivalence classes of any array of this particular type would be not recognized. Moreover exhibiting well-find languages allow to derive negative closure results.

Such techniques were introduced by Hartmanis and Stearns (Hartmanis and Stearns, 1965) to show that the multitape real time Turing machines do not decide languages
$L1 = \{w_1 s \dots s w_n s w \ : \ s \notin \{0,1\}\ ,\ w, w_1, \dots, w_n \in \{0,1\}^*$ and w is the reverse of w_i for some i where $1 \le i \le n\}$ nor
$L2 = \{w_1 s \dots s w_n s w \ : \ s \notin \{0,1\}, w, w_1, \dots, w_n \in \{0,1\}^*$ and $w = w_i$ for some i where $1 \le i \le n\}$. For Iterative Arrays Cole (Cole, 1969) applied these techniques to show limitations concerning real time computations. In particular he proved that $\Sigma^* P$, where $P = \{u \in \Sigma^* \ : \ |u| \ge 3$, u is a palindrome $\}$ is not recognizable by real time IAs. As P and $P\Sigma^*$ are real time IAs languages, it results that real time IAs are not closed under reversal nor under concatenation. Moreover he established that the computing capability of real time IAs increases as the dimensionality of IAs increases. For that purpose he presented languages recognizable in real time by $(n+1)$-dimensional IAs but not by n-dimensional IAs. The language D below marks the boundary between one and two-dimensional IAs.
$D = \{w_{0,0} c w_{0,1} c \dots c w_{0,k_0} d w_{1,0} c w_{1,1} c \dots c w_{1,k_1} d \dots d w_{n,0} c w_{n,1} c \dots c w_{n,k_n}$
$\sharp d^i c^j w \ : \ c, d, \sharp \notin \{0,1\}, w, w_{i,j} \in \{0,1\}^*, 0 \le i \le n, 0 \le j \le k_i$ and w is the

reverse of $w_{n-i,k_{n-i}-j}\}$.
The languages for the larger dimensions have the same kind of structure. Using the same techniques Rosenberg (Rosenberg, 1967) presented several languages not recognizable in real time by multitape Turing machines, one of them is the deterministic context free language
$L = \left\{0^{k_1}10^{k_2}1\ldots10^{k_r}\sharp^s0^{k_{r-s+1}} \ : \ \sharp \notin \{0,1\}, k_i \geq 1 \text{ and } 1 \leq s \leq r\right\}$.
Consequently he has derived negative closure properties. Furthermore he pointed out that real time IAs as real time multitape Turing machines are machines of "polynomialy-limited accessibility". Indeed the equivalence relations exploited in (Hartmanis and Stearns, 1965) (Rosenberg, 1967) for real time multitape Turing machines and in (Cole, 1969) for IAs have the same properties concerning their index and their refinement ability. So the languages presented in (Hartmanis and Stearns, 1965) (Rosenberg, 1967) which are not real time recognizable by Turing machines are not any more real time recognizable by IAs, and the negative results remain valid for IAs. Precisely he set that for any n, the class of real time n-dimensional IAs as the class of real time Turing machines with n-dimensional tapes, are not closed under concatenation, Kleene closure, sequential machine mapping, reversal nor the operations of taking derivatives and quotients. Subsequently we could ask whether it exists languages recognizable in real time by IAs but not by multitape Turing machines. Negative results were also obtained by adapting these techniques to real time OCAs in (Terrier, 1995) (Terrier, 1996). In this way it has been shown that the languages $L = L1.L1$ where
$L1 = \{w \ : \ w = 1^u0^u \text{ or } w = 1^uOy1O^u \text{ with } y \in \{0,1\}^* \text{ and } u > 0\}$ and
$M = \{uvu \ : \ u,v \in \{0,1\}^* \text{ and } |u| > 1\}$ are not real time OCA languages. As L1 is a linear context free language and so recognizable in real time by OCA it follows that real time OCAs are not closed under concatenation and does not contain all context free languages.

2. Definitions

Let recall the basic definitions of cellular automata.

A one dimensional cellular automaton is a linear array of identical finite automata (cells) numbered 1, 2, ... from left to right and working synchronously at discrete time steps. Each cell is directly connected to its left and right neighbors and takes on value from a finite set Q, the set of states. At each step the state of each cell is updated according to the states of its two neighbors and according to a transition function δ. Formally denoting $< c,t >$ the state of the cell c at time t, we have $< c,t > = \delta(< c-1,t-1 >, < c,t-1 >, < c+1,t-1 >)$. The input mode is parallel : at time 1, the i-th bit of the input string w is on the i-th cell. The evolution of a CA is often represented by a time-space diagram : the t-th row corresponds to the configuration of the CA at time t.

A one-way cellular automaton is a CA where the neighborhood is restricted to the right, the evolution of a cell is defined by its own state and the state of its right neighbor : $< c,t >= \delta(< c,t-1 >, < c+1,t-1 >)$. The input mode is parallel.

An iterative array has the same neighborhood than a CA . The restriction is on the input mode which is sequential : the i-th digit of the input is received by the cell 1 at time i.

Here we are interested in CAs as language recognizer. For that purpose we distinguish two subsets of Q : Σ and Q_{accept}; Σ is the alphabet of the input words and Q_{accept} is the subset of accepting states. We specify also a distinguished cell : the cell 1 where the result of the computation is got. A CA (IA or OCA) recognizes a language $L \subset \Sigma^*$ in time $f(n)$ if it accepts words of length n in $f(n)$ steps, i.e. if the cell 1 enters an accepting state at a time less than $f(n)$. The real time corresponds to the minimal time for the cell 1 to read the whole input, that is $f(n) = n$.

To describe the evolution of an IA (or OCA) we will use the following notations. The word $s(c,t,w)$ denotes the state of the cell c at time t of the evolution of the IA (or OCA) on the input string w. The word $s([i,j],t,w)$ of Q^{j-i+1} refers to $s(i,t,w)s(i+1,t,w)\dots s(j,t,w)$, $|w|$ denotes the length of the word w and $|X|$ the cardinality of the set X.

Recall that if an IA (or OCA) A recognizes in real time a language L then on any input u the cell 1 enters an accepting state at time $|u|$ if and only if u belongs to L. So formally $u \in L$ if and only if $s(1,|u|,u) \in Q_{accept}$.

Let precise some definitions relative to equivalence classes.
Let E_1 and E_2 be two equivalence relations on Σ^*. We say that E_1 refines E_2 if for all $u,v \in \Sigma^*$ $(u,v) \in E_1 \Longrightarrow (u,v) \in E_2$.

Observe that if E_1 refines E_2 then the index of E_1 sets an upper bound on the index of E_2.

3. Iterative array

We recall in this section the arguments introduced in (Cole, 1969). For IAs, as the input mode is sequential, the k last letters of any input word are fed during the k last steps of the computation. So these k last letters and the information on the past input history interact on a number of cells independent of the length of the word. That sets an upper bound on the number of distinct behaviors. More precisely let consider the equivalence relation $E_k(A)$ relating to a real time IA A and the equivalence relation $F_k(L)$ relating to a language L (see Figure 1).

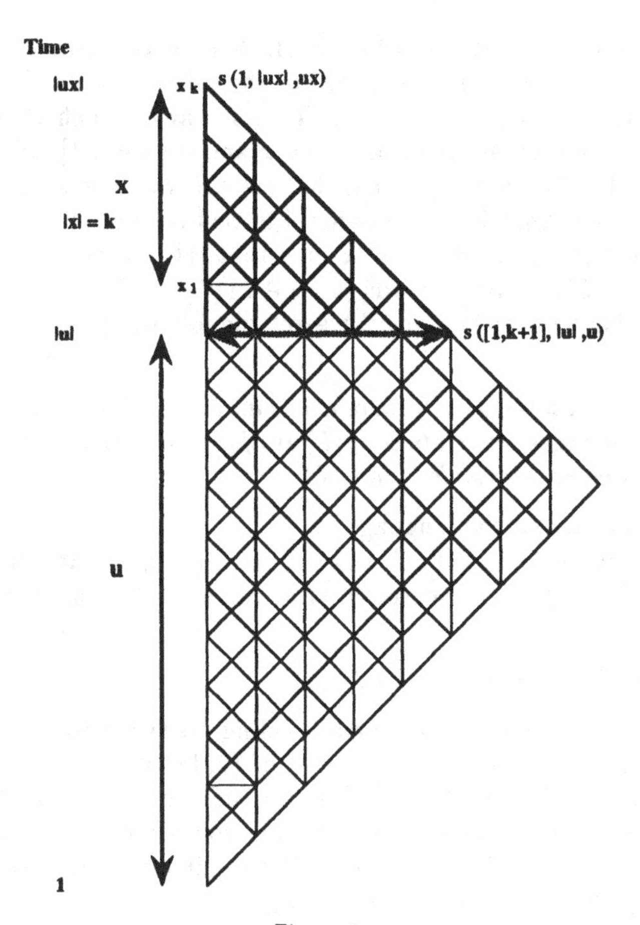

Figure 1.

Let A an IA with input alphabet Σ and k a positive integer be given. The equivalence relation $E_k(A)$ on Σ^* is defined by : $(u, v) \in E_k(A)$ if and only if $s([1, k+1], |u|, u) = s([1, k+1], |v|, v)$.

Fact 1

The index of $E_k(A)$ is at most q^{k+1} where q is the number of states of A.

Proof Indeed the number of distinct $s([1, k+1], |u|, u)$ which are words of Q^{k+1} is at most q^{k+1}. □

Let $L \subset \Sigma^*$ and k a positive integer be given. The equivalence relation $F_k(L)$ on Σ^* is defined by : $(u, v) \in F_k(L)$ if and only if (for all $z \in \Sigma^*$ such that $|z| \leq k$ we have $(uz \in L \Longleftrightarrow vz \in L)$).

Lemma 1

If L is recognized in real time by some IA A then $E_k(A)$ refines $F_k(L)$.

Proof The Figure 1 illustrates this proof. Among the cells of the $|u|$-th row the only ones which have an effect on the cell 1 at times $|u|+h$ with $h=0,\dots,k$ are the cells $1,\dots,k+1$. Thus for any x such that $|x|\leq k$, $s(1,|ux|,ux)$ is completely determined by x and $s([1,k+1],|u|,u)$. Hence if $(u,v)\in E_k(A)$, then for all x such that $|x|\leq k$ we have $s(1,|ux|,ux)=s(1,|vx|,vx)$. Recall that if L is recognized in real time by A , $ux\in L$ when $s(1,|ux|,ux)\in Q_{accept}$. Thus for all x such that $|x|\leq k$ we have ($ux\in L$ if and only if $vx\in L$). In other words $(u,v)\in F_k(L)$. □

According Fact 1 and Lemma 1 we get the following proposition.

Proposition 1

Let $L\subset\Sigma^*$ be a language. If there exists an integer k such that the index of the induced equivalence relation $F_k(L)$ is of greater order than $2^{O(k)}$ then L is not real time recognizable by any IA.

We take as example the language
$L2=\{w_1s\dots sw_nsw\ ;\ s\notin\{0,1\},w,w_1,\dots,w_n\in\{0,1\}^*$ and $w=w_i$ for some i where $1\leq i\leq n\}$ presented in (Hartmanis and Stearns, 1965).

Fact 2

The index of $F_k(L2)$ is at least 2^{2^k}.

Proof Consider the set $\{0,1\}^k$ corresponding to the words on $\{0,1\}$ of length k. To any subset $A=\{w_1,\dots,w_n\}$ of $\{0,1\}^k$ we associate the word $\Psi_A=w_1s\dots sw_ns$ of $\{0,1,s\}^{n(k+1)}$. Observe that for all $z\in\{0,1\}^k$ ($\Psi_Az\in L$ if and only if $z\in A$). Hence if A and B are two distinct subsets of $\{0,1\}^k$, Ψ_A and Ψ_B are not $F_k(L2)$-equivalent. Thus the index of $F_k(L2)$ is at least the number of subsets of $\{0,1\}^k$, which is 2^{2^k}. □

Corollary 1

L2 is not real time recognizable by IAs.

4. One-way cellular automata

In this section we consider two variants to prove limitation on the real time recognition ability of OCAs. As the communication is one-way, a feature of the real time OCAs is that the evolution on any word u contains the evolution on all the subwords of u. The following fact formalizes this property. See Figure 2.

Fact 3

For any word $u=u_1\dots u_n$ of Σ^* and integers i, j, t such that $1\leq i\leq j\leq n+1-t$ we have $s([i,j],t,u)=s([1,1+j-i],t,u_i\dots u_{j-1+t})$.

Proof First observe that at time $t=1$, $s([i,j],1,u)=s([1,1+j-i],1,u_i\dots u_j)=u_i\dots u_j$ for all i,j where $1\leq i\leq j\leq n$. Then for any

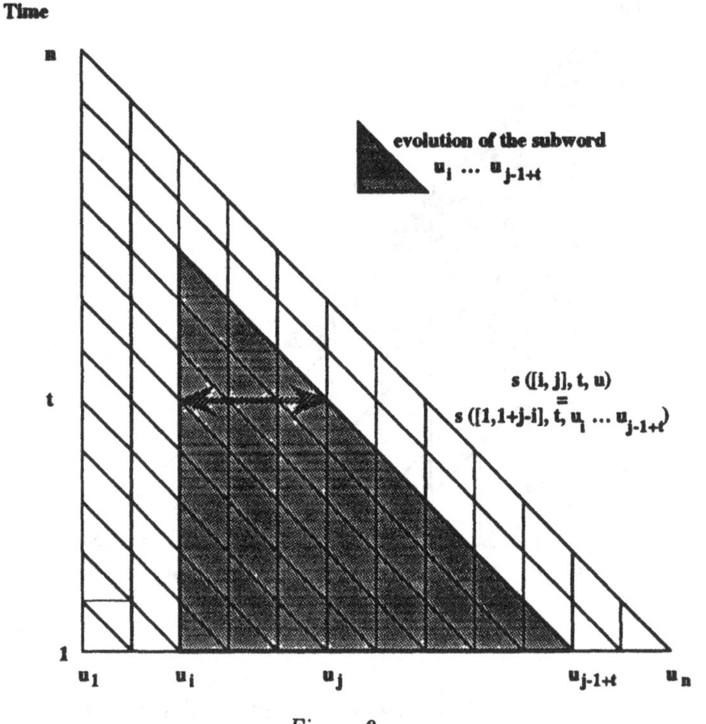

Figure 2.

integer c with $i \leq c < j$ we have $s(c, 2, u) = \delta(s(c, 1, u), s(c+1, 1, u)) = \delta(s(1, 1, u_i \dots u_j), s(2, 1, u_i \dots u_j)) = s(1, 2, u_i \dots u_j)$. Hence $s([i, j], 2, u) = s([1, 1+j-i], 2, u_i \dots u_{j+1})$ for all i, j where $1 \leq i \leq j \leq n-1$. So by induction on t, $s([i, j], t, u) = s([1, 1+j-i], t, u_i \dots u_{j-1+t})$ for all i, j where $1 \leq i \leq j \leq n+1-t$. □

More precisely the k last steps of the computation on an input word w contains the outcome of the $\frac{k(k+1)}{2}$ computations of the subwords of w of length greater than $|w| - k$. This k last steps depend only on the k sites of the row $|w| - k$. That sets an upper bound on the number of different cases which can be discriminated. The first method derives from this observation, where an equivalence relation $G_k(A)$ relating to a real time OCA A and an equivalence relation $H_k(L)$ relating to a language L are defined. See Figure 3.

Let A an OCA with input alphabet Σ and k a positive integer be given. The equivalence relation $G_k(A)$ on Σ^* is defined by : $(u, v) \in G_k(A)$ if and only if $s([1, k], |u| - k + 1, u) = s([1, k], |v| - k + 1, v)$.

Fact 4

The index of $G_k(A)$ is at most q^k where q is the number of states of A.

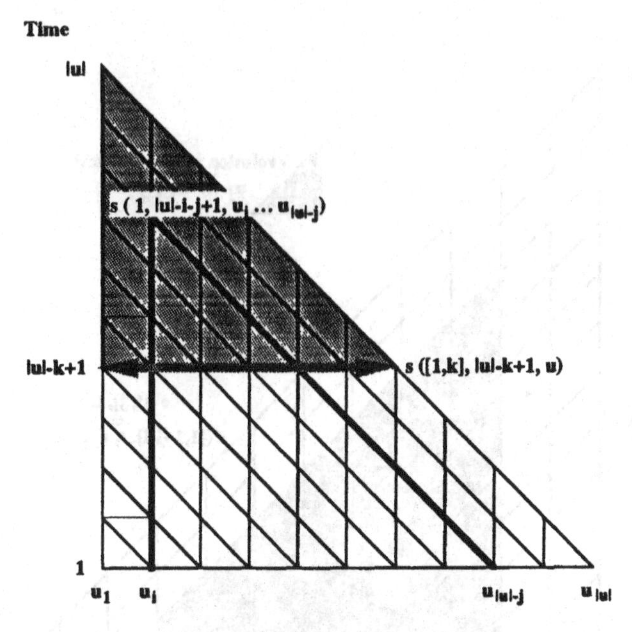

Figure 3.

Let $L \subset \Sigma^*$ and k a positive integer be given. The equivalence relation $H_k(L)$ on Σ^* is defined by : for $u = u_1 \ldots u_m$ and $v = v_1 \ldots v_n$ $(u,v) \in H_k(L)$ if and only if for all i,j such that $1 \leq i \leq i+j \leq k$ we have $(u_1 \ldots u_{m-j} \in L \iff v_1 \ldots v_{n-j} \in L\ )$.

Lemma 2

If L is recognized in real time by some OCA A then $G_k(A)$ refines $H_k(L)$.

Proof Let $u = u_1 \ldots u_m$, $v = v_1 \ldots v_n$ be two words of Σ^* and m,n be their lengths. Suppose that $(u,v) \in G_k(A)$. By definition of $G_k(A)$ $s(i,m-k+1,u) = s(i,n-k+1,v)$ for $i = 1,\ldots,k$.
Then $s(i,m-k+2,u) = \delta(s(i,m-k+1,u), s(i+1,m-k+1,u)) = \delta(s(i,n-k+1,v), s(i+1,n-k+1,v)) = s(i,n-k+2,v)$ for $i = 1,\ldots,k-1$. And by induction on t, $s(i,m-k+t,u) = s(i,n-k+t,v)$ for all t,i such that $1 \leq t \leq k$ and $1 \leq i \leq k+1-t$. Substituting t by $k-i-j+1$ we get $s(i,m-i-j+1,u) = s(i,n-i-j+1,v)$ for all i,j such that $1 \leq i \leq i+j \leq k$. Hence according to Fact 3 we have $s(1,m-i-j+1,u_i \ldots u_{m-j}) = s(1,n-i-j+1,v_i \ldots v_{n-j})$ for all i,j such that $1 \leq i \leq i+j \leq k$. So the real time computations on $u_i \ldots u_{m-j}$ and on $v_i \ldots v_{n-j}$ lead to the same state. Thus for all i,j such that $1 \leq i \leq i+j \leq k$, $u_i \ldots u_{m-j} \in L$ if and only if $v_i \ldots v_{n-j} \in L$. So $(u,v) \in H_k(L)$. □

From the Fact 4 and Lemma 2 we get the following proposition.

Proposition 2

Let $L \subset \Sigma^*$ be a language. If there exists an integer k such that the index of the equivalence relation $H_k(L)$ is of greater order than $2^{O(k)}$ then L is not real time recognizable by any OCA.

Let f be a bijection from $\mathbf{N}^2$ to $\{0,1\}^*$ and $L_f = \{0^i\sharp w_1\sharp \ldots \sharp w_n\sharp 0^j \ : \ w_1 \ldots w_n \in \{0,1\}^*, \sharp \notin \{0,1\}$ and $w_t = f(i,j)$ for some t where $1 \leq t \leq n\}$.

Fact 5

The index of $H_k(L_f)$ is at least $2^{\frac{k(k+1)}{2}}$.

Proof Consider the set $X = \{(x,y) : 1 \leq x, y \leq k$ and $x + y > k\}$ of cardinality $|X| = \frac{k(k+1)}{2}$. To each subset $A = \{(i_1,j_1), \ldots, (i_r,j_r)\}$ of X we associate the word $\Psi_A = \sharp f(i_1,j_1)\sharp \ldots \sharp f(i_r,j_r)\sharp$. As f is a bijection, $0^i\psi_A O^j \in L_f$ if and only if $(i,j) \in A$. Hence if A and B are two distinct subsets of X , Ψ_A and Ψ_B are not $H_k(L_f)$-equivalent. Thus the index of $H_k(L_f)$ is at least the number of subsets of X , which is $2^{\frac{k(k+1)}{2}}$. □

Corollary 2

L_f is not real time recognizable by OCAs.

The second approach uses a more simple language equivalence relation $J_{i,j,X}(L)$ and a more complex OCA equivalence relation $I_{i,j,X}(A)$. See the Figure 4.

Let A an OCA with input alphabet Σ , X a subset of Σ and i,j positive integers be given. The equivalence relation $I_{i,j,X}(A)$ on Σ^* is defined by : $(u,v) \in I_{i,j,X}(A)$ if and only if for all $x,y \in X^*$ such that $|x| = i$ and $|y| = j$ we have $s([1,i+j], 1+|u|, xuy) = s([1,i+j], 1+|v|, xvy)$.

Fact 6

If q is the number of states of A then the index of $I_{i,j,X}(A)$ is at most $q^{|X|^{i+1}+|X|^{j+1}}$ if $|X| > 1$ or q^{i+j} if $|X| = 1$.

Proof First observe that (u,v) is $I_{i,j,X}(A)$-equivalent if and only if (u,v) is $I_{i,0,X}(A)$- and $I_{0,j,X}(A)$- equivalent. Indeed on the input xuy with $|x| = i$ and $|y| = j$, at time $1 + |u|$ the segment $[1,i]$ is independent of y and the segment $[i+1, i+j]$ is independent of x. Formally
$s([1,i+j], 1+|u|, xuy) = s([1,i], 1+|u|, xuy)s([i+1,i+j], 1+|u|, xuy) =$
$s([1,i], 1+|u|, xu)s([1,j], 1+|u|, uy)$
according to Fact 3. So the index of $I_{i,j,X}(A)$ is the product of the indexes of $I_{i,0,X}(A)$ and $I_{0,j,X}(A)$. Now let determine the index of $I_{i,0,X}(A)$. Note that for $x = x_1 \ldots x_i$ we have
$s([1,i], 1+|u|, xu) = s(1, 1+|u|, x_1 \ldots x_i u)s(1, 1+|u|, x_2 \ldots x_i u)\ldots s(1, 1+|u|, x_i u)$.
Actually for integers $t = 1, \ldots, i$ the words u can be viewed as functions from X^{i+1-t} to the set of states Q which associates a word $x_t \ldots x_i$ to a

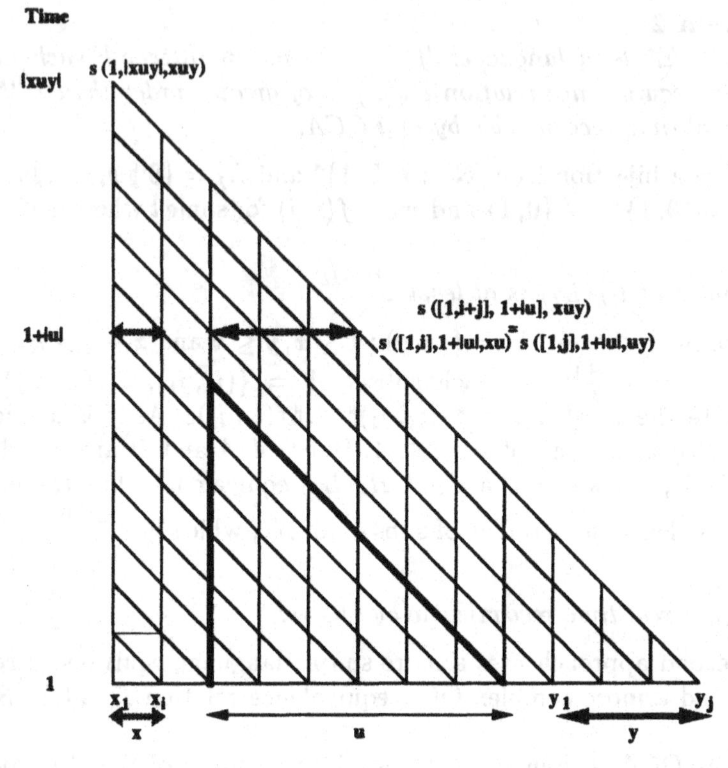

Figure 4.

state $s(1, 1+|u|, x_t \dots x_i u)$. The number of such functions is $q^{|X|^{i+1-t}}$. Thus the number of equivalence classes of $I_{i,0,X}(A)$ is $\prod_{t=1}^{i} q^{|X|^t}$ which is lower than $q^{|X|^{i+1}}$ if $|X| > 1$ or equal to q^i if $|X| = 1$. By symmetry the index of $I_{0,j,X}(A)$ is lower than $q^{|X|^{j+1}}$ if $|X| > 1$ or equal to q^j if $|X| = 1$. □

Let $L \subset \Sigma^*$, X a subset of Σ and i, j positive integers be given. The equivalence relation $J_{i,j,X}(L)$ on Σ^* is defined by : $(u, v) \in J_{i,j,X}(L)$ if and only if for all $x, y \in X^*$ such that $|x| = i$ and $|y| = j$ we have ($xuy \in L \iff xvy \in L$)).

Lemma 3

If L is recognized in real time by some OCA A then $I_{i,j,X}(A)$ refines $J_{i,j,X}(L)$.

Proof Letu, v be two words of Σ^* such that $(u, v) \in I_{i,j,X}(A)$. By definition for all $x, y \in X^*$ such that $|x| = i$ and $|y| = j$ we have $s([1, i+j], 1+|u|, xuy) = s([1, i+j], 1+|v|, xvy)$. Hence $s(1, |xuy|, xuy) = s(1, |xvy|, xvy)$. So the real time computations on xuy and xvy lead to the same state. It follows that for all $x, y \in X^*$ such that $|x| = i$ and $|y| = j$ we have ($xuy \in L \iff xvy \in L$). That is $(u, v) \in J_{i,j,X}(L)$. □

From Fact 6 and Lemma 3 we obtain the following proposition.

Proposition 3

Let $L \subset \Sigma^*$ be a language, X a set of cardinality greater than 1. If there exists an integer k such that the index of the equivalence relation $J_{k,k,X}(L)$ is of greater order than $2^{O(|X|^k)}$ then L is not real time recognizable by any OCA.

Remark 1

With this second approach, the previous proposition remains true even if we know at initial time where the middle of the input word is (observing that $L_{middle} = \left\{x_1 \ldots \tilde{x}_{\lceil \frac{n}{2} \rceil} \ldots x_n \;:\; x_1 \ldots x_n \in L\right\}$ is not also recognizable by any OCA).

Let g be a bijection from $\{0,1\}^* \times \{0,1\}^*$ to $\{0,1\}^*$ and $L_g = \{u\sharp w_1 \sharp \ldots \sharp w_n \sharp v \;;\quad u, v, w_1, \ldots, w_n \in \{0,1\}^*, \sharp \notin \{0,1\}$ and $w_t = g(u,v)$ for some t where $1 \leq t \leq n\}$.

Fact 7

The index of $J_{k,k,\{0,1\}}(L_g)$ is at least $2^{2^{2k}}$.

Proof Consider the set $\{0,1\}^k \times \{0,1\}^k$ of cardinality 2^{2k}. To each subset $A = \{(u_1, v_1), \ldots, (u_n, v_n)\}$ of $\{0,1\}^k \times \{0,1\}^k$ we associate the word $\Psi_A = \sharp g(u_1, v_1)\sharp \ldots \sharp g(u_n, v_n)\sharp$. As g is a bijection we have for all $x, y \in \{0,1\}^k$, $x\Psi_A y \in L$ if and only if $(x, y) \in A$. Hence if A and B are two distinct subsets of $\{0,1\}^k \times \{0,1\}^k$, Ψ_A and Ψ_B are not $J_{k,k,\{0,1\}}(L_g)$ -equivalent. Thus the index of $J_{k,k,\{0,1\}}(L_g)$ is at least the number of subsets of $\{0,1\}^k \times \{0,1\}^k$, which is $2^{2^{2k}}$. □

Corollary 3

L_g is not real time recognizable by OCAs.

References

Choffrut C. and Čulik K. On real time cellular automata and trellis automata. *Acta Informatica* **Vol. no. 21**: 393–407, 1984.

Cole S. Real-time computation by n-dimensional iterative arrays of finite-state machine. *IEEE Trans. Comput.* **Vol. no. C-18**: 349–365, 1969.

Čulik K., Gruska J. and Salomaa A. Systolic trellis automata (for VLSI). *Res. Rep. CS-81-34* Dept. of Comput. Sci., University of Waterloo, **Vol. no. 34**, 1981.

Hartmanis J. and Stearns R. On the computational complexity of algorithms. *Trans. Amer. Math. Soc.* **Vol. 117 no. 5**: 285–306, 1965.

Ibarra O. and Jiang T. On the computational complexity of algorithms. *Theoretical Computer Science* **Vol. no. 57**: 225–238, 1988.

Rosenberg A. Real-time definable languages. *J. ACM* **Vol.14 no. 4**: 645–662, 1967.

Smith A. Cellular Automata and formal languages. *Proc. 11th IEEE Ann. Symp. Switching Automata Theory* **vol. 14 no. 4**: 216–224, 1970.

Smith A. Real-time language recognition by one-dimensional cellular automata. *J. Comput. System Sci* Vol. no. 6: 233–253, 1972.
Terrier V. On real time one-way cellular automata. *Theoretical Computer Science* Vol. no. 141: 331–335, 1995.
Terrier V. Language not recognizable in real time by one-way cellular automata. *Theoretical Computer Science* Vol. no. 156: 281–287, 1996.

Part 4
Dynamics

Dynamics is understood here as the study of phase spaces of cellular automata. The computation field is left for a domain rather close to the modeling one. Typical problems are decision problems such as:

Data: a set of configurations of a cellular automaton,

Question: has the orbits set of these configurations property P?

Most of these problems are undecidable.

It is natural to try to classify cellular automata according to some dynamic properties of their phase spaces. One choice leads to a classification which is effective for the 256 elementary cellular automata.

TOPOLOGICAL DEFINITIONS OF DETERMINISTIC CHAOS

Applications to Cellular Automata Dynamics

G. CATTANEO
Dipartimento di Scienze dell'Informazione,
Via Comelico 39, 20135 *Milano, Italy*

E. FORMENTI
Laboratoire de L'Informatique du Parallélisme,
Allee D'Italie 46, 69364 Lyon, France

AND

L. MARGARA
Dipartimento di Scienze dell'Informazione,
Via Mura Anteo Zamboni 7, Bologna, Italy

Abstract. Various notions of topological chaos for iterated discrete time dynamical systems (DTDS) are introduced and compared.
In particular we consider the Devaney's definition based on the topological properties of transitivity and regularity; the critique and the new proposal of topological chaos discussed by Knudsen, and some original definitions introduced by the authors of the present paper.
The adequacy of these definitions is tested considering the global dynamics induced on the phase space of one-dimensional, bi-infinite configurations cellular automata (CA) local rules of arbitrary radius and alphabet.

1. Autonomous Discrete Time Dynamical Systems

Let us start this section with the general definitions involving (deterministic) *autonomous discrete time dynamical systems* (DTDS) (also *iterated dynamical systems*).

Definition 1.1

An (autonomous) DTDS is a pair $\langle X, g\rangle$ consisting of a nonempty met-

[0]This work has been partially supported by MURST, under the project "Efficienza di Algoritmi e Progetto di Strutture Informative", by CEC ESPRIT BRA contract n. 6317 - ASMICS 2 and by CNR contract n. CT94.00452.CT12.115.26124

rical space (X, d), *called the* phase *or* state space, *and a continuous map* $g : X \mapsto X$, *called the* next state *or* evolution transformation.

For any fixed state $x \in X$, the *positive motion* (or *orbit*) of initial state x is the X–valued sequence $\gamma_x : \mathbf{N} \mapsto X$, which tells us in which state the system is at each time $t \in \mathbf{N}$, if at time 0 it was in state x:

$$\forall t \in \mathbf{N},\ \gamma_x(t) := g^t(x)$$

where $g^t : X \mapsto X$ is the t-times iterated transformation induced by g, recursively defined for any $x \in X$ as $g^t(x) = g\left(g^{t-1}(x)\right)$, and $g^0(x) = x$.

This orbit is the "solution" of the difference equation with initial condition (Cauchy problem for difference equations):

$$\begin{cases} x(t+1) = g(x(t)) & \forall t \in \mathbf{N} \\ x(0) = x \end{cases} \tag{1.1}$$

As to this relation between DTDS and difference equations, we quote the following LaSalle statement (LaSalle, 1976):

> Difference equations are important and significant mathematical models for real phenomena and systems in the physical and non-physical sciences. After all, the observational data we have for real systems is often discrete. [...]
>
> The author wishes to confess that this has been the first time he has looked systematically at the subject of difference equations. What he has done is to do everything by analogy with the more highly developed theory for ordinary differential equations [...] It has been surprising to him to discover new results and to find at this level so interesting a theory and so much of pratical significance.
>
> He is now an advocate for considering difference equations as a prerequisite to the study of differential equations [...], and the theory of systems. Interesting phenomena modeled by difference equations are not too difficult to find. Computations are easy, and it is a good introductory applied mathematics.

Definition 1.2

A DTDS $\langle X, g\rangle$ *is said to be* reversible *iff the next state map* $g : X \mapsto X$ *is a bijective (one-to-one and onto) and bicontinuous (both* $g : X \mapsto X$ *and its inverse* $g^{-1} : X \mapsto X$ *must be continuous).*

In the case of reversible DTDS, we can consider the *orbit* of initial state $x \in X$ as the two-sided X-valued sequence, $\gamma_x : \mathbf{Z} \mapsto X$ defined $\forall t \in \mathbf{Z}$ as follows: if $t \geq 0$, $\gamma_x(t) = g^t(x)$ and if $t < 0$, $\gamma_x(t) = \left(g^{-1}\right)^{-t}(x)$; this positive motion satisfies the difference equation (1.1) extended to all $t \in \mathbf{Z}$.

Since we are going to study the qualitative dynamical behaviour of DTDS, we introduce now some definitions which will be very useful in the sequel.

Definition 1.3

Let $\langle X, g\rangle$ be a DTDS. A point $p \in X$ is said to be periodic *of period k iff $g^k(p) = p$ and $\forall h < k$, $g^h(p) \neq p$. Periodic points of period 1 (i.e., the fixed points of g) are also called* equilibrium *points.*
We denote by $Per_k(g)$ the collection of all periodic points of period $k \in \mathbb{N} \setminus \{0\}$ (sometimes, by $\Phi(g)$ the set of all equilibrium points), and by $Per(g) := \{p \in X \mid \exists k \in \mathbb{N} \setminus \{0\} : g^k(p) = p\}$ the collection of all periodic points.

The following definition is often refered to as an element of *regularity* of the system (Devaney, 1989).

Definition 1.4

A DTDS $\langle X, g\rangle$ is regular *iff the set of periodic points $Per(g)$ is dense in X, i.e.,*

$$\forall x \in X,\ \forall \epsilon > 0,\ \exists p_\epsilon \in Per(g) : d(x, p_\epsilon) < \epsilon\,.$$

Example 1.1

Let $\mathcal{A}$ be a nonempty and finite set, considered as an alphabet *of a finite number of* letters*; this set can be equipped with the Hamming distance: $\forall a_i, a_i \in \mathcal{A}$, $d_H(a_i, a_j) = 1$ if $a_i \neq a_j$, and $= 0$ otherwise. This is a metric for the discrete topology of $\mathcal{A}$ in which any subset of $\mathcal{A}$ is clopen.*

Let us consider the collection $\mathcal{A}^{\mathbb{Z}}$ of all bi-infinite (also two-sided) $\mathcal{A}$-valued sequences $\alpha : \mathbb{Z} \mapsto \mathcal{A}$, $i \to \alpha(i)$. Sometimes, elements from the alphabet $\mathcal{A}$ are called states *and two-sided sequences from $\mathcal{A}^{\mathbb{Z}}$* configurations*; thus, for any* site *$i \in \mathbb{Z}$, the letter $\alpha(i) \in \mathcal{A}$ is the state of the configuration α in this site. The set of all configurations $\mathcal{A}^{\mathbb{Z}}$ can be equipped with the* product *(or* Tychonoff*) metric induced by the Hamming distance d_H of $\mathcal{A}$:*

$$d(\alpha, \beta) := \sum_{i=-\infty}^{+\infty} \frac{1}{|\mathcal{A}|^{|i|}} d_H(\alpha(i), \beta(i)) \tag{1.2}$$

With respect to this metric $\mathcal{A}^{\mathbb{Z}}$ turns out to be a Cantor space *((*Wiggins, 1988*), p. 99), i.e.,*

Compact *In particular sequentially compact, that is, from any sequence in $\mathcal{A}^{\mathbb{Z}}$ it is possible to single out a convergent subsequence.*

Perfect *That is, it is closed and has no isolated points (i.e., it is equal to the set of its own limit points).*

Totally disconnected *That is, every pair of distinct points can be separated by a disconnection (i.e., for every pair of points $x, y \in \mathcal{A}^{\mathbb{Z}}$, with $x \neq y$, there exists two open subsets A and B of $\mathcal{A}^{\mathbb{Z}}$ with $A \cap B = \emptyset$ and $A \cup B = \mathcal{A}^{\mathbb{Z}}$, the* disconnection *of $\mathcal{A}^{\mathbb{Z}}$, such that $x \in A$, $y \in B$).*

The topology induced by the Tychonoff distance on $\mathcal{A}^{\mathbf{Z}}$ is just the product topology induced from the discrete topology of $\mathcal{A}$. Let us recall that this topology is the coarsest one with respect to the pointwise convergence of sequences:
the sequence of configurations $\{\alpha_n\} \subseteq \mathcal{A}^{\mathbf{Z}}$ is convergent to the configuration α, $\lim_{n\to\infty} \alpha_n = \alpha$ in $\mathcal{A}^{\mathbf{Z}}$ iff for any $i \in \mathbf{Z}$, $lim_{n\to\infty}\alpha_n(i) = \alpha(i)$ in $\mathcal{A}$. Of course, since the alphabet is finite, this happens after a finite number of steps: $\forall i \in \mathbf{Z}$, $lim_{n\to\infty}\alpha_n(i) = \alpha(i)$ in $\mathcal{A}$ iff $\exists n_0(i) : \forall n > n_0(i)$, $\alpha_n(i) = \alpha(i)$.

The (two-sided) $\mathcal{A}$-left shift DTDS is the pair $\left\langle \mathcal{A}^{\mathbf{Z}}, \sigma \right\rangle$, based on the next state function $\sigma : \mathcal{A}^{\mathbf{Z}} \mapsto \mathcal{A}^{\mathbf{Z}}$, $\alpha \to \sigma(\alpha)$ defined, $\forall \alpha \in \mathcal{A}^{\mathbf{Z}}$, $\forall i \in \mathbf{Z}$, as follows:

$$[\sigma(\alpha)](i) := \alpha(i+1). \tag{1.3}$$

Analogously, the (two-sided) $\mathcal{A}$-right shift DTDS is the pair $\left\langle \mathcal{A}^{\mathbf{Z}}, \rho \right\rangle$, based on the next state function $\rho : \mathcal{A}^{\mathbf{Z}} \mapsto \mathcal{A}^{\mathbf{Z}}$, $\alpha \to \rho(\alpha)$ defined, $\forall \alpha \in \mathcal{A}^{\mathbf{Z}}$, $\forall i \in \mathbf{Z}$, as follows:

$$[\rho(\alpha)](i) := \alpha(i-1). \tag{1.4}$$

With respect to the Tychonoff metric, both right and left shifts are uniformely continuous maps and bijections such that $\sigma^{-1} = \rho$. Thus they define two reversible DTDS.

A configuration $\underline{p} \in \mathcal{A}^{\mathbf{Z}}$ is a cyclic point of period k with respect to both the shift maps iff it is spatially *periodic of the same period i.e., there exists a nonperiodic word of length k, $\omega_k \in \mathcal{A}^k$, such that $\underline{p} = (\ldots\omega_k\omega_k\omega_k\ldots)$ [also denoted by $\underline{p} = (\overline{\omega}_k)$].*

Now, it is a standard result about Tychonoff topology to prove that the set of all these periodic points is dense in the phase space $\mathcal{A}^{\mathbf{Z}}$, i.e., the two-sided full shift DTDS are regular.

Definition 1.5

Let $\langle X, g\rangle$ be a DTDS. A subset A of the state space X is positively invariant *(resp.,* strictly positively invariant*) iff $g(A) \subseteq A$ (resp., $g(A) = A$).*

If A is positevely invariant, then $\forall t \in \mathbf{N}$, $g^t(A) \subseteq A$. For any positively invariant set A, the pair $\langle A, g\rangle$ can be considered a dynamical system by its own, i.e., a *dynamical sub-system* of $\langle X, g\rangle$, since any positive motion of initial state $a \in A$ is such that $\forall t \in \mathbf{N}$, $\gamma_a(t) = g^t(a) \in A$. In a certain sense, a positively invariant set A is a "*trapping*" set since if a positive motion of initial state $x \in X$ falls into A for the first time at instant t_0

[i.e., $\gamma_x(t_0) \in A$], then from now on the orbit is "trapped" to stay in A [i.e., $\forall t > t_0$, $\gamma_x(t) \in A$].

Some authors prefer to define a dynamical sub-system as a positively invariant set which is also topologically closed; we adopt the above definition, which, owing to the following result, does not preclude this stronger situation.

Proposition 1.1

If A is positively invariant, then also its topological closure $\overline{A}$ is positively invariant.

Standard strictly positively invariant subsets of any DTDS are expressed by the following result:

Proposition 1.2

Let $\langle X, g\rangle$ be any DTDS. Then, for any fixed integer $k = 1, 2, \ldots$ the set of all period k points $Per_k(g)$ is strictly positively invariant. Moreover, the set $Per(g)$ of all periodic points is strictly positively invariant.

Example 1.2

Let us consider the DTDS of Example 1.1 in the particular case of the boolean alphabet $\mathcal{A} = \{0, 1\}$. A configuration $\underline{x} \in \{0, 1\}^{\mathbf{Z}}$ is called 0-finite *iff only a finite number [possibly 0] of sites are in the state 1, i.e., iff it is either of the form*

$$\underline{x} = (\ldots, 0, 0, 1, *, *, \ldots, *, *, 1, 0, 0, \ldots)$$

with $*$ arbitrary element of $\{0, 1\}$, or the configuration $\underline{0} = (\ldots, 0, 0, 0, \ldots)$. The collection of all 0-finite configurations (also, configurations in a background *of 0) will be denoted by $\mathcal{F}_0$.*

Note that $\mathcal{F}_0$ is a dense subset of the whole phase space $\{0, 1\}^{\mathbf{Z}}$ which is strictly positively invariant with respect both shift maps σ and ρ: thus, $\langle \mathcal{F}_0, \sigma\rangle$ and $\langle \mathcal{F}_0, \rho\rangle$ are dynamical sub-systems of the corresponding full shifts.

Analogous definitions can be introduced in the case of the collection $\mathcal{F}_1$ of all configurations from $\{0, 1\}^{\mathbf{Z}}$ in a background of 1s (also 1-finite configurations): $\underline{x} \in \mathcal{F}_1$ iff it is either of the form

$$\underline{x} = (\ldots, 1, 1, 0, *, *, \ldots, *, *, 0, 1, 1, \ldots)$$

or $\underline{1} = (\ldots, 1, 1, 1, \ldots)$. Also in this case, $\mathcal{F}_1$ is dense in $\{0, 1\}^{\mathbf{Z}}$ and strictly positively invariant with respect to both shift maps.

1.1. INDECOMPOSABILITY, TRANSITIVITY AND STRONG TRANSITIVITY

Another topological property of dynamical systems is a condition of minimality.

Definition 1.6

A DTDS $\langle X, g\rangle$ is indecomposable *iff X is not the union of two non empty closed, disjoint, and positively invariant subset.*

Of course, if two closed subsets decompose X, then they must be also open. This means that for an indecomposable DTDS the phase space X cannot be split into two (nontrivial) clopen dynamical sub-systems; indecomposability is in a certain sense an *irreducibility* condition (Ruelle, 1979). Note that in (LaSalle, 1976) this property is also called condition of *invariant connection* and X is said to be *invariantly connected*.

A stronger condition is the so-called transitivity.

Definition 1.7

A DTDS is said to be transitive *iff for any pair A, B of nonempty open subsets of the phase space X, there exists an integer $n \in \mathbf{N}$ such that $g^n(A) \cap B \neq \emptyset$.*

We can prove several conditions equivalent to transitivity.

Proposition 1.3

For any DTDS $\langle X, g\rangle$ the following assertions are equivalent.

1. *$\langle X, g\rangle$ is transitive.*
2. *For any nonempty open set A, we have that*

$$\overline{\bigcup_{n\in\mathbf{N}} g^n(A)} = X$$

3. *For any pair A, B of nonempty open subsets of the phase space X, there exists an integer $n \in \mathbf{N}$ such that $g^{-n}(A) \cap B \neq \emptyset$.*
4. *For any nonempty open set A, we have that*

$$\overline{\bigcup_{n\in\mathbf{N}} g^{-n}(A)} = X$$

5. *For any pair of points $x, y \in X$, for any pair of open balls centered in these points $B(x), B(y)$, there exist $z \in B(x)$ and $t_0 \in \mathbf{N}$ such that $g^{t_0}(z) \in B(y)$.*

Condition 3 is used to discuss not only topological but also measure-theoretic properties of DTDS, for instance ergodicity (Blanchard *et al.*, 1997).

A sufficient condition implying transitivity (and sometimes easier to handle than this latter) is given by the following result.

Proposition 1.4

If the DTDS $\langle X, g\rangle$ is such that there exists a positive motion $\gamma_{x_0} : \mathbf{N} \mapsto X$ whose trajectory is dense in X ($\overline{Im(\gamma_{x_0})} = X$), then it is transitive.

An interesting consequence of transitivity is the following.

Proposition 1.5

Let $\langle X, g\rangle$ be a transitive DTDS, then $\overline{g(X)} = X$.

Proof For any $x \in X$ and for any open ball $B(x)$ of x we prove that, owing to transitivity, there must exist an $y \in B(x) \cap g(X)$. Indeed, if $B(x) \cap g(X) = \emptyset$, let U be any open set disjoint from $B(x)$, then for any $n > 0$, $B(x) \cap g^n(U) \subseteq B(x) \cap g(X) \subseteq \emptyset$ whereas for $n = 0$ we have that $B(x) \cap U = \emptyset$, contrary to transitivity. □

In the case of a compact DTDS we have the following stronger result.

Proposition 1.6

Let $g : X \mapsto X$ be a continuous map on a **compact** *metric space (X, d). Then, the DTDS $\langle X, g\rangle$ is transitive iff there exists an orbit which is dense in X.*
In this case the map g must necessarily be onto: $g(X) = X$, in other words X must be strictly positively invariant.

Note that conditions 1–4 in Proposition 1.3 are global, whereas condition 5 is a local one: a topological "transitive" map has points which eventually move under iteration from one arbitrarily small neighborhood to any other. This allows one to introduce a stronger version of transitivity.

Definition 1.8

A DTDS is strongly transitive *iff one of the following equivalent conditions holds:*

1. *For any nonempty open set A, we have that*

$$\bigcup_{n \in \mathbf{N}} g^n(A) = X\,.$$

2. *For any nonempty open set A, we have that*

$$\bigcup_{n \in \mathbf{N}} g^{-n}(A) = X\,.$$

3. *For any pair of points $x, y \in X$, for any open ball $B(x)$ centered in x there exist $z \in B(x)$ and $t_0 \in \mathbf{N}$ such that $g^{t_0}(z) = y$.*

Another stronger condition for the transivity is the following one:

Definition 1.9

A DTDS is called mixing *iff for any pair A, B of nonempty open subsets of the phase space X, there exists an integer $n_0 \in \mathbf{N}$ such that*

$$\forall n \geq n_0,\; g^{-n}(A) \cap B \neq \emptyset\,.$$

1.2. ATTRACTORS

In the literature about DTDS one can find a lot of different definitions of *attractor*, in general identified with an attracting closed subsystem of the given dynamical system; the attracting condition consisting in a set which "ultimately captures all orbits starting in its domain of attraction". Following Guckenheimer and Holmes (Guckenheimer and Holmes, 1983), the first definition of attractor one can adopt is the following one.

Definition 1.10

A nonempty subset A of the phase space of the DTDS $\langle X, g\rangle$ is called attractor *iff it satisfies the following conditions:*

1. *A is* closed.
2. *A is* positively invariant.
3. *A is* attracting, *in the sense that there exists some neighborhood U of A such that*
   a. *U is positively invariant.*
   b. *U is attracted by A: $\forall u \in U$, $\lim_{t\to\infty} d(g^t(u), A) = 0$.*

Any neighborhood satisfying conditions (3.a,b) is said to be an attracted neighborhood *of A. Note that the existence of such an attracted neighborhood U does not assure its unicity: there could exist many attracted neighborhoods of A. The* basin of attraction $\mathcal{B}(A)$ *of A is the maximal such attracted neighborhood:*
$\mathcal{B}(A) := \cup\{U \in \mathcal{P}(X) : U$ *attracted neighborhood of* $A\}$.

Proposition 1.7

Let A be an attractor and let the domain of attraction *of A be defined as the collection of all initial states whose orbits are ultimately captured by A:*

$$\mathcal{D}_\infty(A) := \left\{y \in X : \lim_{t\to\infty} d(g^t(y), A) = 0\right\}.$$

Then, $\mathcal{D}_\infty(A)$ is the basin of attraction of A:

$$A \subseteq U \subseteq \mathcal{B}(A) = \mathcal{D}_\infty(A). \tag{1.5}$$

Proof $\mathcal{D}_\infty(A)$ is g-positively invariant. Indeed, for any $y \in \mathcal{D}_\infty(A)$ we have that

$$0 = \lim_{t\to\infty} d(g^{t-1}(g(y)), A) = \lim_{t\to\infty} d(g^t(g(y)), A)$$

and so $g(y) \in \mathcal{D}_\infty(A)$.

From the assumption that A is an attractor we have that there exists a neighborhood U of A satisfying condition (3.a,b) of Definition 1.10; in particular, there exists an open set O_U such that $A \subseteq O_U \subseteq U$. Since $U \subseteq \mathcal{D}_\infty(A)$, we can conclude that $\mathcal{D}_\infty(A)$ is a positively invariant neighborhood of A, whose points are attracted by A. □

Remark 1.1

A singleton p is an attractor according to Definition 1.10 iff p is an equilibrium point contained in a positively invariant neighborhood U such that all orbits starting from it converge to p.

Definition 1.11

Let A be an attractor of $\langle X, g\rangle$ and U an attracted neighborhood of A. Then the U-basin of attraction *of A is the subset*

$$\mathcal{B}_U(A) := \bigcup_{n\in\mathbf{N}} g^{-n}(U) = \{b \in X : \exists n \in \mathbf{N}\ s.t.\ g^n(b) \in U\}\ . \tag{1.6}$$

The U-basin of attraction of A is thus the set of all initial states of orbits which enter U after a finite number of time steps [and then converge to A remaining inside U].

Proposition 1.8

The U-basin of attraction of the attractor A is a neighborhood of A which contains U [$A \subseteq U \subseteq \mathcal{B}_U(A)$] and satisfies the following conditions:

1. *$\mathcal{B}_U(A)$ is positively invariant;*
2. *$\mathcal{B}_U(A)$ is attracted by A: $\forall b \in \mathcal{B}_U(A)$, $\lim_{t\to\infty} d(g^t(b), A) = 0$.*

Therefore, $\mathcal{B}_U(A)$ is an attracted neighborhood of A for which the following chain of inclusions holds (compare with (1.5)):

$$A \subseteq U \subseteq \mathcal{B}_U(A) \subseteq \mathcal{B}(A) = \mathcal{D}_\infty(A) \tag{1.7}$$

The particular case of a compact attractor is very interesting.

Proposition 1.9

If K is a **compact** *attractor, for any attracted neighborhood U of K the corresponding U-basin of attraction coincides with the collection of all states ultimately captured by K, i.e., let $b \in X$, then $b \in \mathcal{B}_U(K)$ iff* $\lim_{t\to\infty} d(g^t(b), K) = 0$.

Therefore, in this case $\mathcal{B}_U(K)$ is independent of the choice of U, the basin and the domain of attraction of K coincide and are determined by (1.6). In particular for a compact attractor K one has that for any attracted neighborhood U of K:

$$K \subseteq U \subseteq \mathcal{B}_U(K) = \mathcal{B}(K) = \mathcal{D}_\infty(K) \tag{1.8}$$

Theorem 1.1

Let $\langle X, g\rangle$ be a DTDS and $K \subseteq X$ a **compact** *and g-positively invariant ($g(K) \subseteq K$) subset of X. Then the* limit set *of K:*

$$\Omega_g(K) := \bigcap_{t=0}^{+\infty} g^t(K)$$

is:

1. *nonempty and compact,*
2. *strictly g-positively invariant:* $g(\Omega_g(K)) = \Omega_g(K)$;
3. *the greatest strictly g-positively invariant set contained in* K:

$$g(B) = B \text{ and } B \subseteq K \quad \Rightarrow \quad B \subseteq \Omega_g(K).$$

4. *its domain of attraction is* K: $\mathcal{D}_\infty(\Omega_g(K)) = K$.

The compactness of the phase space is a crucial condition in order to assure that the limit set is nonempty.

Example 1.3

Let us consider the (bounded but not closed) DTDS

$$\left\langle (0,1), g(x) = \tfrac{1}{2}x \right\rangle$$

Then $\forall t \in \mathbf{N}$, $g^t(0,1) = \left(0, \frac{1}{2^t}\right)$, *and so the limit set is empty.*
In the case of the (closed but not bounded) DTDS $\langle [1,\infty), h(x) = 2x \rangle$ *we have that* $\forall t \in \mathbf{N}$, $h^t[1,\infty) = [2^t, \infty)$ *and so also in this case the limit set is empty.*

The results of Theorem 1.1 do not imply that $\Omega_g(K)$ is an attractor having K as basin of attraction.

Example 1.4

Let us consider the DTDS $\langle \mathbf{R}^n, g \rangle$, *where* $g(\underline{x}) = \|\underline{x}\|\,\underline{x}$. *The closed ball centered in* $\underline{0}$ *and radius 1,* $K_1(\underline{0}) = \{\underline{x} \in \mathbf{R}^n : \|\underline{x}\| \leq 1\}$, *is compact, strictly positively invariant, with limit set* $\Omega_g(K_1(\underline{0})) = K_1(\underline{0})$. *But, since* $\forall \underline{y} \in \mathbf{R}^n \setminus K_1(\underline{0})$, $\lim_{t\to\infty} d(g^t(\underline{y}), K_1(\underline{0})) = \infty$, *it is impossible to find a subset* U *of* $\mathbf{R}^n$, *containing an open set* O_U *which in its turn contains* $K_1(\underline{0})$, *and such that all its points are initial states of orbits converging to* $K_1(\underline{0})$.

On the contrary, any closed ball centered in $\underline{0}$ *with radius* $r < 1$, $K_r(\underline{0})$, *is an attractor having* $K_1(\underline{0})$ *as basin of attraction and with limit set* $\Omega_g(K_r(\underline{0})) = \{\underline{0}\}$; *this limit set is a compact attractor (the singleton* $\{\underline{0}\}$*) whose basin of attraction is the unit closed ball:* $\mathcal{B}(\{\underline{0}\}) = K_1(\underline{0})$.

Finally, any closed ball centered in $\underline{0}$ *with radius* $R > 1$ *is not an attractor since it is not positively invariant. The corresponding domain of attraction is the closed unit ball, which is strictly contained in it:* $\mathcal{D}_\infty(K_R(\underline{0})) = K_1(\underline{0}) \subset K_R(\underline{0})$. *Notice that in this case also the limit set is the closed unit ball:* $\Omega_g(K_R(\underline{0})) = K_1(\underline{0})$.

A stronger notion of attractor can be found in Wiggins (Wiggins, 1988), in which condition 2 of Definition 1.10 is substituted by the stronger condition of strictly positive invariance of A; in this case we will say that A is a *W-attractor*.

Another definition of attractor frequently used is the one introduced by Conley (Conley, 1978) (see also Hurley (Hurley, 1982)). Conley and Hurley restrict their attention to compact (continuous time) dynamical systems; we introduce here the corresponding discrete time version in the more general situation of a noncompact phase space.

Definition 1.12

A nonempty subset A of the phase space of the DTDS $\langle X, g\rangle$ is called CH-attractor *iff it satisfies the following conditions:*

1. *A is* closed.
2. *There exists an open neighborhood U of A such that*
   a. $g(\overline{U}) \subseteq U$.
   b. $A = \Omega_g(U) = \bigcap_{t=0}^{\infty} g^t(U)$.

Trivially, from condition (2.b) and the positively invariance of any limit set, it immediately follows that A is positively invariant; moreover, from condition (2.a) we get $g(U) \subseteq g(\overline{U}) \subseteq U$, that is also U is positively invariant.

Condition (2.b) means that for any $a \in A$ there exists a sequence $(u_n)_{n\in\mathbf{N}}$ in U such that $\forall n \in \mathbf{N}$, $g^n(u_n) = a$. In the case of a compact DTDS it is possible to prove that if A is a CH-attractor, then any open neighborhood U satisfying conditions (2.a,b) of Definition 1.12 is attracted by the CH-attractor A. In conclusion we have te following result.

Proposition 1.10

Let $\langle X, g\rangle$ be a **compact** *DTDS, then any CH-attractor is an attractor (according to Definition 1.10).*

Let us stress that many authors make a distinction between *attracting sets* and *attractors*. Precisely, they define as attracting sets what we have introduced in Definition 1.10 as attractor, and then they consider as attractor any attracting set which is *irreducible*: Let us quote for instance Ruelle (Ruelle, 1981): "If an attracting set consists of a number of disjoint invariant pieces, one would like to consider each piece as an attractor, removing irrelevant points like (perhaps) wandering points. Examples of irreducibility conditions are positive transitivity (there is $x \in A$ such that the set of limit points of γ_x is A)".

This position is assumed in (Guckenheimer and Holmes, 1983) where one can find the following definition: "An *attractor* is any attracting set which contains a dense orbit (where this second property implies that the system is indecomposable)".

This definition of attractor can be also found in Holmes (Holmes, 1990):

> "An *attractor* for a [...] map is an *indecomposable, closed, invariant* set for the [...] map, which attracts all orbits starting at points in some

neighborhood. The maximal such neighborhood is the *domain of attraction* or *basin*. [...] Here

indecomposable means that the set cannot be separated into smaller, "basic" pieces: the existence of a dense orbit [...] implies indecomposability.

Invariant means that orbits starting in the set remain in it for all forward [...] time.

The main physical consequence of indecomposability is that typical orbits attracted to the set continually wander about, exploring its entirely, and not settling down to some "simpler" subset".

All the above definitions of attractor mimic to the case of discrete time standard notions introduced in the case of continuous time dynamical systems. However, as pointed out by LaSalle in the statement quoted at the very beginning of this paper, there are specific properties of the discrete time dynamical systems which cannot be recovered in the continuous time theory. One of this interesting notions regards attractors which can be reached by all the orbits starting from a suitable neighborhood after a finite transient.

Definition 1.13

A finite time *attractor A of a DTDS $\langle X, g\rangle$ is any non empty, closed, and positively invariant subset A of X admitting a positively invariant neighborhood U, any point of which is attracted by A in a finite number of steps: $\forall u \in U$, $\exists t_0 \in \mathbb{N}$ such that $g^{t_0}(u) \in A$.*

If in all the above Definitions 1.10, 1.12, and 1.13 of attractor the basin of attraction is the whole phase space, we will say that the corresponding attractor is *global*.

Example 1.5

In the phase space $\mathbb{R}^2$ let us consider the linear DTDS whose next state transformation is the linear mapping induced by the matrix $\left(\begin{smallmatrix}1&1\\1&1\end{smallmatrix}\right)$:

$$\begin{pmatrix} x^1 \\ x^2 \end{pmatrix} \in \mathbb{R}^2 \mapsto \begin{pmatrix} x^1 + x^2 \\ x^1 + x^2 \end{pmatrix} \in \mathbb{R}^2 \tag{1.9}$$

The unstable manifold

$$\mathbb{R}^2_u = \left\{ (x^1, x^2) \in \mathbb{R}^2 : x^1 = x^2 \right\}$$

is a nonempty, closed, and strictly invariant dynamical sub-system which is a global attractor in one time step.
On the contrary, the stable manifold $\mathbb{R}^2_s = \{(x^1, x^2) \in \mathbb{R}^2 : x^1 = -x^2\}$ *is a nonempty, closed, and invariant dynamical sub-system which is not an attractor.*

Example 1.6

Another example of DTDS is the so-called one-sided (left) shift on an alphabet of finite cardinality $\mathcal{A}$: the configuration space of this DTDS is the collection $\mathcal{A}^{\mathbf{N}}$ of all $\mathcal{A}$-valued one-sided sequences $\hat{\alpha} : \mathbf{N} \mapsto \mathcal{A}$, and the next state left-shift function $\hat{\sigma} : \mathcal{A}^{\mathbf{N}} \mapsto \mathcal{A}^{\mathbf{N}}$ is (similarly to the two-sided left shift of Example 1.1) defined for any $\hat{\alpha} \in \mathcal{A}^{\mathbf{N}}$ and any $n \in \mathbf{N}$ as

$$[\hat{\sigma}(\hat{\alpha})](n) := \hat{\alpha}(n+1) \tag{1.10}$$

As a dynamical sub-shift system we consider the collection of all definitively 0 one-sided sequences: $\Sigma_0 := \{\underline{x} \in \mathcal{A}^{\mathbf{N}} : \exists n_0 \in \mathbf{N}\, s.t.\, \forall n \geq n_0, x(n) = 0\}$, i.e., $\underline{x} \in \Sigma_0$ iff it is of the form $\underline{x} = (*, *, \ldots, *, 1, 0, 0, \ldots)$.

Of course, Σ_0 is shift invariant, the corresponding one-sided sub-shift dynamical system $\langle \Sigma_0, \sigma \rangle$ has the null *configuration $\underline{0} = (0, \ldots, 0, \ldots)$ as the unique equilibrium point, and this null configuration is a finite time global attractor.*

1.3. STABILITY, UNSTABILITY, AND SENSITIVE DEPENDENCE ON INITIAL CONDITIONS

The classical studies on DTDS deal with control theory and related questions about stability of systems. Recently, the interest about dynamical systems has been moved to problem of unstability, with an effort to characterize strong forms of this dynamical behaviour.

Definition 1.14 Stable positive motion.

A positive motion $\gamma_x : \mathbf{N} \mapsto X$ is stable iff

$$\forall \epsilon > 0, \exists \delta = \delta(x, \epsilon) : \forall y \in X[d(x, y) < \delta \Rightarrow \forall t, d(\gamma_x(t), \gamma_y(t)) < \epsilon]$$

Associated with this notion involving a single orbit of a DTDS, we have a global notion of stability involving the DTDS as a whole.

Definition 1.15 Stable DTDS.

A DTDS is said to be stable iff any positive motion is stable. Formally iff

$$\forall x \in X, \forall \epsilon > 0, \exists \delta = \delta(x, \epsilon) : \forall y \in X[d(x, y) < \delta \Rightarrow \forall t, d(\gamma_x(t), \gamma_y(t)) < \epsilon]$$

The dual notions of unstable positive motion and of unstable DTDS can be obtained as logical negations of the now introduced notions of local and global stability.

Definition 1.16 Unstable positive motion.

A positive motion $\gamma_x : \mathbf{N} \mapsto X$ is unstable iff it is not stable. Formally, iff

$$\exists \epsilon = \epsilon(x) > 0 : \left\{\forall \delta : \exists \hat{y} \in X, \exists \hat{t} \in \mathbf{N} : [d(x, \hat{y}) < \delta \text{ and } d(\gamma_x(\hat{t}), \gamma_{\hat{y}}(\hat{t})) \geq \epsilon]\right\}$$

So the global notion of unstability is the following.

Definition 1.17 Unstable DTDS.

A DTDS is unstable *if all its positive motions are unstable. Formally, iff* $\forall x \in X, \exists \epsilon = \epsilon(x) > 0$:
$\left\{\forall \delta : \exists \hat{y} \in X, \exists \hat{t} \in \mathbf{N} : [d(x, \hat{y}) < \delta \text{ and } d(\gamma_x(\hat{t}), \gamma_{\hat{y}}(\hat{t})) \geq \epsilon]\right\}$

Note that in the Definition 1.17 of unstable DTDS the quantity ϵ depends on the initial state x. Some stronger forms of the above definition of unstable DTDS can be found in the literature. Let us start with the first of them, which can be found for instance in Devaney (Devaney, 1989)

Definition 1.18 Simply unpredictable DTDS.

A DTDS is said to be sensitive on initial conditions *iff* $\exists \epsilon > 0 : \forall x \in X$,
$\left\{\forall \delta > 0, \exists \hat{y} \in X, \exists \hat{t} \in \mathbf{N} : [d(x, \hat{y}) < \delta \text{ and } d(\gamma_x(\hat{t}), \gamma_{\hat{y}}(\hat{t})) \geq \epsilon]\right\}$.
The constant ϵ*, which in this case is independent of the initial state* x*, can be considered as a constant characterizing the system and is called the* sensitivity constant.

From a general point of view:

"sensitive dependence is a condition which is easily understood by mathematicians and non-mathematicians alike. It has been dubbed 'the butterfly effect' in example of popular literature [...] This condition embodies the essence of chaos - the utter unpredictability of what ought to be simple systems - and so there is something popularly pleasing about requiring sensitive dependence on initial conditions." (Crannell, 1995).

Intuitively, a map is sensitive to initial conditions, or simply sensitive, if for any state x there exist points arbitrary close to x which eventually separate from x under iteration of g by at least a quantity ϵ, independent of x. We emphasize that not all points near x need eventually separate from x, but it is sufficient that at least one such point exists in every neighbourhood of x.

Sensitive dependence on initial conditions is recognized as a central notion in chaos theory since it captures the feature that in chaotic systems small errors in experimental readings lead to large scale divergence, i.e., the system is *unpredictable*. For instance in (Eckmann and Ruelle, 1985) it is claimed that: "Sensitive dependence on initial conditions is also expressed by saying that the system is *chaotic*". If a system is dependent on initial conditions and we are not able to measure with infinite precision the initial state (that is instead of the initial state x, we measure a perturbed one), we cannot predict the dynamical evolution, because

"no matter how small a neighbourhood of x we consider, there is at least one point in this neighbourhood such that after a finite number of iterations, x and this point are separated by some fixed distance" (Wiggins, 1988).

Hence

> "If a map possesses sensitive dependence on initial conditions, then for all practical purposes, the dynamic of the map defy numerical computation. Small errors in computation which are introduced by round-off may become magnified upon iteration. The results of numerical computation of an orbit, no matter how accurate, may bear no resemblance whatsoever with the real orbit" (Devaney, 1989).

A further notion of unpredictability (and thus, a fortiori, of unstability), which in the particular case of *perfect* DTDS is stronger than the above sensitivity to initial conditions, is expansiveness.

> "Sensitive dependence on initial conditions suggests that orbits of chaotic systems corresponding to even very nearby initial conditions *may* separate as time grows. Expansiveness occurs when all initial conditions yield orbits which must, for some time value, be separated by some minimal distance" (MacEachern and Berliner, 1993).

Definition 1.19 Positively Expansive DTDS

A DTDS is (positively) expansive iff

$$\exists \epsilon > 0 : \forall x, y \in X \ (x \neq y), \exists \hat{t} \in \mathbf{N} : \ d(\gamma_x(\hat{t}), \gamma_x(\hat{t})) \geq \epsilon$$

Also in this case the constant ϵ is independent of the initial state and is called the expansivity constant *of the system.*

Example 1.7

Let X be a finite phase space equipped with the trivial metric $d_H(x, y) = 1$ if $x \neq y$, and $= 0$ otherwise. Let $g : X \mapsto X$ be any bijection on X as (d_H-continuous) next state function of a DTDS. Then, this DTDS is expansive with expansivity constant $\epsilon = 1$, but it does not share the sensitive dependence on initial conditions since fixed any $x \in X$, for any $\delta < 1$ there is no $y \in X$ such that $d_H(x, y) < \delta$ and $d_H(g^{t_0}(x), g^{t_0}(y)) \geq \epsilon$ for some t_0.

Since in this paper we are interested in expansivity as a particular strong version of unpredictability (and unstability), from now on we will consider only perfect DTDS, also if not explicitely stated.

Actually many people recognize nonlinearity as a characteristic feature of chaos, but the simple example of the linear one dimensional system $(\mathbf{R}, g_\alpha)$, with $\alpha > 1$, where $\forall x \in \mathbf{R}$, $g_\alpha(x) = \alpha x$, shows an expansive (and thus also unpredictable and unstable) behaviour. If one accepts the point of view that chaos is an intrinsic property of nonlinear systems, then he is obliged to characterize as chaotic those systems satisfying some further conditions besides expansivity.

Another drawback of a definition of chaos based on the unique condition of unpredictability is shown by the following example of an expansive system with a unique globally attracting equilibrium point.

Example 1.8

The right-subshift on the phase space Σ_0 of all definitively null one-sided sequences has the null sequence $\underline{0} = (0,0,0,\ldots)$ as the unique equilibrium point, which is a global (finite steps) attractor. But this DTDS is expansive with expansivity constant equal to 1. Indeed, for any pair of different one-sided sequences $\underline{x} = (x(0), x(1), \ldots, x(m), \overline{0})$ (with $x(m) \neq 0$) and $\underline{y} = (y(0), y(1), \ldots, y(n), \overline{0})$ (with $y(n) \neq 0$), where the site j is the last place in which $x(j) \neq y(j)$ (and so $\forall i > j$, $x(i) = y(i)$), we have that

$$d\left(\sigma^j(\underline{x}), \sigma^j(\underline{y})\right) = d_H(x(j), y(j)) = 1$$

1.4. SOME TOPOLOGIAL DEFINITIONS OF DETERMINISTIC CHAOS

In modern theory of dynamical systems an important role is played by the study of chaotic phenomena. For DTDS there is no universally accepted definition of chaos. Several different quantities have been introduced to single out chaotic phenomena, such as entropy and Lyapunov exponent, and many different properties have been proposed to characterize chaoticity of dynamical systems. The first topological definition of chaos for DTDS we consider in this paper has been introduced by Devaney (Devaney, 1989):

Definition 1.20

A DTDS is Devaney chaotic *(or* D-chaotic*) iff the following conditions are satisfied:*

(D_1) it is regular (densness of periodic points);
(D_2) it is transitive.

In an elegant paper (Banks *et al.*, 1992) it has been shown that, for dynamical systems with infinite phase space, the above two conditions of regularity and transitivity imply sensitivity according to Definition 1.18. "That is, despite its popular appeal, sensitive dependence is mathematically redundant–so that in fact, [Devaney's] chaos is a property relying only on the topological, and not on the metric, properties of the space." (Crannell, 1995).

Knudsen in (Knudsen, 1994) proved the following result.

Theorem 1.2

Let $\langle X, g\rangle$ be a DTDS and $\langle Y, g\rangle$ any sub-DTDS dense in X [i.e., $g(Y) \subseteq Y$ and $\overline{Y} = X$]. Then,

1. *$\langle X, g\rangle$ is sensitive iff $\langle Y, g\rangle$ is sensitive.*
2. *$\langle X, g\rangle$ is transitive iff $\langle Y, g\rangle$ is transitive.*

Quoting Knudsen (Knudsen, 1994):

Let us describe some consequences of the two theorems. Take a dynamical system $[\langle X, g\rangle]$ that is chaotic according to Devaney's definition. Consider the restriction of the dynamics to the set of periodic points which is clearly invariant.

Theorem 1.2 implies that the restricted dynamical system $\langle Per(g), g\rangle$ is topological transitive, and that the system exhibits dependence to initial conditions. Together with the trivial fulfilled condition of denseness of periodic points, this implies that the system is chaotic.

Due to the lack of nonperiodicity this is not the kind of system most people would consider labeling chaotic.

It might, of course, be argued that since for any periodic point there is another periodic point closed by with arbitrary high period, and therefore this is a kind of chaos. Nevertheless, insisting on the distinction between a truly nonperiodic orbit, say, a dense orbit, and a periodic orbit of even arbitrary high period, such a system should not be termed chaotic.

As a consequence of these results, Knudsen proposed "the following definition of chaos which excludes chaos without nonperiodicity" (Knudsen, 1994).

Definition 1.21

A DTDS is Knudsen chaotic *(or* K-chaotic*) iff the following conditions are satisfied*

(K_1) *it has a dense orbit;*
(K_2) *it is sensitive.*

Using the above Proposition 1.6 the following immediately follows.

Proposition 1.11

If a **compact** *DTDS $\langle X, g\rangle$ is chaotic in the sense of Devaney, then it is chaotic in the sense of Knudsen.*

Then from the fact that standard proof of topological D-chaoticity of shift maps is based (see (Devaney, 1989)) on the direct construction of a dense orbit, the following is true.

Corollary 1.1

Both two-sided left-shift $\left\langle \mathcal{A}^{\mathbb{Z}}, \sigma\right\rangle$ and right-shift $\left\langle \mathcal{A}^{\mathbb{Z}}, \rho\right\rangle$ on a finite alphabet $\mathcal{A}$ are chaotic in the sense of Devaney and in the sense of Knudsen.

A stronger definition of chaos can be considered in order to distinguish two-sided *shift-like* chaotic behavior from more complex chaotic ones. We formulate it for general DTDS as follows.

Definition 1.22 (E-chaos)

Let $g : X \mapsto X$, be a continuous map on a metric space (X, d). Then the DTDS $\langle X, g\rangle$ is positively expansive chaotic (E-chaotic) iff

(E_1) *it is transitive,*
(E_2) *it is regular (denseness of periodic points),*
(E_3) *it is positively expansive.*

The one-sided shift dynamical system $\left\langle \mathcal{A}^{\mathbb{N}}, \sigma \right\rangle$ on the finite alphabet $\mathcal{A}$ is a paradigmatic example of E-chaos.

1.5. TOPOLOGICAL SEMI-CONJUGATION, FACTORS, AND TOPOLOGICAL CONJUGATIONS

The notion of isomorphism between DTDS is very crucial to identify dynamics with the same qualitative behavior. In this section we introduce this concept, distinguishing a topological notion of isomorphism from the metrical one.

Definition 1.23

A DTDS $\langle X, g\rangle$ is said to be topologically semi-conjugate *to the DTDS $\langle Y, h\rangle$ iff there exists a surjective and continuous map $\varphi : X \mapsto Y$ such that $g \circ \varphi = \varphi \circ h$, i.e., such that the following diagram commutes:*

$$\begin{array}{ccc} X & \xrightarrow{\;g\;} & X \\ \varphi \downarrow & & \downarrow \varphi \\ Y & \xrightarrow[\;h\;]{} & Y \end{array} \tag{1.11}$$

In this case, the system $\langle Y, h\rangle$ is called a factor *of the system $\langle X, g\rangle$.*

A certain number of qualitative (topological and set theoretic) dynamical properties of the system $\langle X, g\rangle$ are inheredited by any of its factors. In particular, $p \in Per_k(g)$ implies that $\varphi(p) \in Per_j(h)$, with $j \leq k$; moreover, (see (Devaney, 1989) for the proofs),

$\langle X, g\rangle$		$\langle Y, h\rangle$
Regular	$\Rightarrow$	Regular
Transitive	$\Rightarrow$	Transitive
D-Chaotic	$\Rightarrow$	D-Chaotic
K-Chaotic	$\Rightarrow$	K-Chaotic

Definition 1.24

Two DTDS $\langle X, g\rangle$ and $\langle Y, g\rangle$ are said to be topological *[resp.,* metrical*] conjugated iff the map of the above Definition 1.23 is a topological homeomorphism [resp, a surjective isometry].*

Topological [metrical] conjugation states an equivalence relation on the set of all DTDS. If two DTDS are topologically conjugate, then the positive motion of initial state x in X is in a one-to-one correspondence with the positive motion of initial state $\varphi(x)$ in Y; indeed, we have that $\forall t \in \mathbf{N}$, $\gamma_x^{(g)}(t) = \varphi[\gamma_{\varphi(x)}^{(h)}(t)]$ as a consequence of the property $g^t(x) = \varphi[h^t(\varphi(x))]$; in symbols:

$$\gamma_x^{(g)}(t) \in X \underset{\varphi^{-1}}{\overset{\varphi}{\rightleftarrows}} Y \ni \gamma_{\varphi(x)}^{(h)}(t) \tag{1.12}$$

Let us stress that both conditions of "transitivity" and "regularity" (and so also D and K-chaos) are preserved under topological conjugacy, as they are pure topological properties. However, as shown by the following counter example (see (Banks *et al.*, 1992)), there exists a pair of topologically conjugated DTDS such that one of the two is expansive (and so, unstable) and the second one is not sensitive, but stable. In conclusion, "stability", "sensitivity", and "expansiveness" are not preserved by topological conjugations. Of course, these properties are preserved under metrical conjugations.

Example 1.9

The DTDS $g : (1,\infty) \mapsto (1,\infty)$, $g(x) = 2x$ is topologically conjugated with the DTDS $h : (0,\infty) \mapsto (0,\infty)$, $h(x) = x+\log 2$ by the homeomorphism $\varphi : (1,\infty) \mapsto (0,\infty)$, $\varphi(x) = \log(x)$:

$$\begin{array}{ccc} (1,\infty) & \xrightarrow{g(x)=2x} & (1,\infty) \\ {\scriptstyle \varphi(x)=\log(x)}\downarrow & & \downarrow{\scriptstyle \log(x)=\varphi(x)} \\ (0,\infty) & \xrightarrow[h(y)=y+\log 2]{} & (0,\infty) \end{array}$$

Trivially, $\langle(1,\infty), g\rangle$ is expansive (and so also sensitive and globally unstable), but on the contrary $\langle(0,\infty), h\rangle$ is globally stable.

This bad behavior can be avoided in the following two cases:

1. Let $\langle X, g\rangle$ be a compact DTDS. If it is sensitive, then any of its (necessarily compact) factor is sensitive too.
2. Let $\langle X, g\rangle$ and $\langle Y, h\rangle$ be metrically conjugate. If one of the two is sensitive, then the other one is sensitive too.

2. One Dimensional Cellular Automata as DTDS

In this paper we shall consider only bi-infinite (also two-sided), one-dimensional cellular automata (CA). A one dimensional CA is a bi-infinite (two-sided) array of identical elements, called *cells*, located on a straight line and whose position or *site* is labelled by an integer number $i \in \mathbf{Z}$. Each cell can assume a *state* chosen from a finite set $\mathcal{A}$, the *alphabet* of the CA, and changes its state according to a *local rule*, *homogeneously* applied to all cells of the automaton, in a discrete time evolution.

Formally, a one dimensional bi-infinite CA is described as a triple $\langle \mathcal{A}, r, f \rangle$, where $\mathcal{A}$ is the finite set of states, $r \in \mathbf{N}$ is the *radius* of the automaton, and $f : \mathcal{A}^{2r+1} \mapsto \mathcal{A}$ is the *local rule*, on the basis of which each cell is updated. The collection of all local rules, for a fixed alphabet $\mathcal{A}$ and a fixed radius r, will be denoted by $\mathcal{R}(\mathcal{A}, r)$ and called the $(\mathcal{A}, r)$*–space of local rules*.

A *configuration* (*global state*) of a CA is a map $\underline{x} : \mathbf{Z} \mapsto \mathcal{A}$, $i \to x(i)$ which specifies a state for any site of the lattice, and can be represented by an $\mathcal{A}$–valued two-sided sequence:

$$\underline{x} \equiv (\ldots, x(-n), x(-n+1), \ldots, x(-1), x(0), x(+1), \ldots, x(n-1), x(n), \ldots) \tag{2.1}$$

where $\forall i \in \mathbf{Z}$, $x(i) \in \mathcal{A}$.

The updating of the whole automaton is performed by the synchronous application of the local rule; this parallel process is formally described by the *global transition map* induced by f, $F_f : \mathcal{A}^{\mathbf{Z}} \mapsto \mathcal{A}^{\mathbf{Z}}$, and associating with any configuration $\underline{x} \in \mathcal{A}^{\mathbf{Z}}$ the *next time step* configuration

$$F_f(\underline{x}) := (\ldots, [F_f(\underline{x})](i-1), [F_f(\underline{x})](i), [F_f(\underline{x})](i+1), \ldots)$$

whose component functions, $\forall i \in \mathbf{Z}$, $[F_f(\,\cdot\,)](i) : \{0,1\}^{\mathbf{Z}} \mapsto \mathcal{A}$ is expressed by the local rule according to the following law:

$$\forall \underline{x} \in \mathcal{A}^{\mathbf{Z}},\ [F_f(\underline{x})](i) = f(x(i-r), \ldots, x(i), \ldots, x(i+r)) \tag{2.2}$$

Thus, the transition from a configuration to the next time step configuration is obtained by the parallel updating of any cell with the uniform application of the local rule; precisely, the updating of the cell in site i [see (2.2)] is performed by the local rule on the basis of the states of all cells in the radius r neighborhood of this cell.

The set of all configurations $\mathcal{A}^{\mathbf{Z}}$, equipped with the Tychonoff distance [see (1.2)], is in particular a complete and compact metric space. In this metric space, for any local rule f, the global transition map F_f is uniformely continuous, so the pair $\langle \mathcal{A}^{\mathbb{Z}}, F_f \rangle$ is a (compact and complete) DTDS with *phase space* $\mathcal{A}^{\mathbb{Z}}$ and *next state function* F_f.

Example 2.1

Let $\mathcal{A}$ be any finite alphabet and r any radius. Let us consider the CA local rules $f_\sigma : \mathcal{A}^r \mapsto \mathcal{A}$ and $f_\rho : \mathcal{A}^r \mapsto \mathcal{A}$ defined by:

$$f_\sigma(a_{-r}, \ldots, a_{-1}, a_0, a_{+1}, \ldots, a_{+r}) = a_{+1}$$

$$f_\rho(a_{-r}, \ldots, a_{-1}, a_0, a_{+1}, \ldots, a_{+r}) = a_{-1}$$

The global transition functions of these CAs are respectively the left-shift map and the right-shift map already introduced in Example 1.1. Indeed, for any $\underline{x} \in \mathcal{A}^{\mathbb{Z}}$ and any $i \in \mathbb{Z}$,

$$F_\sigma(\underline{x})](i) = f_\sigma(x(i-r), \ldots, x(i-1), x(i), x(i+1), \ldots) = x(i+1)$$

$$F_\rho(\underline{x})](i) = f_\rho(x(i-r), \ldots, x(i-1), x(i), x(i+1), \ldots) = x(i-1)$$

The following theorem of Hedlund (Hedlund, 1969) gives a characterization of those DTDS based on the phase space $\mathcal{A}^{\mathbb{Z}}$ and whose discrete time dynamics is produced by iterations of a generic (Tychonoff) continuous next state transformation $G : \mathcal{A}^{\mathbb{Z}} \mapsto \mathcal{A}^{\mathbb{Z}}$.

Theorem 2.1

Let us consider the state space $\mathcal{A}^{\mathbb{Z}}$ of all bi-infinite sequences in the alphabet $\mathcal{A}$, equipped with the Tychonoff metric. Then, the following two statements are equivalent.

1. *$G : \mathcal{A}^{\mathbb{Z}} \mapsto \mathcal{A}^{\mathbb{Z}}$ is a next state transformation which commutes with the shift map (i.e. $G \circ \sigma = \sigma \circ G$).*
2. *$G : \mathcal{A}^{\mathbb{Z}} \mapsto \mathcal{A}^{\mathbb{Z}}$ is a global function of a suitable CA local rule f, i.e., $G = F_f$.*

A particular subclass of CA dynamics, which we shall investigate in the sequel, is introduced in the following definition.

Definition 2.1

Let $q \in \mathcal{A}$ be a fixed element of the alphabet. A CA rule $f \in \mathcal{R}(\mathcal{A}, r)$ is said to be q-quiescent *iff $f(q, \ldots, q, \ldots, q) = q$. The collection of all q-quiescent $(\mathcal{A}, r)$-CA rules is denoted by $\mathcal{R}_q(\mathcal{A}, r)$. For instance, both the left and the right-shifts are q-quiescent local rules for any q.*

Trivially, if f is a q-quiescent $(\mathcal{A}, r)$-local rule of a CA, then the q-constant configuration $\underline{q} = (\ldots, q, q, q, \ldots) \in \mathcal{A}^{\mathbb{Z}}$ is an equilibrium point of the induced global dynamics.

Example 2.2

In the case of a radius r boolean CA rule $f \in \mathcal{R}(\{0,1\}, r)$ we can consider the 0-quiescent and the 1-quiescent rules with respect to which the null configuration $\underline{0} = (\ldots, 0, 0, 0, \ldots)$ and the 1-identical configuration $\underline{1} = (\ldots, 1, 1, 1, \ldots)$ are equilibrium points respectively.

For any fixed *initial configuration* $\underline{x} \in \mathcal{A}^{\mathbf{Z}}$, the CA evolves through a sequence of configurations by the iteration of the global function. As pointed out in Section 1, the *positive orbit* (*motion*) starting from $\underline{x}$ is then the sequence of configurations $\gamma_{\underline{x}} : \mathbf{N} \mapsto \mathcal{A}^{\mathbf{Z}}$ associating with any time step $t \in \mathbf{N}$ the configuration at time t, $\gamma_{\underline{x}}(t) \in \mathcal{A}^{\mathbf{Z}}$, obtained by the t-times iteration of the global function:

$$\forall t \in \mathbf{N},\ \gamma_{\underline{x}}(t) := [F_f]^t(\underline{x}) \tag{2.3}$$

The *space-time pattern* of initial configuration $\underline{x}$ can be represented by a bi-infinite figure, where for the sake of simplicity we set $x^t(j) := [F_f^t(\underline{x})](j)$:

$t=0$	$\ldots$	$x(-2)$	$x(-1)$	$x(0)$	$x(1)$	$x(2)$	$\ldots$	$=\underline{x}$
$t=1$	$\ldots$	$x^1(-2)$	$x^1(-1)$	$x^1(0)$	$x^1(1)$	$x^1(2)$	$\ldots$	$=F_f(\underline{x})$
$\vdots$	$\vdots$	$\vdots$	$\vdots$	$\vdots$	$\vdots$	$\vdots$	$\vdots$	$\vdots$
t	$\ldots$	$x^t(-2)$	$x^t(-1)$	$x^t(0)$	$x^t(1)$	$x^t(2)$	$\ldots$	$=F_f^t(\underline{x})$
$\vdots$	$\vdots$	$\vdots$	$\vdots$	$\vdots$	$\vdots$	$\vdots$	$\vdots$	$\vdots$

Let us recall (see Example 1.1 of Section 1) that the topology induced from the Tychonoff distance is the coarsest one with respect to the component-wise convergence of sequences: the sequence of configurations $\{F^t(\underline{x})\} \subseteq \mathcal{A}^{\mathbf{Z}}$ is convergent to the configuration $\underline{x}_0 \in \mathcal{A}^{\mathbf{Z}}$ iff for any fixed site $i \in \mathbf{Z}$, $\exists t(i) : \forall t \geq t(i), \quad x^t(i) = x_0(i)$.

This result furnishes an "empirical" criterion to test if a dynamical evolution converges to an equilibrium point (a similar consideration can be stated for cyclic point convergence). During an empirical simulation, let us isolate a finite "window" of observation (for instance all the cells between the site i and the site j, with $i < j$), inside a larger portion of the initial configuration; then apply the local CA rule and look to the dynamical evolution inside the window. If after a finite number of steps all the cells of the window reaches an equilibrium point, then we can conjecture to be in presence of an equilibrium point configuration.

If we use the (one-sided) sequential notation to denote the positive orbit of initial state $\underline{x}$:

$$\gamma_{\underline{x}} \equiv (\underline{x}, F_f(\underline{x}), F_f^2(\underline{x}), \dots, F_f^t(\underline{x}), \dots) \in [\mathcal{A}^{\mathbf{Z}}]^{\mathbf{N}}$$

then, the positive orbit of initial state $F(\underline{x})$ is represented by the left-shifted (one-sided) sequence

$$\gamma_{F_f(\underline{x})} \equiv (F_f(\underline{x}), F_f^2(\underline{x}), F_f^3(\underline{x}), \dots, F_f^{t+1}(\underline{x}), \dots) \in [\mathcal{A}^{\mathbf{Z}}]^{\mathbf{N}}$$

In this way we obtain the following commutative diagram:

$$\begin{array}{ccc} \underline{x} \in \mathcal{A}^{\mathbf{Z}} & \xrightarrow{F_f} & \mathcal{A}^{\mathbf{Z}} \ni F_f(\underline{x}) \\ \Phi \downarrow & & \downarrow \Phi \\ \gamma_{\underline{x}} \in [\mathcal{A}^{\mathbf{Z}}]^{\mathbf{N}} & \xrightarrow[\sigma]{} & [\mathcal{A}^{\mathbf{Z}}]^{\mathbf{N}} \ni \gamma_{F_f(\underline{x})} \end{array}$$

The map Φ, associating with any configuration $\underline{x}$ the orbit starting from $\underline{x}$, is one-to-one but not onto; the map σ is the (one-sided) left-shift map on the alphabet of *infinite* cardinality $[\mathcal{A}^{\mathbf{Z}}]$. Since $\Phi(\mathcal{A}^{\mathbf{Z}})$ is shift-invariant, as a conclusion we can say that any DTDS $\left\langle \mathcal{A}^{\mathbf{Z}}, F_f \right\rangle$ (where the next state map F_f is not necessarily induced by some CA local rule) is topologically conjugated to the subshift $\left\langle \Phi(\mathcal{A}^{\mathbf{Z}}), \sigma \right\rangle$ based on an infinite alphabet.

We now construct families of topological semi-conjugations, based on finite alphabets, considering suitable "windows of observation" of the dynamics produced by a fixed next state map F_f.

Let $i, j \in \mathbf{Z}$, with $i \leq j$. The set of sites $[i, j] = (i, i+1, \dots, j)$ is the *window* of observation of extreme cells i and j. For any configuration $\underline{x} \in \mathcal{A}^{\mathbf{Z}}$ we define its (i, j)-*segment* (or *block*) as the portion of this configuration between the sites i and j: $[\underline{x}]_{i,j} = (x(i), x(i+1), \dots, x(j)) \in \mathcal{A}^{j-i+1}$. This segment is a word of length $j - i + 1$ based on the alphabet $\mathcal{A}$.

We can now introduce the map $\Phi_{i,j}^F : \mathcal{A}^{\mathbf{Z}} \mapsto [\mathcal{A}^{j-i+1}]^{\mathbf{N}}$ associating with any configuration $\underline{x} \in \mathcal{A}^{\mathbf{Z}}$ the one-sided sequence on the *finite* alphabet $[\mathcal{A}^{j-i+1}]$:
$\Phi_{i,j}^F(\underline{x}) =$

$$[\gamma_{\underline{x}}]_{i,j} \equiv ([\underline{x}]_{i,j}, [F_f(\underline{x})]_{i,j}, [F_f^2(\underline{x})]_{i,j}, \dots, [F_f^t(\underline{x})]_{i,j}, \dots) \in [\mathcal{A}^{j-i+1}]^{\mathbf{N}} \tag{2.4}$$

This sequence can be represented by the space-time pattern, restricted to the window of observation $[i, j]$:

$$\begin{array}{|c|cccc|l|} t=0 & x(i) & \dots & x(j) & & = [\underline{x}]_{i,j} \\ t=1 & x^1(i) & \dots & x^1(j) & & = [F_f(\underline{x})]_{i,j} \\ \vdots & \vdots & \vdots & \vdots & & \vdots \\ t & x^t(i) & \dots & x^t(j) & & = [F_f^t(\underline{x})]_{i,j} \\ \vdots & \vdots & \vdots & \vdots & & \vdots \end{array}$$

In this way, we have the commutative diagram:

$$\begin{array}{ccc} \underline{x} \in \mathcal{A}^{\mathbf{Z}} & \xrightarrow{F_f} & \mathcal{A}^{\mathbf{Z}} \ni F_f(\underline{x}) \\ \Phi_{i,j} \downarrow & & \downarrow \Phi_{i,j} \\ [\gamma_{\underline{x}}]_{i,j} \in [\mathcal{A}^{j-i+1}]^{\mathbf{N}} & \xrightarrow[\sigma]{} & [\mathcal{A}^{j-i+1}]^{\mathbf{N}} \ni [\gamma_{F_f(\underline{x})}]_{i,j} \end{array}$$

The two DTDS $\left\langle \mathcal{A}^{\mathbf{Z}}, F_f \right\rangle$ and $\left\langle \Phi_{i,j}(\mathcal{A}^{\mathbf{Z}}), \sigma \right\rangle$ are now only semi-conjugated since the homomorphism $\Phi_{i,j}$ is not one-to-one. It will be interesting to investigate under what conditions for some i and j it is possible to have a topological conjugation. The following is easily proved.

Proposition 2.1

Let $\left\langle \mathcal{A}^{\mathbf{Z}}, F \right\rangle$ be the DTDS based on the next state map F, non necessarily induced from a 1-dimensional CA local rule. The following are equivalent.

1. *$\left\langle \mathcal{A}^{\mathbf{Z}}, F \right\rangle$ is positively expansive.*
2. *$\exists i, j \in \mathbf{Z} : \Phi_{i,j}(\mathcal{A}^{\mathbf{Z}})$ is injective.*
3. *$\left\langle \mathcal{A}^{\mathbf{Z}}, F \right\rangle$ is topologically conjugated to a one-sided subshift on a finite alphabet.*

In the particular case of CA dynamics we have the following results.

Theorem 2.2

Let $\left\langle \mathcal{A}^{\mathbf{Z}}, F_f \right\rangle$ the DTDS induced from a 1-dimensional CA local rule f. The following are equivalent.

a. *$\left\langle \mathcal{A}^{\mathbf{Z}}, F_f \right\rangle$ is positively expansive.*

b. *$\left\langle \mathcal{A}^{\mathbf{Z}}, F_f \right\rangle$ is topologically conjugated to the one-sided subshift $\langle \Phi_{0,2k-1}, \sigma \rangle$, which in its turn is topologically conjugated to a one-sided full shift on a finite alphabet (*Nasu, 1995*).*

c. $\left\langle \mathcal{A}^{\mathbb{Z}}, F_f \right\rangle$ *is topologically conjugated to a one-sided full shift on a finite alphabet (*Kůrka, 1997*).*

In the sequel we widely discuss the case of radius 1 boolean CA, the so-called *elementary CA (ECA)*, whose local rule is a mapping $f : \{0,1\}^3 \mapsto \{0,1\}$; the rule space of all ECA is $\mathcal{R}(\{0,1\}, r)$, also denoted by $\mathcal{R}(ECA)$. Thus, there are exactly 256 elementary CA rules, each of them represented by a lookup table:

x, y, z	$f(x, y, z)$
0,0,0	$f(0,0,0)$
0,0,1	$f(0,0,1)$
0,0,0	$f(0,1,0)$
0,1,1	$f(0,1,1)$
1,1,0	$f(1,0,0)$
1,0,1	$f(1,0,1)$
1,1,0	$f(1,1,0)$
1,1,1	$f(1,1,1)$

or by a boolean vector of length 8: $(f(1,1,1), \dots, f(0,0,0))$; to each of the $2^{2^3} = 256$ different ECA rules f we assign the integer number n_f whose binary representation is the vector:

$$n_f = f(0,0,0) \cdot 2^0 + f(0,0,1) \cdot 2^1 + \cdots + f(1,1,0) \cdot 2^6 + f(1,1,1) \cdot 2^7.$$

Coming back to the general case, any finite alphabet $\mathcal{A}$ of cardinality greater than two from now on is identified with the ring of integers modulo m, $\mathcal{A} = \{0, 1, \dots, m-1\}$, equipped with the usual operations of addition and multiplication modulo m, and the *conjugation* operation defined $\forall a \in \mathcal{A} \setminus \{m-1\}$, as $a^* = 1 + a$, and $(m-1)^* = 0$. Now, it is possible to take into account, for any fixed local rule $f \in \mathcal{R}(\mathcal{A}, r)$, at least three other local rules whose global dynamics cannot be distinguished since they are all mutually isometrically conjugate.

Definition 2.2

Let $f : \mathcal{A}^{2r+1} \mapsto \mathcal{A}$ be a CA local rule, we denote by

1. $f^* : \mathcal{A}^{2r+1} \mapsto \mathcal{A}$ *the* conjugate *rule of f, defined by:*
   $\forall (x_{-r}, \dots, x_0, \dots, x_r) \in \mathcal{A}^{2r+1}$
   $f^*(x_{-r}, \dots, x_0, \dots, x_r) = f(x^*_{-r}, \dots, x^*_0, \dots, x^*_r)^*$
2. $f^o : \mathcal{A}^{2r+1} \mapsto \mathcal{A}$ *the* reflected *rule of f, defined by:*
   $\forall (x_{-r}, \dots, x_0, \dots, x_r) \in \mathcal{A}^{2r+1}$
   $f^o(x_{-r}, \dots, x_0, \dots, x_r) = f(x_r, \dots, x_0, \dots, x_{-r})$

3. $f^{o*} : \mathcal{A}^{2r+1} \mapsto \mathcal{A}$ $(= f^{*o})$ *the* reflected conjugate *rule of* f, *defined by:*
$\forall(x_{-r}, \dots, x_0, \dots, x_r) \in \mathcal{A}^{2r+1}$
$f^{o*}(x_{-r}, \dots, x_0, \dots, x_r) = f(x_r^*, \dots, x_0^*, \dots, x_{-r}^*)^*$

To each rule f we can associate the set $C(f) = \{f, f^*, f^o, f^{o*}\}$; we observe that in some cases this set contains less than four rules since some of them could be invariant with respect to the introduced transformations [for instance, in the case of ECA local rule $f_{150} : \{0,1\}^3 \mapsto \{0,1\}$, such that $f_{150}(x_{-1}, x_0, x_1) = 1$ only for the triples $\{001, 011, 100, 110\}$, we have that $f_{150}^* = f_{150}^o = f_{150}^{*o} = f_{150}$].

The transformations introduced in the above Definition 2.2 give rise to isometrical conjugacy between dynamical systems, as expressed in the following result.

Proposition 2.2

Let $f : \mathcal{A}^{2r+1} \mapsto \mathcal{A}$ *be a local rule and* f^* *its conjugate, the dynamical systems* $\langle \mathcal{A}^{\mathbb{Z}}, F_f \rangle$ *and* $\langle \mathcal{A}^{\mathbb{Z}}, F_{f^*} \rangle$ *are isometrically conjugated. This means that the following diagram commutes,*

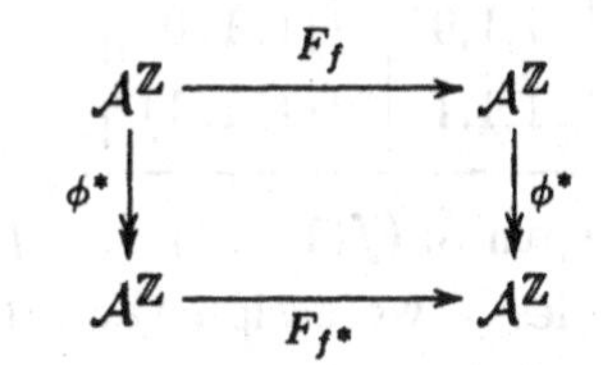

where $\phi^* : \mathcal{A}^{\mathbb{Z}} \mapsto \mathcal{A}^{\mathbb{Z}}$ *is the surjective isometry*

$$\underline{x} \in \mathcal{A}^{\mathbb{Z}} \to \phi^*(\underline{x}) = (\dots, \phi_{i-1}^*(\underline{x}), \phi_i^*(\underline{x}), \phi_{i+1}^*(\underline{x}), \dots) \in \mathcal{A}^{\mathbb{Z}}$$

component-wise defined by the law:

$$\forall i \in \mathbb{Z}, \quad \phi_i^*(\underline{x}) = x(i)^*.$$

An analogous result holds for f^o and f^{o*}, defining the surjective isometries $\phi^o : \mathcal{A}^{\mathbb{Z}} \mapsto \mathcal{A}^{\mathbb{Z}}$ and $\phi^{o*} : \mathcal{A}^{\mathbb{Z}} \mapsto \mathcal{A}^{\mathbb{Z}}$ as follows: $\forall \underline{x} \in \mathcal{A}^{\mathbb{Z}}$, $\forall i \in \mathbb{Z}$, as

$$\phi_i^o(\underline{x}) = x(-i) \quad \text{and} \quad \phi_i^{o*}(\underline{x}) = x(-i)^*.$$

2.1. WOLFRAM'S CLASSIFICATION OF CA

In the case of one-dimensional CA, there have been many attempts of classification according to their dynamical behaviour (see for example (Wolfram,

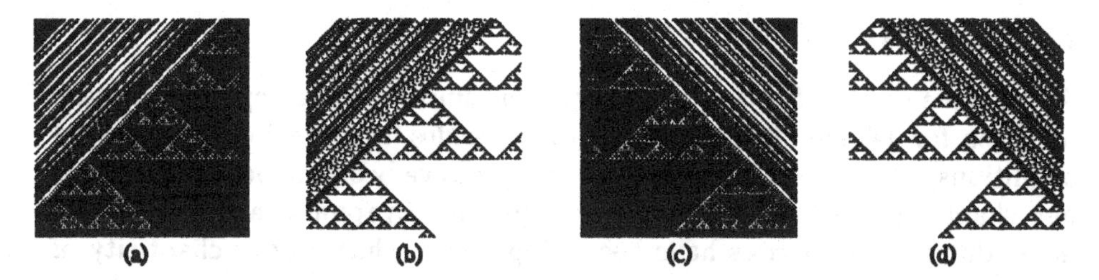

Figure 1. Example of space-time patterns generated by ECA: (a) rule 154 (f), (b) rule 166 (f^*), (c) rule 210 (f°) and (d) rule 180 ($f^{\circ *}$). If $\underline{x}$ is the initial configuration of the simulation involving rule 154, then the patterns of the other rules start from the initial configurations $\phi^*(\underline{x})$, $\phi^\circ(\underline{x})$ and $\phi^{*\circ}(\underline{x})$, respectively.

1986), (Čulik and Yu, 1988), (Culik *et al.*, 1990),(Gutowitz, 1990), (Sutner, 1990)). The classification introduced by Wolfram ((Wolfram, 1984)) has become the starting point of many approaches. This classification is based on the experimental observation of space-time patterns produced during the dynamical evolution of one dimensional CA starting from random generated initial configurations, and looking at a fixed finite window. Wolfram, after a certain number of experimental simulations, divided the one dimensional CA in four classes:

W_1 evolution leads to a homogeneous state, after a finite transient;
W_2 evolution leads to a set of separated simple stable or periodic (space-time) structures;
W_3 evolution leads to aperiodic ("chaotic") space-time patterns;
W_4 evolution leads to complex localized structures, sometimes long-lived.

It seems that none of the elementary CA is in class W_4. There are many problems related to this classification. First, the only way to assign an automaton to a class is to look at the space-time patterns it creates inside a fixed window; the dynamical evolution is observed during a finite number of time steps, starting from many (but finite) randomly generated initial configurations. The decision depends on the particular (finite) set of configurations we have empirically obtained; if we perform the same method another time with different initial configurations, we are not sure to obtain the same results. Furthermore the choice of the class which a rule belongs to is in some sense "subjective" since the definition of the classes is not formal and thus one must decide what does the class description mean.

An attempt of formalization of Wolfram's classification scheme has been done by Culik and Yu (Čulik and Yu, 1988) who split CA into three classes of increasing complexity; unfortunately membership in each of these classes is shown to be undecidable.

2.2. TOPOLOGICAL CHAOS IN CA

As we have seen in Section 1.4, there is no universally accepted definition of chaos for DTDS. Let us notice that besides the previously discussed definitions, several other different quantities have been introduced to single out chaotic phenomena, such as entropy and Lyapunov exponents, and many different properties have been proposed to characterize chaoticity of dynamical systems.

As stressed above, in the context of CA, the question of chaos was first faced by Wolfram, which recognized "chaotic" phenomena in those CA generating aperiodic space-time patterns. This characterization of chaos is not founded on a formal definition nor on a computable quantity that objectively individuate the set of chaotic CA.

Afterwards, Packard (Packard, 1985) gave an algorithm to calculate a numerical approximation of Lyapunov exponent for those CA for which it is well-defined. The empirical results obtained for a finite very narrow number of CA seem to agree with Wolfram's description of chaos: that is rules with positive Lyapunov exponent seem to be the ones in Wolfram class 3. Anyway, in defining chaotic those CA with positive Lyapunov exponent, empirical results, also in this case obtained for a finite number of initial configurations and for a finite number of iterations, cannot guarantee that *all* CA in Wolfram class 3 are chaotic according to this definition.

A third possibility is to make reference to rigorous definitions of chaos based on topological properties of DTDS. Then one can consider CA as DTDS trying to single out the class of those CA satisfying these properties, possibly by some theorems linking the (finite) values of the local rule to the chaotic behaviour of the dynamics produced by the global transition rule on the (infinite) configuration space. In this paper we follow this approach and refer to the definition of chaotic DTDS discussed in Section 1.4. In this context, we have to face another question related to the notion of "chaotic" CA. In fact in Wolfram's classification the subshift CA belong to class 2 and are considered very simple dynamical systems, while, on the basis of the definition of deterministic chaos given in (Devaney, 1989), they are chaotic dynamical systems. This oblige to try to distinguish in the case of CA dynamics chaos of shift type from chaos which is not of this kind.

We recall the following result, which holds for the general case of any finite alphabet (eventually of prime cardinality) and any radius r CA local rule.

Theorem 2.3 (Codenotti and Margara, 1996)
If $\left\langle \mathcal{A}^{\mathbb{Z}}, F_f \right\rangle$ the global dynamics induced by the local rule f is transitive, then it is also unpredictable (sensitive to initial conditions).
That is in CA dynamics, transitivity is equivalent to Knudsen chaoticity.

We give now the definition of permutive local rule and that of leftmost and rightmost permutive local rule, respectively.

Definition 2.3 *(*Hedlund, 1969*)*

The CA local rule f is permutive *in x_i, $-r \leq i \leq r$, if and only if, for any given sequence*
$\overline{x}_{-r}, \ldots, \overline{x}_{i-1}, \overline{x}_{i+1}, \ldots, \overline{x}_r \in \mathcal{A}^{2r}$*, we have*

$$\{f(\overline{x}_{-r}, \ldots, \overline{x}_{i-1}, x_i, \overline{x}_{i+1}, \ldots, \overline{x}_r) : \; x_i \in \mathcal{A}\} = \mathcal{A}.$$

Definition 2.4

f is leftmost *[resp.,* rightmost*] permutive if and only if there exists an integer $i : -r \leq i < 0$ [resp., $i : 0 < i \leq r$] such that*

- *f is permutive in the ith variable, and*
- *f does not depend on x_j, $j < i$, [resp., $j > i$].*

The following result give an interesting link (as sufficient condition) between an easy to test property of CA local rules and the involved chaotic global dynamics.

Theorem 2.4

*Let f be any leftmost or rightmost permutive 1-dimensional CA local rule defined on a finite alphabet $\mathcal{A}$. Then the induced global dynamics $\left\langle \mathcal{A}^{\mathbb{Z}}, F_f \right\rangle$ is topologically transitive (Theorem 3.1 of (*Cattaneo et al., 1997*)) and satisfies the denseness of period point condition (Theorem 3.3 of (*Cattaneo et al., 1997*)), i.e., it is Devaney chaotic.*

Taking into account Propositions 1.6 and 1.11, we can summarize this result by the following chain of implications:

$$\boxed{\text{L or R Permutive}} \Longrightarrow \boxed{\text{D-Chaos}} \Longrightarrow \boxed{\text{K-Chaos}} \Longrightarrow \boxed{\text{Surjective}} \tag{2.5}$$

Let us stress that the results of Theorem 2.2 involve properties of the global dynamics of a 1-dimensional CA; the following result gives a very interesting link between the local behavior of a 1-dimensional CA local rule and the global positively expansive dynamics.

Theorem 2.5 *(*Margara, 1995*)*

Let $\left\langle \mathcal{A}^{\mathbb{Z}}, F_f \right\rangle$ be the DTDS based on the 1-dimensional CA local rule f. If f is leftmost and *rightmost permutive, then $\left\langle \mathcal{A}^{\mathbb{Z}}, F_f \right\rangle$ is topologically conjugated to a one-sided full shift on a finite alphabet.*

As an immediate consequence of this result, taking into account the above Theorem 2.4, we can state that leftmost *and* rightmost permutive CA produce global dynamics which are expansively chaotic:

$$\boxed{\text{L and R Permutive}} \stackrel{CA}{\Longrightarrow} \boxed{\text{E-Chaos}} \tag{2.6}$$

The main results of this Section can be summarized as follows.

(a) Every 1-dimensional CA based on a local rule f which is permutive either in the first (leftmost) or in the last (rightmost) variable is Devaney (and then Knudsen) chaotic.

(b) All the 1-dimensional CA based on a local rule f which is both rightmost and leftmost permutive are expansively chaotic.

(c) There exists a Devaney chaotic CA based on a local rule f with radius 1 which is not permutive in any variable (Cattaneo *et al.*, 1997).

(d) There exists a Devaney chaotic CA defined on a boolean local rule f which is not permutive in any variable (Cattaneo *et al.*, 1997).

In Figure 2 we present the space-time patterns of some leftmost or/and rightmost permutive Boolean radius 2 CA rules.

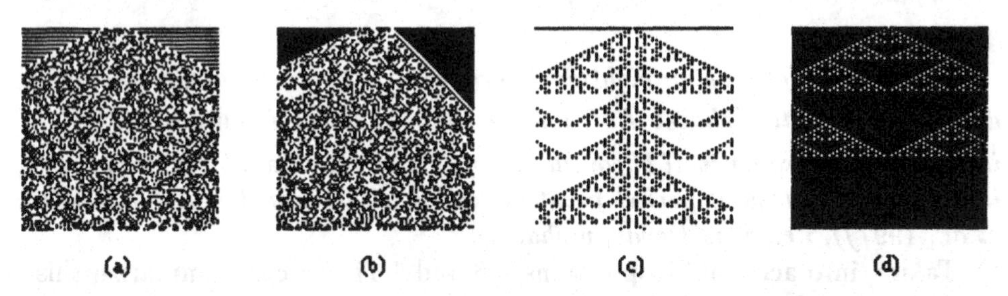

Figure 2. Leftmost [R] or Rightmost [R] permutive radius 2 Boolean CA rules: (a) rule 1706203725 [L], (b) rule 2577029478 [R], (c) rule 2796181845 [LR], (d) rule 2845136490 [LR].

2.3. EFFECTIVE CLASSIFICATION OF ECA BY TOPOLOGICAL CHAOS

In this subsection we deal with the class of ECA, i.e., CA based on a local rule $f : \{0,1\}^3 \mapsto \{0,1\}$. In this case the rule is leftmost permutive iff

$$\forall x_0, x_1 : \quad f(0, x_0, x_1) \neq f(1, x_0, x_1).$$

Similarly, it is rightmost permutive iff

$$\forall x_{-1}, x_0 : \quad f(x_{-1}, x_0, 0) \neq f(x_{-1}, x_0, 1).$$

The following result has been proved in (Cattaneo *et al.*, 1997).

Theorem 2.6

Let $\left\langle \{0,1\}^{\mathbb{Z}}, F_f \right\rangle$ *be a non trivial ECA based on the local rule* f : $\{0,1\}^3 \to \{0,1\}$.

If F_f *is surjective, then* f *is either leftmost or rightmost permutive.*

The next Corollary is a direct consequence of Theorems 2.5 and 2.6 and gives a characterization of D-chaos of the global dynamics of a ECA in terms of the easy to state leftmost or rightmost properties of the local rule.

Corollary 2.1

Let $\left\langle \{0,1\}^{\mathbb{Z}}, F_f \right\rangle$ *be the global DTDS based on the ECA local rule* f. *Then, the following statements are equivalent.*

1. f *is either leftmost or rightmost permutive (or both).*
2. F_f *is Devaney-chaotic.*
3. F_f *is Knudsen-chaotic.*
4. F_f *is surjective and non-trivial.*

All this can be summarized by the following scheme.

$$\boxed{\text{L or R Permutive}} \overset{ECA}{\Longleftrightarrow} \boxed{\text{D-Chaos}} \overset{ECA}{\Longleftrightarrow} \boxed{\text{Non-trivial Surjective}} \tag{2.7}$$

Another characterization involves E-chaos and the condition about the ECA local rule of being simultaneously leftmost *and* righmost permutive (see Table 1).

Corollary 2.2

Let $\left\langle \{0,1\}^{\mathbb{Z}}, F_f \right\rangle$ *be the global DTDS based on the ECA local rule* f. *Then, the following statements are equivalent.*

1. f *is leftmost* and *rightmost permutive.*
2. F_f *is Expansively-chaotic.*

$$\boxed{\text{L and R Permutive}} \overset{ECA}{\Longleftrightarrow} \boxed{\text{E-Chaos}} \tag{2.8}$$

We wish to emphasize that in this way we obtain a complete classification of the ECA rule space based on a rigorous mathematical definition of topological chaos.

In order to make a distinction between shift-like chaotic dynamics and chaotic dynamics characterized by complex space-time patterns, in (Cattaneo *et al.*, 1997) we introduce a notion of *rule entropy* (RE) depending uniquely from the structure of the local rule. Informally, it measures how much the initial uncertainty on the values of some cells influences the knowledge about the future configurations of the CA and propagates during its

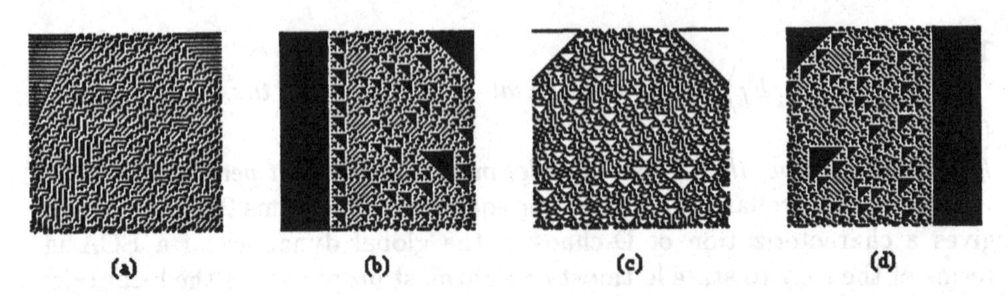

Figure 3. **Leftmost [L] or Rightmost [R] permutive ECA: (a) ECA rule 45 [L], (b) ECA rule 60 [LC], (c) ECA rule 149 [R], (d) ECA rule 102 [RC].**

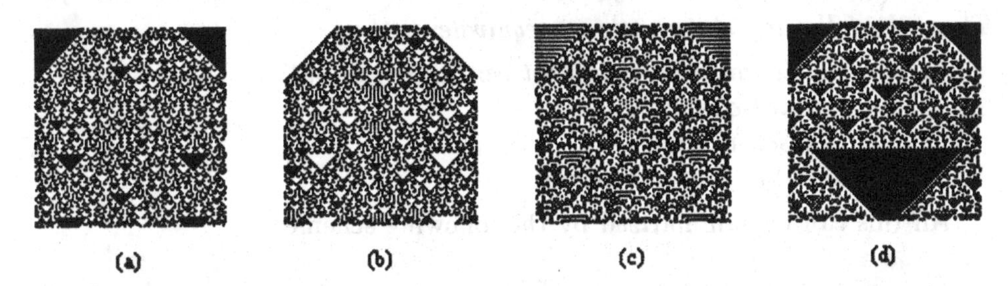

Figure 4. **Leftmost *and* Rightmost [LR] permutive ECA: (a) (b) rule 150, (c) rule 105, (d) rule 90.**

evolution. In other words, the RE expresses the inherent tendency of a CA to increase disorder in presence of uncertainty. Furthermore, it is effectively computable since it depends from the values assumed by the local rule.

In order to formally introduce this notion of RE, let us define the quantities (we adopt the usual convention for entropy: $0 \times \infty = 0$):

$$H_l^{(1)}(f) = \sum_{y,z\in\{0,1\}} \left[\left(\sum_{a\in\{0,1\}} \frac{f(a,y,z)}{2} \right) \log \left(\frac{2}{\sum_{a\in\{0,1\}} f(a,y,z)} \right) + \right.$$
$$\left. + \left(1 - \sum_{a\in\{0,1\}} \frac{f(a,y,z)}{2} \right) \log \frac{1}{1 - \sum_{a\in\{0,1\}} \frac{f(a,y,z)}{2}} \right]$$

$$H_l^{(2)}(f) = \sum_{z\in\{0,1\}} \left[\left(\sum_{a,b\in\{0,1\}} \frac{f(a,b,z)}{2} \right) \log \left(\frac{2}{\sum_{a,b\in\{0,1\}} f(a,b,z)} \right) + \right.$$
$$\left. + \left(1 - \sum_{a,b\in\{0,1\}} \frac{f(a,b,z)}{2} \right) \log \frac{1}{1 - \sum_{a,b\in\{0,1\}} \frac{f(a,b,z)}{2}} \right]$$

In an analogous way, let us introduce $H_r^{(1)}(f)$ and $H_r^{(2)}(f)$. Then, the *left-*

RE and *right-RE* are defined respectively as

$$E_l(f) := H_l^{(1)}(f) + H_l^{(2)}(f) \quad \text{and} \quad E_r(f) := H_r^{(1)}(f) + H_r^{(2)}(f).$$

Let us say that an ECA local rule f is *independent* of the variable x_{-1} (of the left variable) iff

$$\forall x_0, x_1 \in \{0,1\}: \quad f(0, x_0, x_1) = f(1, x_0, x_1).$$

Analogously, it is *independent* of the variable x_1 (of the right variable) iff

$$\forall x_{-1}, x_0 \in \{0,1\}: \quad f(x_{-1}, x_0, 0) = f(x_{-1}, x_0, 1).$$

The following result can be found in (Cattaneo *et al.*, 1997).

Proposition 2.3
In ECA rule space the following hold:

- *Both rules f and $1 \oplus f$ have the same left/right rule entropies.*
- *An ECA local rule f has the maximum left/right rule entropy iff it is left/rightmost permutive.*
- *An ECA local rule has zero left/right rule entropy iff it is independent on the left/right variable.*

In Table 1 we collect the rules entropies for the case of all left or rightmost permutive ECA, with also the computation of the central entropy (whose definition is a straightforward extension of the above introduced entropies). In this table, we also take into account ECA rules which are central permutive.

As shown by this table (in agreement with Proposition 2.3)

(En_1) the expansive chaos of leftmost *and* rightmost permutive ECA local rules is characterized by the maximum of RE values ($E_l = 6$ and $E_r = 6$);

(En_2) the dynamics of leftmost or rightmost permutive (1,1)-shift [full shift rules 240 and 170] and (2,2)-shift [full double alternate shift rules 15 and 85] are characterized by the RE values $(6,0)$ and $(0,6)$;

(En_3) for all the other leftmost and rightmost permutive ECA, any intermediate value of the local rule entropies [(6,2), (2,6) and (6,3.6), (6,3.6)] can be considered as a "degree" of chaoticity between the shift-like chaotic dynamics and the "stronger" expansive chaos.

On the contrary, the central permutive ECA local rules 204 and 51 show a very small entropy values (2,2); the global dynamics of these rules is characterized by the property that all points are fixed for F_{204} and all points are cyclic of period two for F_{51}.

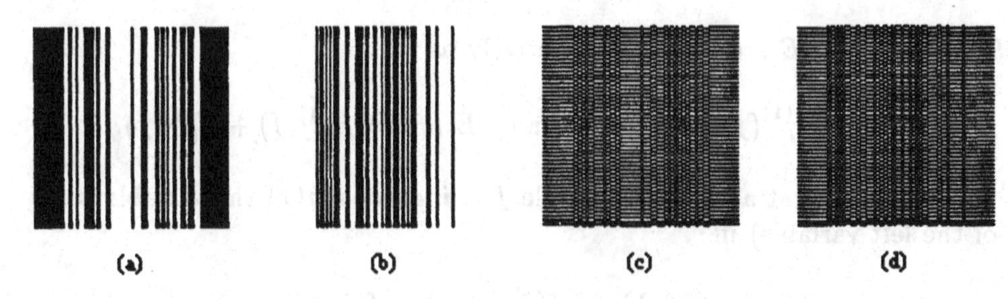

Figure 5. Central permutive ECA rules: (a) (b) rule 204, (c) (d) rule 51; (a) (c) in background of 0s, (b) (d) background of 1s

3. ECA: An Effective Classification of 0–Quiescent Rules

Recently Braga et al. (Braga *et al.*, 1995) gave a complete and effective classification of 0-quiescent elementary cellular automata (ECA) according to their dynamical global behavior. The aim was: on the basis of the finite number of information from any 0-quiescent ECA [i.e., CA based on a local rule $f : \{0,1\}^3 \mapsto \{0,1\}$, with the constraint $f(0,0,0) = 0$] to deduce, by theorems, information about the qualitative behaviour of the corresponding global dynamic [i.e., about the dynamics generated by iterations of the map $F_f : \{0,1\}^{\mathbb{Z}} \mapsto \{0,1\}^{\mathbb{Z}}$.]

In (Braga *et al.*, 1995) the rule space $\mathcal{R}_0(ECA)$ of all 0-quiescent ECA has been classified according to the mutually disjoint classes described by Table 2.

Let us call $\mathcal{S}_{right}$ the class $\mathcal{S}_r \cup \mathcal{S}_r' \cup \mathcal{S}_r''$ and $\mathcal{S}_{left}$ the class $\mathcal{S}_l \cup \mathcal{S}_l' \cup \mathcal{S}_l''$. These classes are effective, that is, there is a simple algorithm to establish if a given 0-quiescent ECA local rule belongs to one of them: it is enough to look at the (finite) transition table of the local ECA rule (and not to the space-time patterns produced by the global evolution starting from various initial configurations).

From the point of view of global dynamics, we will consider the set $\mathcal{F}_0$, dense in $\{0,1\}^{\mathbb{Z}}$, consisting of all 0-*finite* configurations, i.e., configurations of the form

$$\underline{x} = (\ldots, 0, 0, x(m), \ldots, x(p), 0, 0, \ldots)\,,$$

with m and p the minimun and the maximum site respectively in which $x(m) = x(p) = 1$, plus the 0-*quiescent* configuration $\underline{0} = (\ldots, 0, 0, 0, \ldots)$. To any such 0-finite configuration $\underline{x} \in \mathcal{F}_0$ we assign its *length* $\ell(\underline{x}) := p - m + 1$ [$\ell(\underline{0}) = 0$]; the ***support or active region*** of such a finite configuration $\underline{x}$ is the set of sites $\{m, m+1, \ldots, p-1, p\}$ and the ***finite pattern*** is the block (i.e., finite sequence) of boolean states $\{x(m), x(m+1), \ldots, x(p-1), x(p)\}$.

Let us note that for any 0-quiescent ECA local rule $f \in \mathcal{R}_0(ECA)$, $\mathcal{F}_0$ is positively invariant with respect to the corresponding global next state function (i.e., $F_f(\mathcal{F}_0) \subseteq \mathcal{F}_0$); thus $\mathcal{F}_0$ is a *trapping region* with respect to F_f and the pair $\langle \mathcal{F}_0, F_f \rangle$ defines a dynamical sub-system of the DTDS $\left\langle \{0,1\}^{\mathbf{Z}}, F_f \right\rangle$, dense in this latter.

From the point of view of the global dynamics on the phase space $\mathcal{F}_0$, the following classification has been introduced in (Braga *et al.*, 1995):

Definition 3.1

A quiescent ECA local rule $f \in \mathcal{R}_0(ECA)$, with induced global next state function F_f on the phase space $\mathcal{F}_0$ of all 0-finite configurations, is in class:

$$\begin{aligned} \mathcal{C}_1 &\Leftrightarrow \forall \underline{x} \in \mathcal{F}_0 : \lim_{n \to \infty} \ell(F_f^n(\underline{x})) = 0, \\ \mathcal{C}_2 &\Leftrightarrow \forall \underline{x} \in \mathcal{F}_0 : sup_{n \in \mathbf{N}} \ell(F_f^n(\underline{x})) < \infty, \\ \mathcal{C}_3 &\Leftrightarrow \exists \underline{x} \in \mathcal{F}_0 : sup_{n \in \mathbf{N}} \ell(F_f^n(\underline{x})) = \infty . sup_{n \in \mathbf{N}} \end{aligned}$$

It follows immediately from the above definitions that the following relations between the classes hold:

$$\mathcal{C}_1 \subsetneq \mathcal{C}_2 = \mathcal{R}_0(ECA) \setminus \mathcal{C}_3 .$$

For a rule of class $\mathcal{C}_2$ the length of each configuration stays limited during the dynamical evolution, but the support region can move under iterations of the global function.

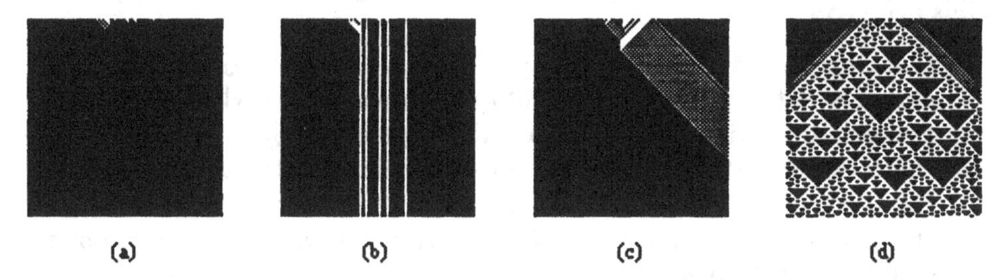

Figure 6. 0-quiescent ECA rules in background of 0s: (a) rule 224 $\in \mathcal{C}_1$, (b) rule 228 $\in \mathcal{C}_2$, (c) rule 184 $\in \mathcal{C}_2$, (d) rule 122 $\in \mathcal{C}_3$

The link between the finite informations on ECA local rules condensed in Table 2 and the global dynamical qualitative behavior expressed in Definition 3.1 are given by the following proposition whose proof can be found in (Braga *et al.*, 1995):

Proposition 3.1

The following equalities hold:

$$\begin{aligned} \mathcal{C}_1 &= \mathcal{L} \cap (\mathcal{S}_l \cup \mathcal{S}_r), \\ \mathcal{C}_2 &= \mathcal{L} \cup \mathcal{S}_{right} \cup \mathcal{S}_{left}, \\ \mathcal{C}_3 &= \mathcal{R}_0(ECA) \setminus (\mathcal{L} \cup \mathcal{S}_{right} \cup \mathcal{S}_{left}). \end{aligned}$$

The qualitative dynamical behavior generated by global next state functions induced by various local rules of $\mathcal{C}_1$ and $\mathcal{C}_2$ classes can be theoretically described on the basis of the properties of the local rules introduced in Table 2. In particular, it has been shown in (Braga *et al.*, 1995) that the following kinds of dynamics exhaust the above class $\mathcal{C}_2$:

Class $\mathcal{C}_1$

1. the null $\underline{0}$ configuration is the unique equilibrium point, and it is a finite time global attractor $[W_1]$.

Class $\mathcal{L} \setminus \mathcal{C}_1$

2. the set of all equilibrium points is a finite time global attractor containing the null configuration $\underline{0}$ as a reachable state (i.e., there exists at least an orbit which is attracted by it) $[W_2]$;
3. the set of all equilibrium points is a finite time global attractor containing the null configuration $\underline{0}$ as an unreachable state (i.e., there is no orbit which is attracted by it) $[W_2]$;
4. the ECA 108, which shows both equilibrium and period two cyclic points whose collection is a finite time global attractor with the null configuration as unreachable state $[W_2]$.

Class $\mathcal{S}_1 \cup \mathcal{S}_2 \setminus \mathcal{L}$

5. there exists a sub-shift finite time global attractor Σ_0 (i.e., Σ_0 is a closed, strictly invariant subset of the phase space on which the global transition function behaves as a shift mapping) $[W_2]$;
6. there exists a (2,2) sub-shift finite time global attractor $\Sigma_{(2,2)}$ (i.e., $\Sigma_{(2,2)}$ is a closed, strictly invariant subset of the phase space on which the square of the global transition function behaves as the square of a shift mapping) $[W_2]$;
7. The ECA local rule 180 (and its topological complex conjugate local rule 166) which shows a "pathological" shift-like behaviour, called *generalized shift dynamics* completely studied in (Cattaneo *et al.*, 1997) $[W_3]$.

Rules of class $\mathcal{C}_1$ are all classified as $[W_1]$ by Wolfram; rules of class $\mathcal{L} \setminus \mathcal{C}_1$ are $[W_2]$ and all rules of class $\mathcal{S}_1 \cup \mathcal{S}_2 \setminus \mathcal{L}$ are of class $[W_2]$ except rule 180 (and its complex conjugate 166) which are $[W_3]$ (also in Wolfram classification the "strangeness" of rules 180 and 166 is recognized). In Table 3.1 we summarize the results about classes $\mathcal{C}_1$ and $\mathcal{C}_2$, and compare them with the original Wolfram's classification.

The following dynamics exhaust all the 0-quiescent ECA of class C_3.

Class C_3

8. The set of all cyclic points is an asymptotical global attractor. We can collect these ECA rules in classes consisting of period 1 till period 4 attractors. Moreover, this gathering can be further on specified in ECA which are in Wolfram $[W_1]$ and ECA which are in $[W_2]$ classes.
   We have to stress that in the case of period 3 and 4 the results are only empirical.
9. The ECA rule 250 in which the global asymptotic attractor consists of all period 1 *and* period 2 cyclic points (see (c1,2) of Figure 7) $[W_1]$.

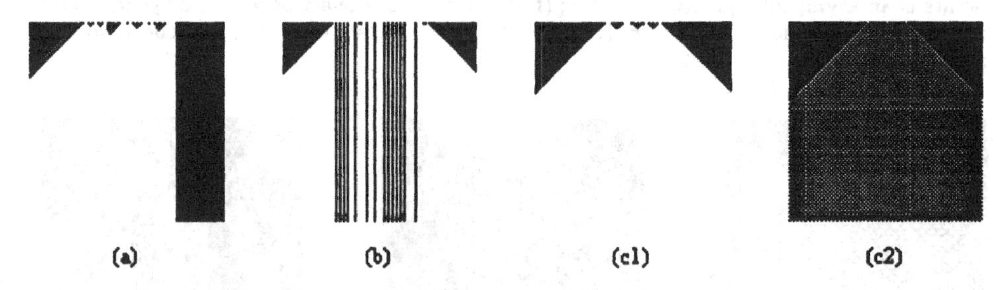

Figure 7. 0-quiescent ECA rules in class C_3: (a) ECA 234 and (b) ECA 222, the set of all fixed points is an asymptotic global attractor; (c1,2) ECA rule 250 in which the asymptotic global attractor consists of period 1 and 2 cyclic points.

10. ECA rules which behave exactly as the leftmost *and* rightmost (expansively chaotic) rule 90 on a closed strictly invariant subset of the phase space (i.e, sub 90 ECA rules) $[W_2$ and $W_3]$.
11. ECA rules which are leftmost/rightmost permutive (D-chaotic); two of them (210 and 154) are also sub 90 rules $[W_3]$.
12. ECA rules which are leftmost *and* rightmost permutive.
13. ECA rules 54, 110, 124 which are outside our theoretical classification, but sharing "complex" space-time patterns $[W_3]$.

In Table 4 we list all the 0-quiescent ECA of class C_3 classified according to their dynamical behavior and compared with Wolfram classification.

Let us stress the interesting case of ECA rules 54, 111, and 124 which empirically share "complex" space-time patterns (see Figure 9) but of which we are not able to give any theoretical result.

3.1. ECA CLASSIFICATION OF 1-QUIESCENT RULES

The corresponding classification of 1-quiescent ECA rules can be similarly obtained in the case of the phase space $\mathcal{F}_1$ of all finite configurations in a background of 1s. In particular, some results can be trivially proved from

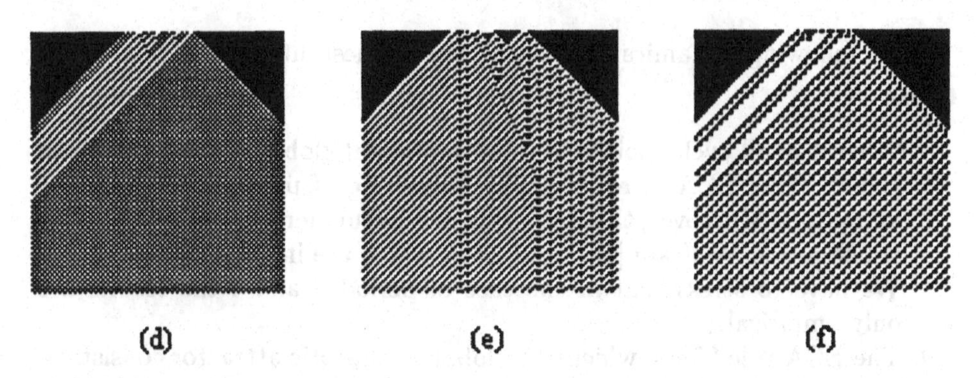

Figure 8. 0-quiescent ECA rules in class $\mathcal{C}_3$: (e) ECA rule 58, the set of period 2 cyclic points is an asymptotic global attractor; (f) ECA rule 62, the set of period 3 cyclic points is an asymptotic global attractor; (e) ECA rule 158, the set of period 4 cyclic points is an asymptotic global attractor;.

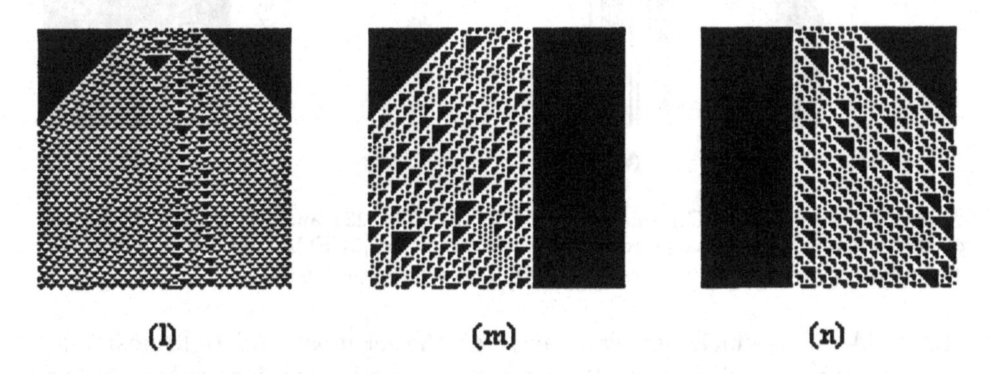

Figure 9. 0-quiescent ECA rules in class $\mathcal{C}_3$ which show a "complex" space-time pattern, but without any theoretical results: (a) ECA rule 54; (b) ECA rule 110: (c) ECA rule 124(= 110°).

the previous results about 0-quiescent ECA rules, once noticed that f is a 1-quiescent ECA rule iff its conjugate ECA rule f^* is a 0-quiescent ECA rule.

In Tables 5 and 6, we list all the odd 1-quiescent ECA rules gathered according to the same qualitative dynamical behavior. Table collects all the rules which are both 0-quiescent and 1-quiescent; this case is very interesting since it shows some "mixed" situations depending on the kind of background one adopts. Indeed, we can gather 0 and 1-quiescent ECA rules according to the global dynamics with respect to the pair of phase spaces $\mathcal{F}_0 - \mathcal{F}_1$ respectively; in particular we stress as particularly interesting the $\mathcal{C}_1 - \mathcal{C}_3$ or $\mathcal{C}_2 - \mathcal{C}_3$ cases.

In particular, in Figure 10 it is shown the space-time patterns of two ECA rules, 160 and 162, which are in classes $\mathcal{C}_1$-$\mathcal{C}_3$ and $\mathcal{C}_2$-$\mathcal{C}_3$ respectively.

In both cases the simulations correspond to an initial state which is in a background of 0s [(a) and (c)] and a background of 1s [(c) and (d)].

The case of ECA rules which show a class $\mathcal{C}_2$ behavior in background of 0s [patterns (e) and (f) in Figure 11] and a class $\mathcal{C}_3$ behavior in background of 1s [patterns (g) and (h) in Figure 11], is well represented by rule 180. This rule has been classified by Wolfram has a W_2 rule: "the evolution leads to a set of simple stable or periodic structure"; but this is true in the case of initial configurations in background of 0s, whereas for the other case of configurations in a background of 1s the behavior of this rule is locally the same as the expansively chaotic ECA rule 165, which is both leftmost and rightmost permutive. This confirm us on the conjecture that Wolfram classification is essentially based on configurations in background of 0s (or finite configuration with 0 as fixed boundary conditions).

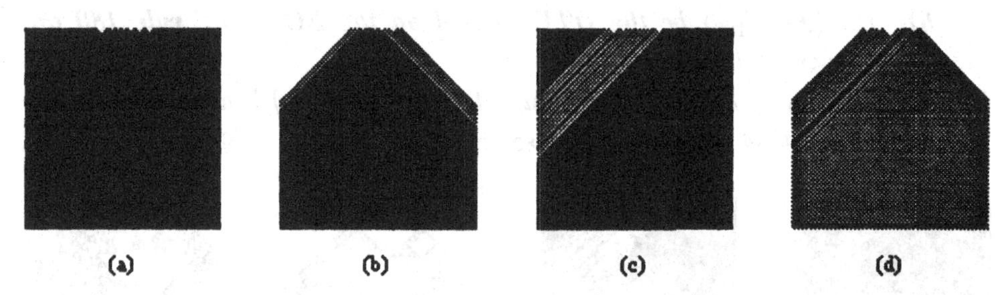

Figure 10. 0 and 1-quiescent ECA rules: (a) and (b) ECA rule 160 $\in \mathcal{C}_1$-$\mathcal{C}_3$; (c) and (d) ECA rule 162 $\in \mathcal{C}_2$-$\mathcal{C}_3$.

The following results proved in (Cattaneo and Margara, 1998) summarize the main properties of ECA rule 180:

Theorem 3.1

The global dynamics induced on the phase space $\{0,1\}^{\mathbb{Z}}$ from ECA local rule 180 is Devaney chaotic.

(a) Let $\langle \mathcal{F}_0, F_{180} \rangle$ be the DTDS based on the ECA local rule 180 on the phase space of 0-finite configurations (finite configurations in a background of 0s).

(i) the null configuration $\underline{0}$ is an asymptotic global attractor, i.e., for any configuration $\underline{x} \in \mathcal{F}_0$,

$$\lim_{n\to\infty} d(F_{180}^n(\underline{x}), \underline{0}) = 0,$$

(ii) the only periodic orbit for F_{180} is the null configuration,

(iii) F_{180} does not have any dense orbit.

(iv) Notwithstanding the very "regular" qualitative dynamics of above points (i)–(iii), the global dynamics of rule 180 on the phase space $\mathcal{F}_0$ preserve some component of chaoticity; to be precise, it is transitive and

sensitive to initial conditions [note that, since $\mathcal{F}_0$ is not compact, transitivity is not equivalent to the existence of a dense orbit].

(*v*) F_{180} *is a generalized subshift, i.e., there exists a map* $M : \{0,1\}^* \mapsto \mathbb{N}$, *which satisfies the following property.*

$$\forall x \in \{0,1\}^*,\ F_{180}^{M(x)}(\underline{x}) = \sigma^{M(x)}(\underline{x}),$$

where $\underline{x} \in \mathcal{F}_0$ *denotes any bi-infinite extension of block* x *in a background of 0s.*

In particular, this map is such that for every $n \in \mathbb{N}$

$$M(1^n) = 2^{\lfloor \log(n+1) \rfloor},$$

where 1^n *is the 1-constant block of length n.*

(b) Let $\langle \mathcal{F}_1, F_{180} \rangle$ *be the DTDS based on the ECA local rule 180 on the phase space of 1-finite configurations.*

(*vi*) F_{180} *locally (on an invariant subset* $D_{180,165}$ *of* $\mathcal{F}_1$*) behaves as (in the sense that it is identical on* $D_{180,165}$ *to) ECA local rule 165.*

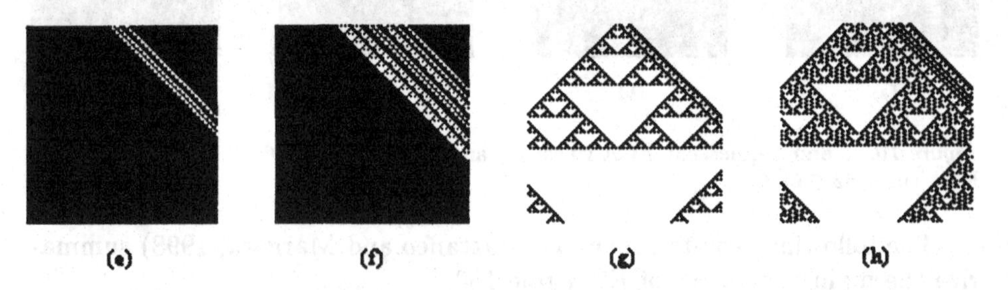

Figure 11. 0 and 1-quiescent ECA rule 180 in class $\mathcal{C}_2 - \mathcal{C}_3$; (e) and (f) in background of 0s, (c) and (d) in background of 1s.

References

Banks J., Brooks J., Cairns G., Davis G. and Stacey P. On Devaney's definition of chaos. *American Mathematical Monthly* **Vol. no. 99**: 332–334, 1992.

Blanchard F., Kůrka P. and Maas A. Topological and measure-theoretic properties of one-dimensional cellular automata. *Physica D* **Vol. no. 103**: 86–99, 1997.

Braga G., Cattaneo G., Flocchini P. and Vogliotti C. Pattern growth in elementary cellular automata. *Theoretical Computer Science* **Vol. no. 145**: 1–26, 1995.

Cattaneo G., Finelli M. and Margara L. Investigating topological chaos by elementary cellular automata dynamics. Submitted to *Theoretical Computer Science*, 1997.

Cattaneo G. and Margara L. Generalized sub-shifts in elementary cellular automata: the strange case of chaotic rule 180. Accepted in *Theoretical Computer Science*, 1998.

Codenotti B. and Margara L. Transitive cellular automata are sensitive. *American Mathematical Monthly* **Vol. no. 103**: 58–62, 1996.

Conley C. Isolated invariant sets and the Morse index. *American Mathematical Society, CBMS Regional Conference*, **Ser. 38**, 1978.

Crannell A. The role of transitivity in Devaney's definition of chaos. *American Mathematical Monthly* **Vol. no. 102**: 768–793, 1995.

Čulik K., Hurd L. and Yu S. Computation theoretic aspects of CA. Physica, **Vol. no. 45**: 357–378, 1990.

Čulik K. and Yu S. Undecidability of CA classification schemes. *Complex Systems* **Vol. no. 2**: 177–190, 1988.

Devaney R. *Introduction to chaotic dynamical systems.* Addison-Wesley, second ed., 1989.

Eckmann J-P. and Ruelle D. Ergodic theory of chaos and strange attractors. *Rev. Modern Physics* **Vol. no. 57**: 617–656, 1985.

Guckenheimer J. and Holmes P. *Nonlinear oscillations, dynamical systems, and biforcations of vector fields.* Springer Verlag, Berlin, 1983.

Gutowitz H. A hierarchical classification of cellular automata. *Physica D* Special issue, **Vol. no. 45**: 130–156, 1990.

Hedlund G. Endomorphism and automorphism of the shift dynamical system. *Mathematical System Theory* **Vol. no. 3**: 320–375, 1969.

Holmes P. Poincaré, celestial mechanics, dynamical system theory and chaos. *Physics Report (Review Section of Physics Letters)* **Vol. no. 193**: 137–163, 1990.

Hurley M. Attractors, persistence and density of their basin. *Trans. Amer. Math. Soc.* **Vol. no. 269**: 247–27, 1982.

Knudsen C. Chaos without nonperiodicity. *American Mathematical Monthly* **Vol. no. 101**: 563–565, 1994.

Kůrka P. Languages, equicontinuity and attractors in cellular automata. *Ergod. Th. & Dynam. Sys* **Vol. no. 17**: 417–433, 1997.

LaSalle J-P. Stability Theory for Difference Equations. MAA Studies in Math., *American Mathematical Society*, 1976.

MacEachern S. and Berliner L. Aperiodic chaotic orbits. *American Mathematical Monthly* **Vol. no. 100**: 237–241, 1993.

Margara L. Cellular Automata and Chaos. PhD dissertation, Università di Pisa, Dipartimento di Informatica, 1995.

Nasu M. Textile Systems for Endomorphisms and automorphisms of the shift. Memoires, *American Mathematical Society*, 1995.

Packard N. Complexity of growing patterns in cellular automata. in *Dynamical Systems and Cellular Automata*, Demongeot M. T. J., Goles E., ed., Academic Press, 1985.

Ruelle D. Strange attractors. *Math. Intelligencer*, **Vol. no. 2**: 126–137, 1980.

Ruelle D. Small random perturbations of dynamical systems and the definition of attractors. *Commun. Math. Phys.* **Vol. no. 82**: 137–15, 1981.

Sutner K. Classifying circular cellular automata. *Physica D* **Vol. no. 45**: 386–395, 1990.

Wiggins S. *Global Bifurcations and Chaos.* Springer, Berlin, 1988.

Wolfram S. Universality and complexity in cellular automata. *Physica D* **Vol. no. 10**: 91–125, 1984, also in (Wolfram, 1986).

Wolfram S. *Theory and Applications of Cellular Automata.* World Scientific, Singapore, 1986.

Permutivity	Rule	E_l	E_r	E_c
C	204, 51	2	2	4
L	240, 15	6	0	2
R	170, 85	0	6	2
LC	60, 195	6	2	6
CR	102,153	2	6	6
L	30, 45, 75, 120, 135, 180, 210, 225	6	3.6	4
R	86, 89, 101, 106, 149, 154, 166, 169	3.6	6	4
LR	90, 165	6	6	2
LCR	150, 105	6	6	4

TABLE 1. Left entropy (E_l), right entropy (E_r), and central entropy (E_c) for leftmost (L), rightmost (R), and central (C) permutive ECA.

$\mathcal{L}$	$\Leftrightarrow$	$f(0,0,1) = f(1,0,0) = 0$
$\mathcal{S}_r$	$\Leftrightarrow$	$f(0,0,1) = f(0,1,0) = f(0,1,1) = 0$
$\mathcal{S}'_r$	$\Leftrightarrow$	$f(0,0,1) = 0,\ f(0,1,0) = 1,\ f(0,1,1) = 0$
$\mathcal{S}''_r$	$\Leftrightarrow$	$f(0,0,1) = 0,\ f(0,1,1) = 1,\ f(0,1,0) = 0,$ $f(1,1,0) \cdot (f(1,1,1) \vee f(1,0,1)) = 0$
$\mathcal{S}_l$	$\Leftrightarrow$	$f(1,0,0) = f(0,1,0) = f(1,1,0) = 0$
$\mathcal{S}'_l$	$\Leftrightarrow$	$f(1,0,0) = 0,\ f(0,1,0) = 1,\ f(1,1,0) = 0$
$\mathcal{S}''_l$	$\Leftrightarrow$	$f(1,0,0) = 0,\ f(1,1,0) = 1,\ f(0,1,0) = 0,$ $f(0,1,1) \cdot (f(1,1,1) \vee f(1,0,1)) = 0$

TABLE 2. Classification of all 0-quiescent ECA rules in mutually disjoint classes.

Our Class	Basic Class	Dynamics	Rule number	Wolfram's Class
$\mathcal{C}_1$	$\mathcal{L} \cap (\mathcal{S}_r \cup \mathcal{S}_l)$	$\underline{0}$ unique global attracting equilibrium state	0,8,32,40,64,96,128 136,160,168,192,224	W_1
$\mathcal{C}_2 \setminus \mathcal{C}_1$	$\mathcal{L} \setminus \mathcal{C}_1$	Per_1 is a global attractor containing reachable $\underline{0}$	72,104,200,232 4,36,132,164	W_2
		Per_1 is a global attractor containing unreachable $\underline{0}$	12,44,68,100,140,172 196,228, 76,204,236	
		global attractor consisting of fixed and period two cyclic points	108	
	$\mathcal{S}_r \setminus \mathcal{L}$	right subshift	16,48,80,112,144	W_2
		Right full shift and L-permutive	176,208, 240	W_2
	$\mathcal{S}'_r \setminus \mathcal{L}$	right subshift	84,116,212,244	W_2
		right (2,2)-subshift	20,52,148	W_2
		right shift-like dynamics	180	W_3
	$\mathcal{S}''_r \setminus \mathcal{L}$	right subshift	24,56,152,184	W_2
		right (2,2)-subshift	88	W_2
	$\mathcal{S}_l \setminus \mathcal{L}$	left subshift	2,10,34,42,130	W_2
		Left full shift and R permutive	138,162, 170	W_2
	$\mathcal{S}'_l \setminus \mathcal{L}$	left subshift	14,46,142,174	W_2
		left (2,2)-subshift	6,38,134	W_2
		left shift-like dynamics	166	W_3
	$\mathcal{S}''_l \setminus \mathcal{L}$	left subshift	66,98,194,226	W_2
		left (2,2)-subshift	74	W_2

TABLE 3. All the 0-quiescent ECA rules in classes $\mathcal{C}_1$ and $\mathcal{C}_2$.

Class	Dynamic	Rule Number	Wolfram's class
C_3	Per_1 asymptotic global attractor	234,238,248,252,254	W_1
	$Per_{1,2}$ asymptotic global attractor	250	W_1
	Per_1 asymptotic global attractor	78,92,202,206,216,220,222	W_2
	Per_2 asymptotic global attractor	28,50,58,70,114,156 178,186,198,242	W_2
	$(Per_3)^\dagger$ asymptotic global attractor	62,118	W_2
	$(Per_4)^\dagger$ asymptotic global attractor	158,188,190,214,230,246	W_2
	sub 90	94,218	W_2
		18,22,26,82	W_3
		122,126,146,182	W_3
	left permutive	30,60,120	W_3
	left perm. and sub 90	210 (= 180*)	W_3
	right permutive	86,102,106	W_3
	right perm. and sub 90	154 (= 166*)	W_3
	left and right permutive	90,150	W_3
	???	54, 110, 124	W_3

TABLE 4. All the 0-quiescent ECA rules in class C_3. Classes marked by † correspond to empirical classifications without theoretical results.

Our Class	Dynamic	Rule Number	Wolfram's class
$\mathcal{C}_1$	1 unique global attratting equilibrium state	235,239,249,251,253,255	W_1
$\mathcal{C}_2 \setminus \mathcal{C}_1$	globally attratting equilibrium configurations	203,205,207,217,219 221,223,233,237	W_2
	globally attratting periodic configurations of period $k \leq 2$	201	W_2
	right subshift	241,243,245,247	W_2
	right subshift	209,213	W_2
	$(2,2)$-right subshift	211,215	W_2
	right subshift	227,231	W_2
	(2-2)-right subshift	229	W_2

TABLE 5. All the odd 1-quiescent elementary rules (Rule Number = odd *and* > 128), part a).

Our Class	Dynamic	Rule Number	Wolfram's class
$C_2 \setminus C_1$	left subshift	171,175,187,191	W_2
	left subshift	139,143	W_2
	(2, 2)-left subshift	155,159	W_2
	left subshift	185,189	W_2
	(2, 2)-left subshift	173	W_2
C_3	$Per_k(F_f) : k \leq 4$ is a global asymptotical attractor	131,141,145,157 163,177,179,197,199	W_2
	sub 165	133	W_2
		129,151,161	W_3
		167,181,183	W_3
	left permutive	135,195,225	W_3
	right permutive	149,153,169	W_3
	left and right permutive	165	W_3
	???	137,147,193	W_3

TABLE 6. All the odd 1-quiescent elementary rules (Rule Number = odd *and* > 128), part b).

$\mathcal{F}_0$	$\mathcal{F}_1$	Rule number	Wolfram
$\mathcal{C}_1$	$\mathcal{C}_3$	128 (*= 254), 136 (*= 238), 160 (*= 250), 168 (*= 234) 192 (*= 252), 244(*= 248)	W_1
$\mathcal{C}_2$	$\mathcal{C}_2$	138 (*= 174), 142 (*= 142), 170 (*= 170), 184 (*= 226), 200 (*= 236), 204 (*= 204), 208 (*= 244), W_2 212 (*= 212), 232 (*= 232), 240 (*= 240)	
$\mathcal{C}_2$	$\mathcal{C}_3$	130 (*= 190), 132 (*= 222), 134 (*= 158), 140 (*= 206) 144 (*= 246),148 (*= 214), 152 (*= 230), 162 (*= 186), 164 (*= 218) 172 (*= 202),176 (*= 242) 194 (*= 188), 196 (*= 220), 228 (*= 216)	W_2
$\mathcal{C}_2$	$\mathcal{C}_3$	166 (*= 154), 180 (*= 210)	W_2
$\mathcal{C}_3$	$\mathcal{C}_3$	178 (*= 178), 198 (*= 156)	W_2
$\mathcal{C}_3$	$\mathcal{C}_3$	146 (*= 182), 150 (*= 150)	W_3

TABLE 7. Dynamical classification of all ECA rules which are both 0 *and* 1 quiescent.

Part 5
Modeling

Cellular automata are used in many fields in order to model complex phenomena. Actually, their local and discrete features allow them to formalize phenomena which lead to nonlinear differential equations, and which find applications in physics, chemistry, biology. Examples can be found in journals like *Complex Systems*, *Physica D*, the *Journal of Theoretical Biology*. The following contributions belong to less standard domains. The first one is about innovation diffusion resting on probabilistic cellular automata. The second one is an example of a breakthrough in theoretical cardiology.

MODELING DIFFUSION OF INNOVATIONS WITH PROBABILISTIC CELLULAR AUTOMATA

N. BOCCARA AND H. FUKŚ
Department of Physics, UIC,
845 W. Taylor, Chicago, IL 60607 – 7059, *USA*

Abstract. We present a family of one-dimensional cellular automata modeling the diffusion of an innovation in a population. Starting from simple deterministic rules, we construct models parameterized by the interaction range and exhibiting a second-order phase transition. We show that the number of individuals who eventually keep adopting the innovation strongly depends on connectivity between individuals.

1. Introduction

Diffusion phenomena in social systems such as spread of news, rumors or innovations have been extensively studied for the past three decades by social scientists, geographers, economists, as well as management and marketing scholars. Traditionally, ordinary differential equations have been used to model these phenomena, beginning with the Bass model (Bass, 1969) and ending with sophisticated models that take into account learning, risk aversion, nature of innovation, etc. (Mahajan *et al.*, 1990; Mahajan and Peterson, 1985; Rogers, 1995). Models incorporating space and spatial distribution of individuals have been also proposed, although most research in this field has been directed to refining of the discrete Hägerstrand models (Hägerstrand, 1952; Hägerstrand, 1965) and to constructing partial differential equations similar to diffusion equations known to physicists (Haynes *et al.*, 1977).

Diffusion of innovations (we will use this term in a general sense, meaning also news, rumors, new products, etc.) is usually defined as a process by which the innovation "is communicated through certain channels over time among the members of a social system" (Rogers, 1995). In most cases, these communication channels have a rather short range, i.e. in our decisions we are heavily influenced by our friends, family, coworkers, but not that much by unknown people in distant cities. This local nature of social interactions makes cellular automata (CA) a well-adapted tool in modeling diffusion phenomena. In fact, epidemic models formulated in terms of automata networks have been successfully constructed in recent years (Boccara and Cheong, 1992; Boccara and Cheong, 1993; Boccara *et al.*, 1994).

2. Simple deterministic models

We will construct models of diffusion of innovations based on elementary (radius-1) cellular automata (Wolfram, 1986). If $s(i,t)$ denotes the state of lattice site i at time t, the function $(i,t) \mapsto s(i,t)$ is a mapping from $\mathbb{Z} \times \mathbb{N}$ to $\mathbb{Z}_2 = \{0,1\}$. Given a function $f : \{0,1\}^{2r+1} \mapsto \{0,1\}$, the discrete dynamical system

$$s(i,t+1) = f(s(i-r,t), s(i-r+1,t), \dots, s(i+r,t)) \tag{1}$$

is called a cellular automaton (CA) of radius r, and f is called its local (function) rule.

In the simplest version of our model, the sites of an infinite lattice are all occupied by individuals. The individuals are of two different types: ***adopters*** $\big(s(i,t) = 1\big)$ and *neutrals* $\big(s(i,t) = 0\big)$. Moreover, we will assume that an individual can get information only from its two nearest neighbors. To simplify our model, we will also assume that once an individual becomes an adopter, he remains an adopter, i.e. his state cannot change. This condition fixes four entries in the rule table: $\{0,1,0\} \to 1$, $\{0,1,1\} \to 1$, $\{1,1,0\} \to 1$ and $\{1,1,1\} \to 1$, and since the information comes from nearest neighbors, if both are neutral, then the individual will stay neutral, i.e., $\{0,0,0\} \to 0$. This leaves three entries in the rule table to be determined, and this can be done in $2^3 = 8$ ways, as shown in Table 1. The first Rule 204 (for rule codes cf.(Wolfram, 1986)) listed in this table is just the identity, and it will be excluded from further considerations. Rules 220 and 252 can be obtained respectively from 206 and 238 by spatial reflection, therefore they will be excluded too. This leaves us with five distinct rules.

Rule 254: An individual adopts if, at least, one of his neighbors is an adopter.

Rule 238: An individual adopts only if his right neighbor is an adopter.

Rule 222: An individual adopts only if exactly one of his neighbors is an adopter.

Rule 206: An individual adopts only if his right neighbor is an adopter and its left neighbor is neutral.

Rule 236: An individual adopts only if both his neighbors are adopters.

code	1,1,1	1,1,0	1,0,1	1,0,0	0,1,1	0,1,0	0,0,1	0,0,0
204	1	1	0	0	1	1	0	0
206	1	1	0	0	1	1	1	0
220	1	1	0	1	1	1	0	0
222	1	1	0	1	1	1	1	0
236	1	1	1	0	1	1	0	0
238	1	1	1	0	1	1	1	0
252	1	1	1	1	1	1	0	0
254	1	1	1	1	1	1	1	0

TABLE 1. Eight possible elementary rules for diffusion of an innovation.

We will now demonstrate that in all five cases, the density of adopters $\rho(t)$ at time t can be exactly computed, assuming that we start from a disordered initial configuration with $\rho(0) = \rho_0$.

(i) In order to understand the dynamics of Rule 254, we can view the initial configuration as clusters of ones separated by clusters of zeros. Only neutral sites adjacent to a cluster of ones change their state, while all other neutral sites remain neutral, as shown in the example below (sites that will change their state to 1 in the next time step are underlined):

$$\cdots \; | \; 1 \; | \; \underline{0} \; | \; 0 \; | \; \underline{0} \; | \; 1 \; | \; 1 \; | \; \underline{0} \; | \; 0 \; | \; 0 \; | \; \underline{0} \; | \; 1 \; | \; 1 \; | \; \cdots$$

This implies that the length l of every cluster of zeros decreases by two every time step, i.e.

$$M(l, t+1) = M(l+2, t), \tag{2}$$

where $M(l, t)$ denotes a number of clusters of zeros of size s per site at time t, and therefore

$$M(l, t) = M(l+2t, 0). \tag{3}$$

For a random initial configuration with initial density ρ_0, the cluster density is given by $M(l, 0) = (1-\rho_0)^l \rho_0^2$, hence

$$M(l, t) = (1-\rho_0)^{l+2t} \rho_0^2. \tag{4}$$

The density of zeros at time t, denoted by $\eta(t)$, is given by

$$\eta(t) = \sum_{l=1}^{\infty} l M(l, t) = (1-\rho_0)^{2t+1}, \tag{5}$$

and finally the density of ones $\rho(t)$ is equal to

$$\rho(t) = 1 - \eta(t) = 1 - (1 - \rho_0)^{2t+1}. \tag{6}$$

(ii) For Rule 238, the derivation is similar, except that now every cluster of zeros decreases by one every time step, which means that we have to replace $2t$ by t in the previous result, i.e.

$$\rho(t) = 1 - (1 - \rho_0)^{t+1}. \tag{7}$$

(iii) In Rule 222, individuals "dislike overcrowding", and they do not become adopters if their two neighbors are already adopters, adopt if both neighbors adopted, i.e. $\{1,0,1\} \to 0$. As a consequence, clusters of size $l > 1$ decrease their size by two units every time step, while clusters consisting of a single isolated zero ($l = 1$) do not change their size. Clusters of size 1 are thus created from clusters of size 3, as well as those of size 1. For the density of ones, this yields (using a similar reasoning as before)

$$\rho(t) = 1 - \rho_0^2(1 - \rho_0) - \rho_0 \frac{(1 - \rho_0)^3}{2 - \rho_0} - \left[1 - \rho_0^2 + \frac{\rho_0(1 - \rho_0)^2}{\rho_0 - 2}\right](1 - \rho_0)^{2t+1}, \tag{8}$$

and

$$\lim_{t \to \infty} \rho(t) = \rho(\infty) = 1 - \rho_0^2(1 - \rho_0) - \rho_0 \frac{(1 - \rho_0)^3}{2 - \rho_0}. \tag{9}$$

(iv) Rule 206 is just like Rule 222, except that clusters of zeros decrease their length by one. Therefore, we obtain

$$\rho(t) = 1 - \rho_0(1 - \rho_0) - (1 - \rho_0)^{t+2}, \tag{10}$$

and $\rho(\infty) = 1 - \rho_0(1 - \rho_0)$.

(v) Rule 236 is the simplest because, after the first iteration, all clusters of zeros of size 1 disappear while all other clusters remain unchanged, therefore

$$\rho(t) = 1 - (1 - \rho_0)^2(1 + \rho_0). \tag{11}$$

In summary, in all cases (except for Rule 236), the density of ones approaches the fixed point $\rho(\infty)$ exponentially. In a real social system, however, this is not the case. The density of adopters usually follows an S-shaped or logistic curve. The model discussed in the next section eliminates this shortcoming.

3. Probabilistic model

In order to generalize the simple model discussed previously, consider a 2-state probabilistic cellular automaton, with a dynamics such that $s(i, t+1)$ depends on $s(i,t)$ and $\sigma(i,t)$, where

$$\sigma(i,t) = \sum_{n=-\infty}^{\infty} s(i+n,t)p(n), \tag{12}$$

and p is a nonnegative function satisfying

$$\sum_{n=-\infty}^{\infty} p(n) = 1. \tag{13}$$

(deterministic automata networks of this type have been studied in details by (Boccara *et al.*, 1997)).

Our generalized model is defined as follows. At every time step, a neutral individual located at site i at time t can become an adopter at time $t+1$ with a probability depending on $\sigma(i,t)$ (in this section we will simply assume that this probability is equal to $\sigma(i,t)$). As before, once an individual becomes an adopter, he remains an adopter, i.e. his state cannot change.

The model can be viewed as a probabilistic cellular automaton with the probability distribution

$$P\Big(s(i,t+1)=0\Big) = \Big(1-s(i,t)\Big)\Big(1-\sigma(i,t)\Big) \tag{14}$$

$$P\Big(s(i,t+1)=1\Big) = 1-\Big(1-s(i,t)\Big)\Big(1-\sigma(i,t)\Big) \tag{15}$$

The *transition probability* $P_{b\leftarrow a}$ is defined as

$$P_{b\leftarrow a} = P\Big(s(i,t+1)=b \,|\, s(i,t)=a\Big), \tag{16}$$

and represents the probability for a given site of changing its state from a to b in one time step. In our case, the transition probability matrix has the form

$$\mathbf{P} = \begin{bmatrix} P_{0\leftarrow 0} & P_{0\leftarrow 1} \\ P_{1\leftarrow 0} & P_{1\leftarrow 1} \end{bmatrix} = \begin{bmatrix} 1-\sigma(i,t) & 0 \\ \sigma(i,t) & 1 \end{bmatrix}. \tag{17}$$

As a first approximation we we consider $\sigma(i,t)$ defined by

$$\sigma(i,t) = \frac{1}{2R}\left(\sum_{n=-R}^{-1} s(i+n,t) + \sum_{n=1}^{R} s(i+n,t)\right). \tag{18}$$

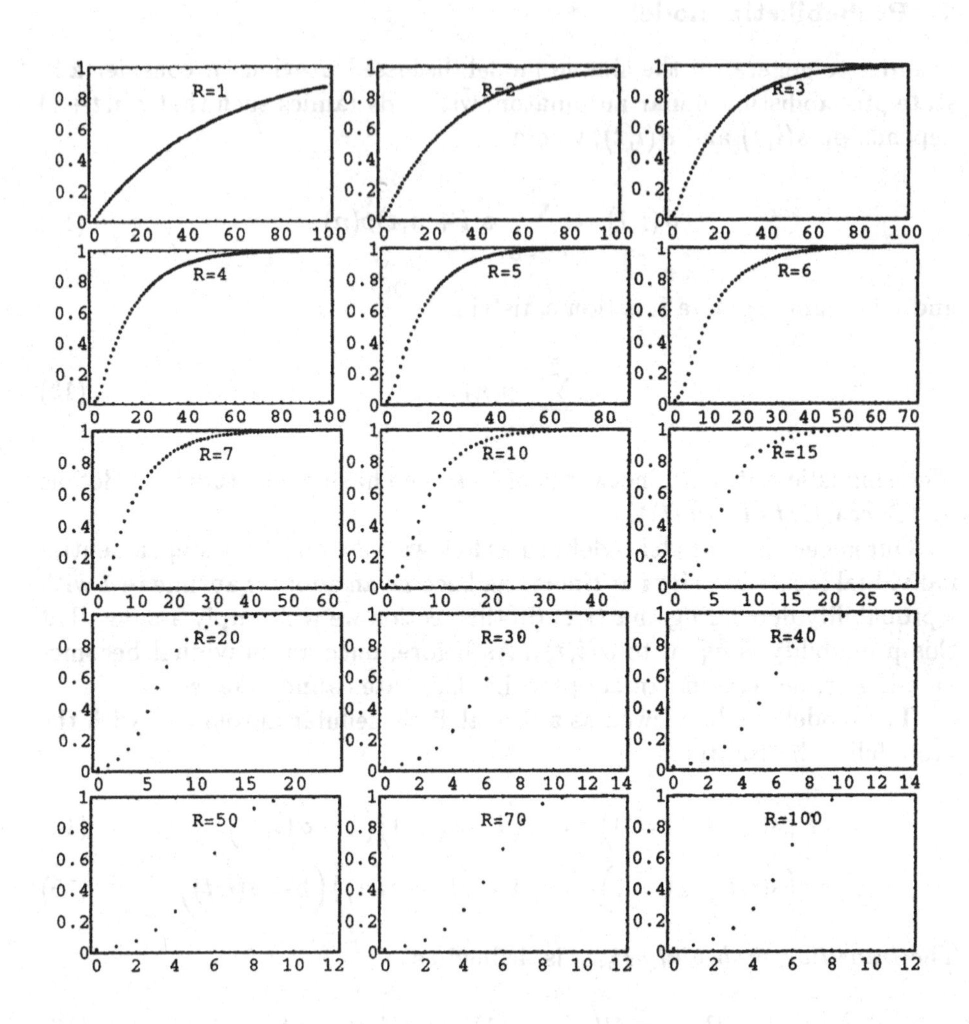

Figure 1. Density of adopters (vertical axis) as a function of time (horizontal axis) for several values of radius R.

$\sigma(i,t)$ is then the local density of adopters at time t over the $2R$ closest neighbors of site i. This choice of σ, although somewhat simplistic, captures some essential features of a real social system: the number of influential neighbors is finite and these neighbors are all located within a certain finite radius R. Opinions of all neighbors have equal weight here, which is maybe not realistic, but good enough as a first approximation. Let $\rho(t)$ be the global density of adopters at time t (i.e. number of adopters per lattice site), and $\eta(t) = 1 - \rho(t)$. Since $P_{1\leftarrow 1} = 1$, the number of adopters increases with time, and $\lim_{t\to\infty} \rho(t) = 1$. If we start with a small initial density of randomly distributed adopters ρ_0, $\rho(t)$ follows a characteristic S-shaped curve, typical of many growth processes. The curve becomes steeper when R increases, and if R is large enough it takes only a few time steps to reach a high density of ones (e.g. 0.99). Figure 1 shows some examples of curves obtained in a computer experiment with a lattice size equal to 10^5 and $\rho_0 = 0.02$.

4. Exact Results

Average densities $\rho(t+1)$ and $\eta(t+1)$ can be obtained from previous densities $\rho(t)$ and $\eta(t)$ using the transition probability matrix ($\langle\rangle$ denotes a spatial average)

$$\begin{bmatrix} \eta(t+1) \\ \rho(t+1) \end{bmatrix} = \begin{bmatrix} \langle P_{0\leftarrow 0}\rangle & \langle P_{0\leftarrow 1}\rangle \\ \langle P_{1\leftarrow 0}\rangle & \langle P_{1\leftarrow 1}\rangle \end{bmatrix} \begin{bmatrix} \eta(t) \\ \rho(t) \end{bmatrix}, \tag{19}$$

hence

$$\rho(t+1) = \rho(t)\langle P_{1\leftarrow 1}\rangle + (1-\rho(t))\langle P_{1\leftarrow 0}\rangle, \tag{20}$$

and using (17) we have

$$\rho(t+1) = \rho(t) + (1-\rho(t))\langle\sigma(i,t)\rangle. \tag{21}$$

This difference equation can be solved in two special cases, $R = 1$ and $R = \infty$.

If $R = 1$, only three possible values of local density σ are allowed: 0, $\frac{1}{2}$ and 1. The initial configuration can be viewed as clusters of ones separated by clusters of zeros. Only neutral sites adjacent to a cluster of ones can change their state, and they will become adopters with probability $\frac{1}{2}$. All other neutral sites have a local density equal to zero, therefore they will remain neutral. This implies that the length of a cluster of zeros will on the average decrease by one every time step, just like in the case of Rule 238. The density of adopters, therefore, is, as for Rule 238, given by

$$\rho(t) = 1 - (1-\rho_0)^{t+1}. \tag{22}$$

Using the above expression, we obtain

$$\rho(t+1) = \rho(t) + (1-\rho(t))\rho_0. \tag{23}$$

Comparing with (21) this yields $\langle\sigma(i,t)\rangle = \rho_0$, i.e. the average probability that a neutral individual adopts the innovation is time-independent and equal to the initial density of adopters, which was *a priori* not obvious.

When $R \to \infty$, the local density of ones becomes equivalent to the global density, thus in (21) we can replace $\langle \sigma(i,t) \rangle$ by $\rho(t)$. Hence

$$\rho(t+1) = \rho(t) + (1 - \rho(t))\rho(t), \tag{24}$$

or

$$1 - \rho(t+1) = (1 - \rho(t))^2, \tag{25}$$

that is

$$\rho(t) = 1 - (1 - \rho_0)^{2^t}. \tag{26}$$

Note that this case corresponds to the mean-field approximation, as we neglect all spatial correlations and replace the local density by the global one.

For $1 < R < \infty$, we may assume that the density of adopters can be written in the form

$$\rho(t) = 1 - (1 - \rho_0)^{f(t,R)}, \tag{27}$$

where f is a certain function satisfying $f(t,1) = t + 1$ and $f(t,\infty) = 2^t$. Computer simulations suggest that for a finite R, $f(t,R)$ becomes asymptotically linear (when $t \to \infty$), and the slope of the asymptote increases with R. Moreover, for large R, f satisfies the following approximate equation

$$f(t+k, 2^k R) = 2^k f(t,R), \tag{28}$$

where k is a positive integer. Detailed discussion of $f(t,R)$ and its properties can be found in (Fukś and Boccara, 1996).

5. Generalization

The model presented in the previous chapter was rather crude. One of its assumptions is that once the individual accepts the innovation, he will never change his mind. In practice, every technology or product has a finite life span. For some products, as TV sets, this time is relatively long, while for other items, like computer software, it is much shorter. One way to incorporate this phenomenon in our model is to assume that at every time step, any adopter can drop the innovation with a given probability p. Therefore, the average time during which an individual is an adopter is $1/p$ (geometric distribution).

To make the model even more realistic, we will also assume that the adoption probability is not equal but proportional to the local density of adopters

$$P_{1\leftarrow 0} = q\sigma(i,t), \tag{29}$$

where $q \in [0,1]$. Hence, the new transition probability matrix is

$$P_{mn} = \begin{bmatrix} P_{0\leftarrow 0} & P_{0\leftarrow 1} \\ P_{1\leftarrow 0} & P_{1\leftarrow 1} \end{bmatrix} = \begin{bmatrix} 1 - q\sigma(i,t) & p \\ q\sigma(i,t) & 1-p \end{bmatrix}. \tag{30}$$

The difference equation for the average density of adopters

$$\rho(t+1) = \rho(t)\langle P_{1\leftarrow 1}\rangle + \Big(1-\rho(t)\Big)\langle P_{1\leftarrow 0}\rangle \tag{31}$$

now becomes

$$\rho(t+1) = (1-p)\rho(t) + q\Big(1-\rho(t)\Big)\langle\sigma(i,t)\rangle. \tag{32}$$

The previous model is recovered for $p = 0$ and $q = 1$. When $q = 0$, the solution is $\rho(t) = \rho_0(1-p)^t$. When $R = 1$, the model can also be considered as a discrete version of the contact process, an irreversible lattice model involving nearest-neighbor interactions, used to study catalytic reactions (Harris, 1974; Liggett, 1985). In the contact process, a particle desorbs spontaneously with rate p and adsorbs at a given unoccupied site at a rate proportional to the number of neighboring occupied sites.

Although it is not possible to solve exactly equation (32), the general nature of the solution can be understood using the mean-field approximation (MFA), in which it is assumed that the average local density $\langle\sigma(i,t)\rangle$ is the same as the global density $\rho(t)$ (this is actually true for $R = \infty$). Our equation becomes

$$\rho(t+1) = (1-p)\rho(t) + q\rho(t)\Big(1-\rho(t)\Big). \tag{33}$$

The substitution

$$\rho(t) = \frac{1-p+q}{q}x(t) \tag{34}$$

yields

$$x_{t+1} = (1-p+q)x(t)\Big(1-x(t)\Big). \tag{35}$$

For $R = \infty$, therefore, the dynamics of the model can be understood as an iteration of the logistic map $Q_\lambda : x \mapsto \lambda x(1-x)$ with $\lambda = 1-p+q$. Note that $0 \leq \lambda \leq 2$, which excludes stable periodic points and chaos.

Q_λ has always two fixed points $x^{(1)}(\infty) = 0$ and $x^{(2)}(\infty) = 1 - 1/\lambda$. Only one, however, is stable, depending on λ, namely $x^{(1)}(\infty)$ when $\lambda < 1$ and $x^{(2)}(\infty)$ when $\lambda > 1$. In terms of $\rho(t)$ we obtain

$$\lim_{t\to\infty} \rho(t) = \begin{cases} 0 & \text{if } p > q, \\ 1 - p/q & \text{otherwise.} \end{cases} \tag{36}$$

Since $|Q'_\lambda(x_\infty)| < 1$ if $\lambda \neq 1$, the stable fixed point is hyperbolic (strongly attracting). At $\lambda = 1$, however, it becomes nonhyperbolic (weakly attracting), and Q_λ exhibits a *transcritical bifurcation* (Bardos and Bessis, 1980) with exchange of stability.

6. Phase Transition

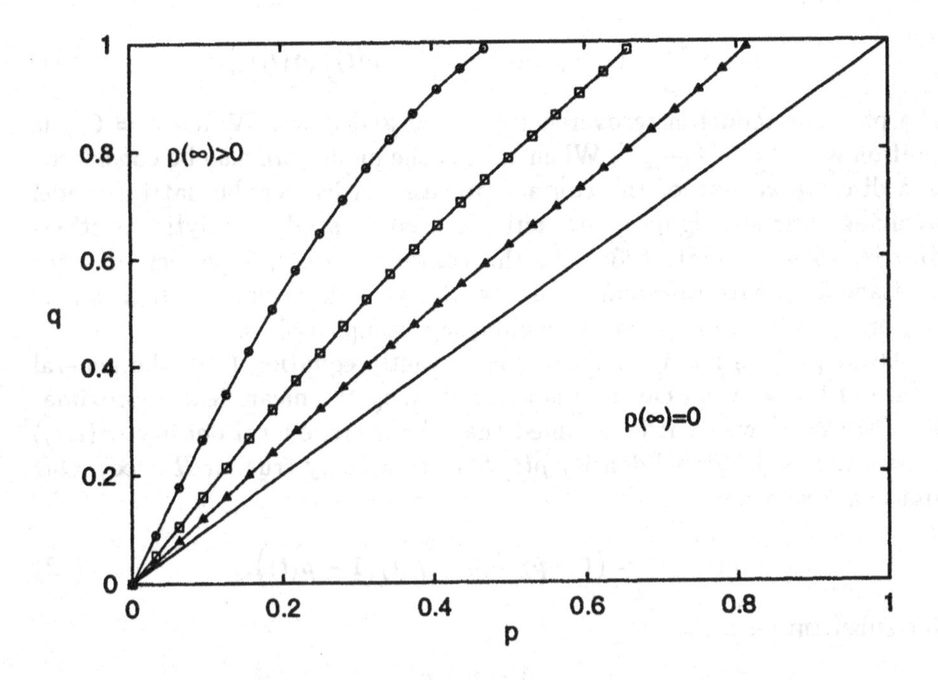

Figure 2. (p, q) phase diagram obtained using Monte Carlo simulations. Lines separating regions where $\rho(\infty) = 0$ and $\rho(\infty) \neq 0$ are shown for $R = 1$ (○), $R = 4$ (□), and $R = 16$ (△). The dotted line represents the mean-field approximation. Lattice size is equal to 10000.

The mean-field approximation discussed in the previous section becomes correct only when $R \to \infty$. For small values of R, like in the basic model, strong correlations are created and substantial deviations from MFA can be expected. Figure 2 represents the (p, q) phase diagram for different values of R obtained in computer simulations. The smaller R is, the larger the deviation from MFA (dotted line) becomes.

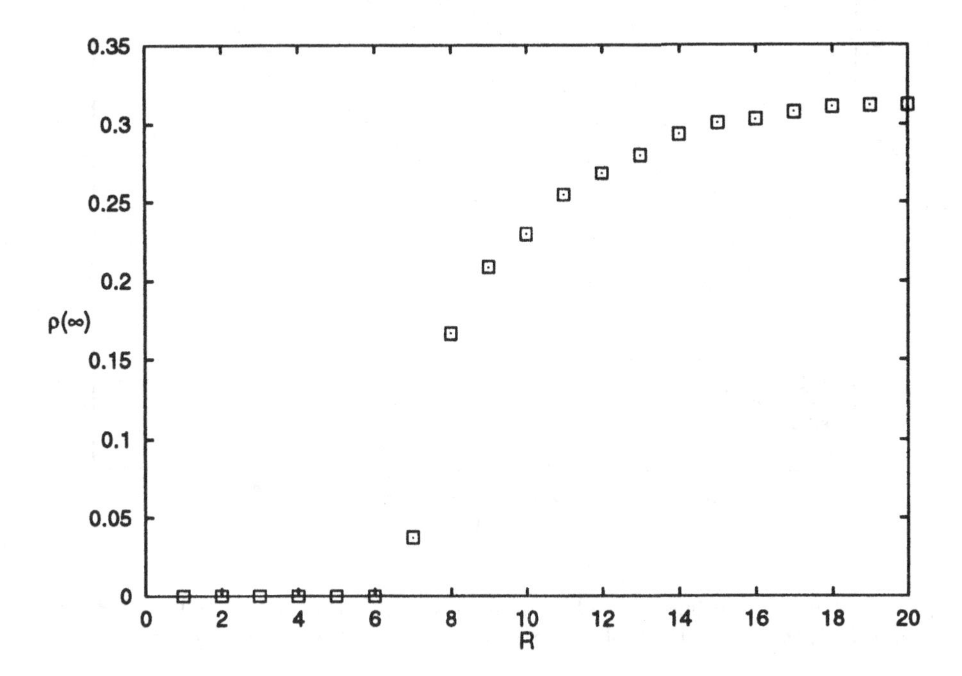

Figure 3. Phase transition of $\rho(\infty)$ with R playing the role of the control parameter for $p = 0.2$ and $q = 0.3$.

As we can see, the line separating $\rho(\infty) \neq 0$ and $\rho(\infty) = 0$ shifts to the left (toward larger p) when R increases. This means that the connectivity between sites increases the robustness of the innovation, i.e. even if p is large, it can be compensated by a large R. To see it, consider a point on the (p, q) phase diagram which is located between the $R = 1$ and the mean-field lines (e.g. $(0.2, 0.3)$). If we plot $\rho(\infty)$ as a function of R, we obtain a bifurcation (or phase-transition) diagram, as shown in Figure 3. For $R \leq 6$ the asymptotic density of adopters goes to zero, but if $R > 6$, $\rho(\infty)$ becomes positive. Three distinct regions of the (p, q) phase space, therefore, can be distinguished:

1. In the region bounded by the q axis and the $R = 1$ critical line, $\rho(\infty) > 0$ regardless of R.
2. In the region bounded by the $R = 1$ and the mean-field critical lines, $\rho(\infty)$ is either equal to zero or positive, depending on the value of R. In general, $\rho(\infty)$ is positive for a sufficiently large R.

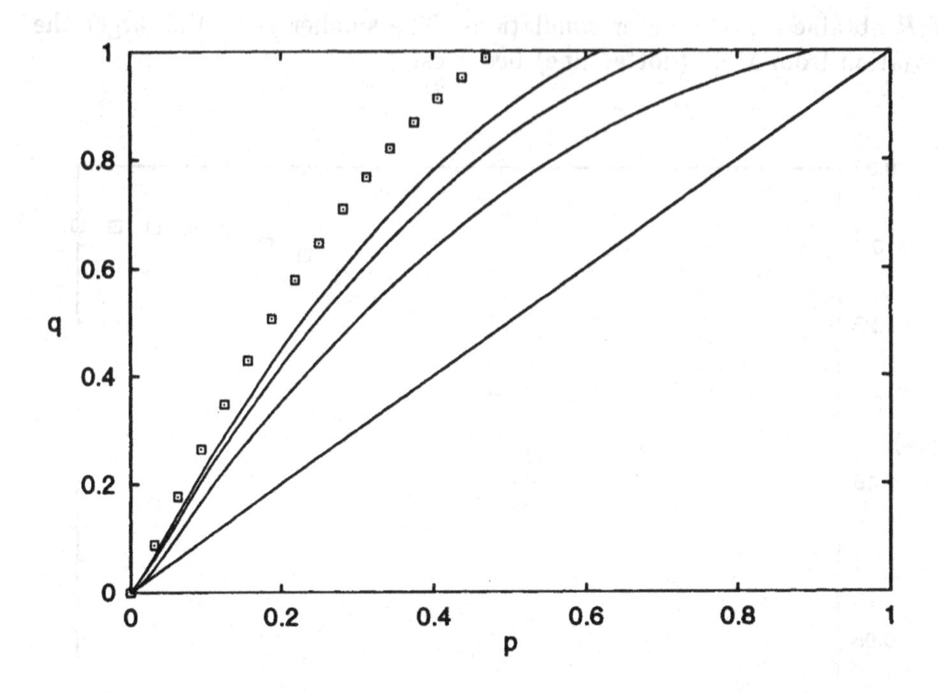

Figure 4. $R = 1$ critical line (□) and its local structure theory approximations. Starting from the right, the consecutive solid lines correspond to $n = 1$, 2, 3 and 4.

3. In the region bounded by the mean-field critical line, $\rho(\infty) = 0$ regardless of R.

7. $R = 1$ critical line

We studied the boundary between regions 1 and 2, i.e., the $R = 1$ critical line, using the local structure theory (LST) up to order 4 (Gutowitz, 1987; Gutowitz *et al.*, 1987), as shown in Figure 4.

As we increase the order n of the approximation, we obtain an approximate critical line closer to the experimental one. This can be also seen if we plot $\rho(\infty)$ obtained from LST as a function p, keeping q constant. An example of such curves is shown in Figure 5, where six consecutive orders of LST are presented. For this particular value of p, the sixth-order critical q-value q_c differs by about 11% from the numerical result, as shown below.

n	1	2	3	4	5	6	simulation
q_c	0.200	0.362	0.430	0.462	0.479	0.490	0.549

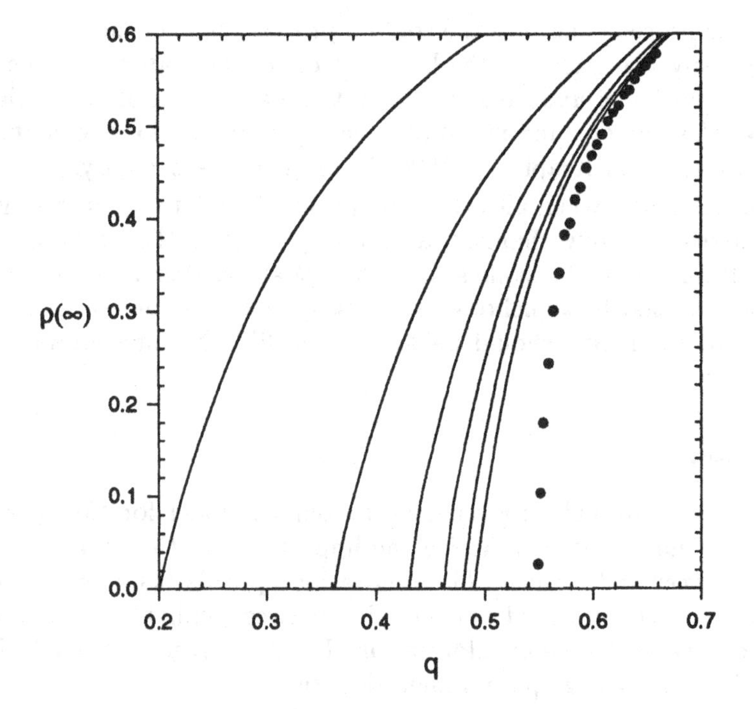

Figure 5. Local structure approximation for $p = 0.2$. Solid lines represent consecutive orders of the local structure approximations, starting with $n = 1$ (the leftmost curve), and ending with $n = 6$ (the rightmost solid curve). The dots represent results of computer simulations ($p = 0.2$, $R = 1$).

As mentioned earlier, for $q = 1$ and $R = 1$, our model can be considered as a discrete realization of the contact process. For the contact process, theoretical studies as well as Monte Carlo simulations suggest that the phase transition occurs at p_c=0.3032 (Dickman and Jensen, 1991; Jensen and Dickman, 1993). The table below compares this result with our numerical simulations and local structure theory approximate results.

n	1	2	3	4	5	simulation	CP
p_c	1.00	1.00	0.70	0.62	0.55	0.48	0.3032

The discrepancy between discrete and continuous time versions of the contact process is understandable, since the synchronous dynamics of CA allows, for example, for one site to change its state from 1 to 0, and, at *exactly* the same time, for its neighbor to change its state from 0 to 1, something that does not occur in the continuous time model. We should stress at this point that although there are many analytical techniques developed to treat stochastic processes such as the contact process, none of them, in general, can be easily translated to the language of cellular automata theory, and the synchronicity of updating is usually the main source of difficulties. For example, the cluster approximation recently proposed for the contact process (Ben-Naim and Krapivsky, 1996) belongs to this category.

On the other hand, when all transition probabilities are small, a single site of the lattice is rarely changed, and during most of the iterations of the CA it will remain in the same state. The updating, therefore, becomes more and more similar to asynchronous updating, and as a consequence, we can expect that the model should behave almost like the time-continuous contact process.

8. Conclusion

We have studied a probabilistic cellular automata model for the spread of innovations. Our results emphasize the importance of the range of the interaction between individuals. The innovation spreads faster when the range increases since increased connectivity between individuals reduces constraints on the exchange of information. Larger connectivity could be also achieved by increasing space dimensionality.

The range of interaction R not only affects the growth rate, but in the region of parameter space (p, q) bounded by the $R = 1$ critical line and the mean-field line it can be a decisive factor for the asymptotic density of adopters $\rho(\infty)$. In this region, if R is too small, $\rho(\infty) = 0$, while for a sufficiently large connectivity $\rho(\infty) > 0$.

References

Bardos C. and Bessis D., editors, *Bifurcation Phenomena in Mathematical Physics and Related Topics.* D. Reidel, Dordrecht, 1980.

Bass F.M. A new product growth for model consumer durables. *Management Sciences* Vol. no.15:215, 1969.

Ben-Naim E. and Krapivsky P.L. Cluster approximation for the contact process. *J. Phys. A: Math. Gen* Vol. no.27:L481, 1994.

Boccara N. and Cheong K. Automata network sir models for the spread of infectious diseases in populations of moving individuals. *J. Phys. A: Math. Gen.* Vol. no.25:2447, 1992.

Boccara N. and Cheong K. Critical behaviour of a probabilistic automata network sis model for the spread of an infectious disease in a population of moving individuals. *J.Phys. A: Math. Gen.* Vol. no.26: 3707, 1993.

Boccara N., Cheong K. and Oram M. Probabilistic automata network epidemic model with births and deaths exhibiting cyclic behaviour. *J. Phys. A: Math. Gen.* Vol. no. 27:1585, 1994.

Boccara N. Fukś H. and Geurten S. A new class of automata networks. *Physica D* Vol. no.103, 1997. To appear.

Dickman R. and Jense, I. Time-dependent perturbation theory for nonequilibrium lattice models. *Phys. Rev. Lett.* Vol. no. 67:2391, 1991.

Fukś H. and Boccara N. Cellular automata models for diffusion of innovations. in *Instabilities and Nonequlibrium Structures VI*, editor E. Tirapegui, Kluwer, Dordrecht, to appear.
Gutowitz H.A. Local Structure Theory for Cellular Automata. Ph.D. thesis, Rockefeller University, New York, New York, 1987.
Gutowitz H.A., Victor J.D. and Knight B.W. Local structure theory for cellular automata. *Physica D* **Vol. no. 28**:18, 1987.
Hägerstrand T. The Propagation of Innovation Waves. Lund Studies in Geography. Gleerup, Lund, Sweden, 1952.
Hägerstrand T. On Monte Carlo simulation of diffusion. *Arch. Europ. Sociol.* **Vol. no. 6**:43, 1965.
Harris T.E. *Ann. Prob.* **Vol. no. 2**:969, 1974.
Haynes K.E. Mahajan V. and White G.M. Innovation diffusion: A deterministic model of space-time integration with physical analog. *Socio-Econ. Plan. Sci.* **Vol. no. 11**:23, 1977.
Jensen I. and Dickman R. Nonequilibrium phase transitions in systems with infinitely many absorbing states. *J. Stat. Phys.* **Vol. no. 48**:1710, 1993.
Liggett T.M. Interacting particle systems. Springer-Verlag, New York, 1985.
Mahajan V., Muller E. and Bass F.M. New product diffusion models in marketing: A review and directions for research. *Journal of Marketing* **Vol. no. 54**:1, 1990.
Mahajan V. and Peterson R.A. Models for Innovation Diffusion. Number 07-048 in Quantitative Applications in the Social Sciences. Sage Publications, Beverly Hills, 1985.
Rogers E.M. *Diffusion of Innovations.* The Free Press, New York, 1995.
Wolfram S. *Theory and applications of cellular automata.* World Scientific, Singapore, 1986.

CELLULAR AUTOMATA MODELS AND CARDIAC ARRHYTHMIAS

An approach of sudden cardiac death prevention

A.L. BARDOU
Cardiologie Théorique, CHU Pitié-Salpêtrière,
91 Bd. de l'Hôpital, 75634 Paris Cedex 13.

AND

P.M. AUGER, R. SEIGNEURIC AND J-L CHASSÉ
UMR CNRS 5558, Université Claude Bernard Lyon 1,
43 Bd. du 11 Novembre, 69622 Villeurbanne Cedex.

1. Introduction

Cardiac Modeling may at present be considered as an important component of basic research in Cardiology. Theoretical studies have progressively been developed in multiple directions, from mechanical models in order to simulate hemodynamics or myocardial mechanics to electrophysiological models. A large part of cardiac action potential modelizations developed till now (Noble, 1975) (Beeler and Reuter, 1977) (Van Capelle and Durrer, 1980) (Luo and Rudy, 1991) have been derived from the model based on differential equations described by Hodgkin and Huxley on the giant squid axon in 1952 (Hodgkin and Huxley, 1952) (Hodgkin and Huxley, 1990).

Nevertheless, attempts have been parallely done to develop cellular automata models, as simpler models than differential equations ones, in order to study propagation phenomenon in myocardial tissue without taking account of the modelization of cellular mechanisms responsible of the genesis of action potential. These cellular automata models have permitted to study patho-physiological behaviour of myocardium associated to modifications of physiological properties of cardiac fibers during ischemia, myocardial infarction, etc... In these models only the propagation of myocardial activation wave (which can be described by different rules) and the state of excitability or refractoriness of the cellular elements are considered.

The first of these cellular automata models, has been developed by Moe et al (Moe *et al.*, 1964) and is of historical interest as it allowed, for the first time, to partially describe mechanisms of atrial flutter. During the last ten years, cellular automata models have mainly been used to try to understand mechanisms of electrophysiological diseases in ventricular myocardium activation wave.((Smith and Cohen, 1984) (Saxberg and Cohen, 1991) (Auger *et al.*, 1987a) (Auger *et al.*, 1987b)).

The cellular automata model presented in this paper has been, during the ten last years, widely applied to the study of cardiac arrhythmias, fibrillation and defibrillation basic mechanisms. It allowed to understand how numerous patho-physiological effects at the cellular electrophysiology level, mainly due to ischemia or ventricular infarction, may have important consequences on propagation in the myocardium, inducing cardiac arrhythmias. Results presently obtained provide important information about different possible reentry mechanisms able to degenerate in sudden cardiac death. Moreover, recent investigations show possible pharmacological applications of this model, mainly on antiarrhythmic drugs effects.

2. Methods

We mainly developed this model in order to systematically study the mechanisms of cardiac arrhythmias and ventricular fibrillation. These diseases are essentially induced by pathological perturbations at the level of propagation of the depolarization wave (activation wave) inside the myocardium. The needs were more to exactly know the position of activation wave at any time, and the refractory or excitable state of the different cells of the simulated myocardium, than the exact potential of each element at any moment. In these conditions, development of a cellular automata model seemed more adequate to this kind of study than ionic models based on differential equations.

Recall of the model

The originality of the present cellular automata model is based on the propagation law in the lattice ((Auger *et al.*, 1987a) (Auger *et al.*, 1987b) (Auger *et al.*, 1989)).

This model is a time discrete model in which the states of the different groups of cells are calculated at equal time intervals Δ t. Knowing the position of the depolarization wave at time t, we chose to apply Huyghens' principle to precisely describe its position at time t+Δ t. It consists in defining the position of wave front at time t+Δ t as the envelope of half circus of radius C.Δ t, centered on each point of the wave front at time t, C being the conduction velocity of the depolarization wave in the myocardium. This value being perfectly known from experimental measurements in physiological and pathological conditions (mainly during ischemia), such a method allows to reproduce the propagation of activation wave in the simulated myocardium in accordance with conditions in actual myocardium.

In this model, a square surface element of the ventricle is represented by a network of 2500 (N=50) elements in the initial version and by 65 536 elements (N=256) in the more recent versions (this increase in N, for example, leading to the possible study of very local ischemic effects at the whole myocardium level). Each point (i,j) of the surface element, where i and j $\in$ [1, 256] corresponds to a group of few cardiac cells which are in a discrete set of cellular states. To each point (i,j) is associated a state matrix consisting of integer values S(i,j) varying with the cellular excitability status. Cells can be at rest (excitable) (S(i,j)=0) or depolarized (S(i,j)>0). The refractory period, R(i,j) at point (i,j) can differ from one point to another but in general it is fixed at the beginning of each simulation. When cells are depolarized, the state matrix varies suddenly from 0 to R(i,j). Then, at each time interval Δ t, the state matrix decreases to give R(i,j)-1 after time Δ t, R(i,j)-2 after 2Δ t, . . . , 1 after {n-1}Δ t, and finally becomes again equal to 0 after nΔ t, n being the number of time steps to reach complete recovery. In this situation, the corresponding cells are again excitable and can be depolarized another time. There are R(i,j)+1 possible state transitions 0 $\rightarrow$ R $\rightarrow$ R-1 $\rightarrow$ R-2 $\rightarrow$ . . . $\rightarrow$ 1 $\rightarrow$ 0, (the states being represented in simulated mappings by different colours), and thus the total refractory period (absolute+relative) is equal to the time necessary for them, i.e. {R(i,j)+1}Δ t. The possible stimulation of a cell in relative refractory period is estimated as a function of stimulus intensity according to stimulation threshold variability as a function of this parameter during the terminal part of action potential (rule mainly applied for simulation of defibrillation shocks).

Application of Huyghens' construction method consists to assume that during the time interval Δ t, each newly depolarized group of cardiac cells at a point (i,j) is able to depolarize any excitable cardiac cell which can be found in a small neighborhood defined by a circle centered on this point (i,j) with a radius R(i,j) = c(i,j).Δ t. This radius R(i,j) corresponds to the distance covered by the wave from this point during the time interval Δ t. Huyghens' principle also means that each point of the wave front at time t is a source of a circular wave and that the wave front at time t+Δ t is obtained by the composition of all these circular waves.

3. Results

In this section, we will briefly summarize the very interesting results that have globally been obtained on different basic mechanisms of induction of arrhythmias and ventricular fibrillation by means of this cellular automata model before describing its present applications to antiarrhythmic drugs.

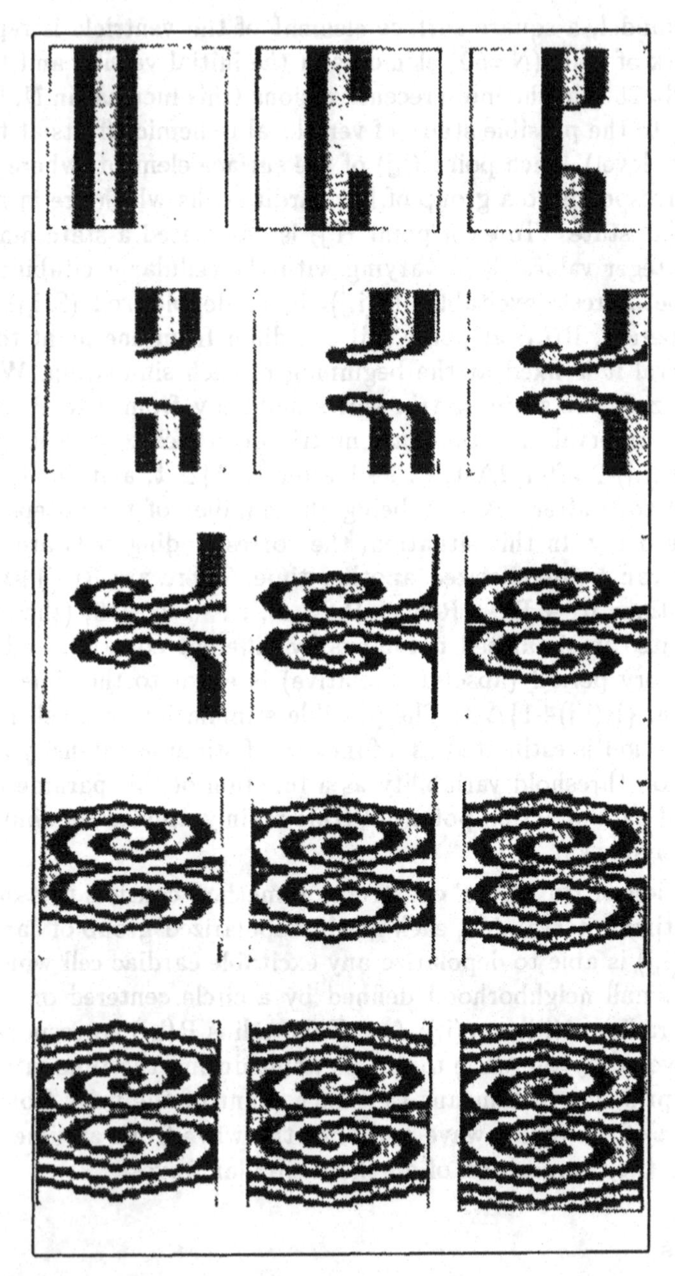

Figure 1. **Unidirectional block smaller than critical size in anisotropic conditions. Occuring of a reentry in these conditions evidence the favorizing role of anisotropy in arrhythmias genesis.**

In a first step, induction of one limb of double limb reentries by unidirectional blocks has been found again by use of this model ((Auger *et al.*, 1987a) (Auger *et al.*, 1987b)). The vortices so induced were durably maintained even if conduction laws were re-established as normal after induction of reentry, showing that the self-maintained property of these reentries was totally compatible with spontaneous irreversibility of ventricular fibrillation when induced.

An interesting notion about unidirectional conduction blocks critical size has been evidenced: there exists a critical size l_{cri} such as $l_{cri} = 2C\tau$ occur (C=conduction velocity, τ =refractory period) (Auger *et al.*, 1989) . This critical size equal to 20cm in physiological conditions (possible explanation of non fibrillation of small size hearts) is able to fall to 2cm in ischemic conditions allowing then a reentry to occur in the myocardium.

Simulation of anisotropic propagation (wavefront considered as the envelope of half-ellipses whose longest axis is three times the smallest one) evidenced the fact that, in these anisotropic propagation conditions, reentries inducing spiral waves were able to occur in the myocardium (Auger *et al.*, 1989) , even for blocks smaller than the critical size precedently defined for isotropic conditions (figure 1). These theoretical results have recently been experimentally validated by Jalife *et al.* by use of optical mapping which directly showed that, in anisotropic conditions, spiral waves were able to underly reentrant activity in isolated cardiac muscle (Pertsov *et al.*, 1993) .

An important problem related to internal defibrillation and implantable defibrillator technique was the possible existence of a defibrillation myocardial critical mass. Simulated defibrillation shocks realized by means of this model showed that there were no critical mass, practically all the cardiac fibers necessiting to be stimulated in order to obtain defibrillation ((Auger *et al.*, 1989) (Auger *et al.*, 1990b) (Auger *et al.*, 1990a)) (figure 2). These theoretical results have been experimentally confirmed by Zhou et al. (Zhou *et al.*, 1989) , who showed in dogs that critical mass of defibrillation, if existing, corresponds at least to 90% of the ventricular mass. An interesting notion of possible defibrillation by increasing simultaneously the refractory period of all the elements of the ventricular sheet has been shown in this theoretical study of defibrillation (figure 3) (Auger *et al.*, 1990a) . We presently deeply study this very important preliminary result which seems to be possibly short-dated clinically validated on atrial fibrillations.

Incidence of ischemia on propagation has been studied by means of this model, mainly through its two major cellular effects that are a modification of the dV/dt of the action potential (leading to a decrease in propagation velocity) and modifications of action potential duration inducing a wide dispersion of refractoriness. This second effect has been found as able to induce "percolation-like" rentries as previously described by Smith et al. (Smith

Figure 2. **Simulation of a defibrillation shock by the cellular automata model. 95% of the cardiac cells are stimulated by the electric shock in this simulation. The small number of non depolarized cells appear nevertheless to be able to initialize again multiple circus movements (different from initial vortex): The critical mass for defibrillation can so be considered as over 95% of the ventricular myocardium (no critical mass for defibrillation).**

et al., 1984) .

One original result has been obtained by studying the effect of this dispersion on very local regions of the myocardium (figure 4). This variability of refractoriness appears as the most dangerous consequence of ischemia, as it seems able to induce reentries even when occuring in very limited regions. These results permit to evidence one of the possible mechanisms by which minor myocardial infarctions may degenerate in ventricular fibrillation (Bardou *et al.*, 1995) (Bardou *et al.*, 1996) .

Some effects of antiarrhythmic drugs have been simulated by means

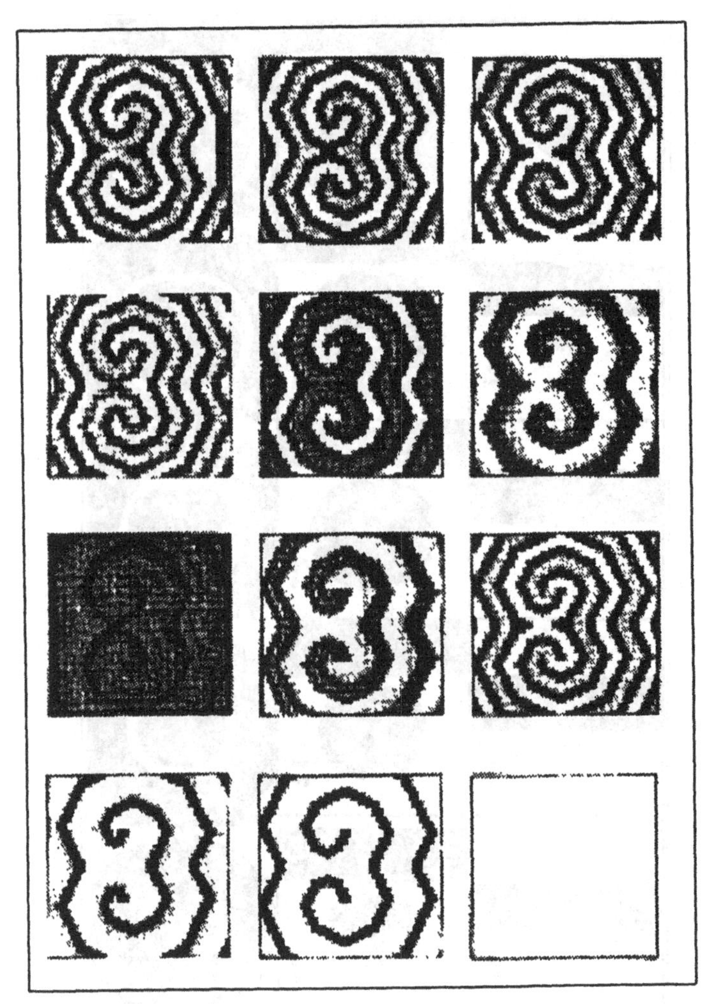

Figure 3. Increase of refractory period of cardiac cells after induction of a reentry in cellular automata model. This prolongation of repolarization of action potential appears able sto stop the initialized reentry.

of this model. These drugs can have multiple and complex effects on the refractory period, the conduction velocity, the excitability threshold or still the ratio between longitudinal and transversal conduction velocities.

The arrhythmogenic effect of some of them has been experimentally evidenced ((Velbit *et al.*, 1982) (Bigger and Sahar, 1987) (Horowitz *et al.*, 1987)) and, in USA, the CAST investigators, mandated by NIH, showed, on a randomized population of post infarction patients, that mortality was more elevated in patients treated with some class 1B and 1C antiarrhythmic drugs than in patients without any treatment (CAST89, 1989) .

In these conditions, it appeared of first importance to try to understand

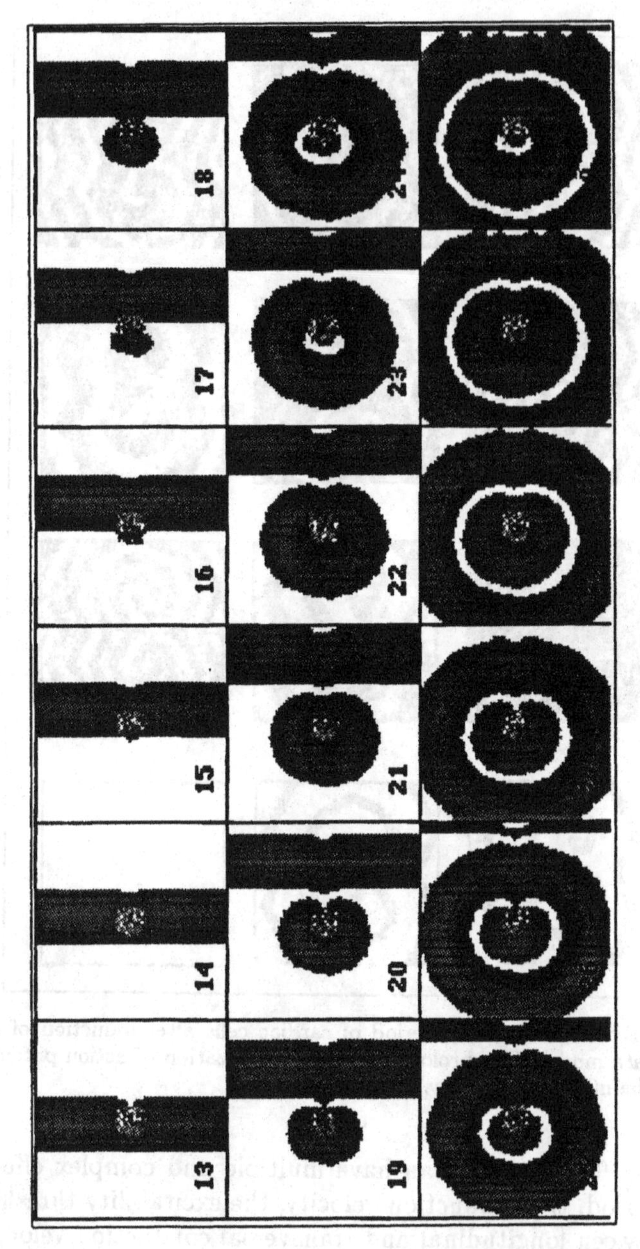

Figure 4. Effect of local ischemia modelized as dispersion in refractoriness in a very limited zone of a 2-D myocardium sheet simulated by the cellular automata model. This dispersion in a local zone appears able to induce a reentry invading the whole element.

the mechanisms corresponding to arrhythmias aggravation or induction by administration of these drugs. Shortening of mean refractory period, appearing as a secondary effect of some of them, has been shown as particularly dangerous by decreasing the wavelength and as a consequence the critical size (Auger *et al.*, 1992) .

Nevertheless, the most dangerous arrhythmogenic effect evidenced by the theoretical studies based on this model has been to show that some of these antiarrhythmic drugs were able to induce transient blocks by depressing sodium channels, these transient blocks being able to degenerate in particular conditions in ventricular fibrillation (Bardou *et al.*, 1993) . These results have been experimentally validated by similar results obtained by means of direct recording of ventricular activation wave using epicardial cardiac mapping in the dog (Bardou *et al.*, 1993) , showing that these molecules (for example lidocaïne) possibly provoked transient conduction blocks triggering vortex in the myocardium.

These results led us to presently develop intensive studies of transient conduction blocks in order to determine their possible conditions of reentries and ventricular fibrillation induction (Chassé *et al.*, 1995) .

4. Discussion

One of the main element of effectiveness of this model has been to stay as close as possible of experimental physio-pathological activation wave propagation conditions by use of Huyghens' principle. It is to be noted that very accurate results were obtained by Solomon et al (Solomon and Selvester, 1973) , to simulate total heart activation sequence, by associating a proper geometrical shape of the ventricles and an activation transmission based on Huyghens' principle. More recently, Szathmary and Osvald, (Szathmary and Osvald, 1994) use the same methodology to define a model of propagated activation with analytically defined geometry of the ventricles.

The totally new and original way introduced in cardiac modelling with cellular automata models development offers the possibility to perform specific investigations in the field of action potential propagation in the myocardium and to diseases associated to propagation troubles, mainly arrhythmias and fibrillation in the study of which they appear as more and more accurate. They open the way to the systematic research of different mechanisms of myocardial reentry and of parameters involved in this phenomenon (unidirectionnal blocks, variability of refractory period induced by ischemia, variability of anisotropy ratio, etc..). A recent and interesting development of these models is to possibly use them on a predictive mode applied to drug screening in order to determine through their multiple and various effects which one appear as being the most important in stopping the reentry process.

The recent studies of Belousov-Zhabotinsky reactions kinetics on chemical reactors and the studies of control or breakdown of so induced spiral waves open a new way to try to control the development of such spiral waves in myocardium which possibly induce ventricular fibrillation.

One of the most important fact to underline is the excessively powerfulness of coupling modelization to experimental methods. This intervalidation between these two approaches have mainly been realized by coupling different methods of cardiac modelling to experimental electrical cardiac mapping on whole heart or ventricular bidimensional sheets ((Auger *et al.*, 1989) (Auger *et al.*, 1990b) (Auger *et al.*, 1992)(Szathmary and Osvald, 1994) (Hélie *et al.*, 1995)(Hélie *et al.*, 1995)) or to optical mapping on sheets of cardiac tissue ((Pertsov *et al.*, 1993)(Chiaso *et al.*, 1990)).

A significant example of the efficiency of such a coupling may be given by the similar results obtained by our cellular automata model and experimental electrical cardiac mapping showing the possible triggering of ventricular fibrillation by transitory blocks induced by Class 1 antiarrhythmic drugs such as Lidocaine (Bardou *et al.*, 1993) , this phenomenon being one of the possible origin of arrhythmogenic effects of these molecules. The present use of a cellular network model based on Van Capelle and Durrer equations applied to systematic study of the effects of transient blocks (Chassé *et al.*, 1995) may lead to potential results of first importance at the clinical level.

An other important field, where significant advances could be expected from modeling, is cardiac pharmacology. An example of contribution of theoretical methods to pharmacology has been developed in this paper about arrhythmogenic effects of antiarrhythmic drugs. When one looks experimentally to drugs effects, they appear to be multiples, molecules acting generally on different parameters. In such a case, models represent a tool permitting to study separately the different effects on the different parameters. From such information, it becomes possible to determine the parameters involved as the most significant targets to reduce the disease concerned by the drug. Important applications of these theoretical methods will surely be, in the future, related to development of new antarrhythmic drugs and/or to their possible application to chemical defibrillation.

At present, results obtained from models initialize the understanding of basic mechanisms of cardiac arrhythmias and diseases based on activation wave propagation troubles in myocardial tissue. At the clinical level, this kind of results about ventricular fibrillation and defibrillation basic mechanisms, coupled to non invasive methods such as cardiomagnetography, is expected to lead, probably in the next 10 years, to the development of a wide prevention of sudden cardiac death.

As a general conclusion, electrophysiological cardiac modeling has been and still remains a very important and potentially very valuable research field.

References

Auger P., Bardou A., Coulombe A. and Degonde J. Computer simulation of different mechanisms of ventricular fibrillation. *Proceedings of the 9th IEEE/EMBS International Conference* 171–174, 1987.

Auger P., Bardou A., Coulombe A. and Degonde J. Computer simulation of different electrophysiological mechanisms of ventricular fibrillation. *Comput. in cardiol.* **Vol. no. 87**: 527–530, 1987.

Auger P., Coulombe A., Govaere M.-C., Chesnais J.-M., Von Euw D. and Bardou A. Computer simulation of mechanisms of ventricular fibrillation and defibrillation. *Innov. Technol. Biol. Med.* **Vol. no. 10**: 299–312, 1989.

Auger P., Bardou A., Coulombe A., Govaere M.-C., Rochette L., Schreiber M.-C., Von Euw D. and Chesnais J.-M. Computer simulation of ventricular fibrillation and of defibrillating electric shocks. Effects of antiarrhythmic drugs. *Mathl. Comput. Modelling* **Vol. no. 14**: 576–581, 1990.

Auger P., Coulombe A., Dumée P., Govaere M.-C., Chesnais J.-M. and Bardou A. Computer simulation of cardiac arrhythmias and of defibrillating electric shocks. Effects of antiarrhythmic drugs. in *Non linear coherent structures in Physics and Biology.* Lecture Notes in Physics **Vol. no. 393**, Remoissenet, M. and Peyrard, M. Eds. Heidelberg: Springer Verlag, 133–140, 1990.

Auger P., Cardinal R., Bril A., Rochette L. and Bardou A. Interpretation of cardiac mapping by means of computer simulations: Applications to calcium, lidocaïne and to brl 34915. *Acta Biotheoretica* **Vol. no. 40**: 161–168, 1992.

Bardou A., Auger P., Cardinal R., Dumée P., Birkui P., Von Euw D. and Govaere M.-C. Theoretical study by means of computer simulation of conditions in which extrasystoles can trigger ventricular fibrillation. validation by epicardial mappings. *Journal of Biological Systems* **Vol. no. 1**: 147–158, 1993.

Bardou A., Auger P., Achour S., Dumée P., Birkui P. and Govaere M.-C. Effect of myocardial infarction and myocardial ischemia on induction of cardiac reentries and ventricular fibrillation. *Acta Biotheoretica* **Vol. no. 43**: 363–372, 1995.

Bardou A., Auger P., Chassé J.-L. and Achour S. Effects of local ischemia and transient conduction blocks on the induction of cardiac reentries. *International Journal of Bifurcation and Chaos* **Vol. no. 6**: 1657–1664, 1996.

Beeler G. and Reuter R. Reconstruction of the action potential of ventricular myocardial fibres. *J. Physiol.* **Vol. no. 268**:177–210, 1977.

Bigger T. and Sahar D. Clinical types of proarrhythmic response to antiarrhythmic drugs. *Amer. J. Cardiol.* **Vol. no. 59**: 2E–9E, 1987.

Cardinal R., Vermeulen M., Shenasa F., Roberge P., Page P., Hélie F. and Savard P. Anisotropic conduction and functional dissociation of ischemic tissue during reentrant ventricular tachycardia in canine myocardial infarction. *Circulation.* **Vol. no. 77**: 1162–1176, 1988.

Chassé J.-L., Auger P., Bardou A. and Achour S. Some mechanisms of initiation of ventricular fibrillation: Transient blocks and ischemic zones. *IEEE/Engin. Med. Biol., Soc. 17th Conf. Proceed.* : 1755–1758, 1995.

Chiaso D., Michaels D. and Jalife J. Supernormal excitability as a mechanism of chaotic dynamics of activation in cardiac purkinje fiber. *Circ. Res.* **Vol. no. 90**: 525–545, 1990.

Hélie F., Cossette F., Vermeulen M. and Cardinal R. Differential effects of lignocaine and hypercalcemia on anisotropic conduction and reentry in the ischemically damaged canine ventricle. *Cardiovascular research.* **Vol. no. 29**: 359–372, 1995.

Hodgkin A. and Huxley A. A quantitative description of membrane current and its application to conduction and excitation in nerve. *J. Physiol.* **Vol. no. 117**: 500–544, 1952.

Hodgkin A. and Huxley A. A quantitative description of membrane current and its application to conduction and excitation in nerve. *Bul.Math.Bio.* **Vol.no. 52**: 25–71,

1990.

Horowitz L., Zipes D., Campbell R., Morganroth J., Podrid B., Rosen M. and Woosley R. Proarrhythmia, arrhythmogenesis or aggravation of arrhythmia-a status report. *Amer. J. Cardiol.* **Vol. no. 59**: 54E–56E, 1987.

Luo C. and Rudy Y. A model of the ventricular action potential. depolarization, repolarization and their interaction. *Circ. Res.* **Vol. no. 68**: 1501–1526, 1991.

Moe G., Rheinboldt W. and Abildskov J. R. A computer model of atrial fibrillation. *Am. Heart J.* **Vol. no. 67**: 200–220, 1964.

Noble D. *The initiation of the heart beat.* Oxford University Press, 1975.

Pertsov A., Davidenko J., Salomonsz R., Baxter W. and Jalife J. Spiral waves of excitation underlie reentrant activity in isolated cardiac muscle. *Circ. Res.* **Vol. no. 72**: 631–650, 1993.

Saxberg B. and Cohen R. Cellular automata models of cardiac conduction. in *Theory of the heart.*, Glass L., Hunter P. and McCulloch A., Eds, Springer Verlag, New-York: 437–476, 1991.

Smith J. and Cohen R. Simple computer model predicts a wide range of ventricular dysrhyythmias. *P.N.A.S.* **Vol. no. 81**: 233–237, 1984.

Smith J., Ritzenberg R. and Cohen R. Percolation theory and cardiac conduction. *Computers in Cardiology*, Washington, DC: IEEE Computer Society Press: 201–204, 1984.

Solomon J. and Selvester R. Simulation of measured activation sequence in the human heart. *Am. Heart J.* **Vol. no. 85**: 518–523, 1973.

Szathmary V. and Osvald R. An intercative computer model of propagated activation with analytically defined geometry of ventricles. *Computers and Biomed. Res.* **Vol. no. 27**: 27–38, 1994.

The Cardiac Arrhythmia Suppresion Trial Investigators [CAST89]: Preliminary report: effect of encaidine and flecaidine on mortality in a randomized trial of arrhythmia suppression after myocardial infarction. *New Engl. J. Med.* **Vol. no. 321**: 406–412, 1989.

Van Capelle F. and Durrer D. Computer simulation of arrhythmias in a network of coupled excitable elements. *Circ. Res.* **Vol. no. 47**: 454–466, 1980.

Velbit V., Podrid P., Lown B., Cohen B. and Graboys T. Aggravation and provocation of ventricular arrhythmias by antiarrhythmic drugs. *Circulation* **Vol. no. 65**: 886–894, 1982.

Zhou X., Daubert J., Lown B., Wolf P., Smith W. and Ideker R. Size of the critical mass for defibrillation. *Circulation* **Vol. no. 80**: II–531, 1989.

Part 6
Particular techniques examples

The additive cellular automata class lends itself well to mathematical treatment, and brings into play algebraic and arithmetical techniques. Two of the following contributions illustrate this and especially show connections between cellular automata and combinatorics on infinite words.

Some cellular automata can be expressed in a non-cellular, more compact form, as is the case of automata with memory. Their study implies threshold functions and proves the efficiency of new techniques.

If one wants to free oneself from the net regularity, one could use the notion of graph of automata, which is very general and includes, in some sense, cellular automata and neural nets. The difficulty is then to define tools that allow local mastering of the graph: local decoding, compass, maps.

Cellular automata are made up of locally interacting components. A way of understanding these interactions better is to make them vary. That is done in Čulik's contribution where the neighborhood is neither finite nor uniform but depends on some "regular global context".

DYNAMIC PROPERTIES OF AN AUTOMATON WITH MEMORY

M. COSNARD
LORIA - INRIA
615 rue du Jardin Botanique
54602 Villers lès Nancy France

Abstract. We review the dynamic properties of a single automaton with memory.

1. Introduction

Automata theory is often used to modelize elementary properties of the nervous system (Caianiello and de Luca, 1966; Kitagawa, 1973; Yamaguti and Hata, 1982; Yoshizawa, 1982). Most of these models discretize the nervous signal and then treat only boolean information. In spite of the simplicity of such an assumption, the dynamical behaviour of these models is extremely rich.

In this paper we review the dynamic property of an automaton with memory introduced by (Caianiello and de Luca, 1966) in order to take into account the refractary character of the neural response. The evolution of the automaton depends upon the current state and some previous states. It has been proven to have a very complex dynamic which is far from being completely known. The automaton is defined by the following equation:

$$x_{n+1} = \mathbf{1}[\sum_{i=0}^{k-1} a_i x_{n-i} - \theta] \qquad (1)$$

The following review is based on a series of papers (Cosnard *et al.*, 1988a), (Cosnard *et al.*, 1988b), (Cosnard *et al.*, 1998) . In the next section, we shall discuss some properties that relate this one-dimensional recurrence relation with k-dimensional automata network and introduce some of the questions to be studied later. Section 3 will be devoted to the presentation of the properties of (1) for positive coefficients. In section 4, we completely handle the case of reversible systems. Section 5 will present the results for symmetric weights, section 6 for geometric coefficients and section 7 will be devoted to the unbounded memory case. This review paper will be concluded by some general results and conjectures.

2. Direct properties and general results

Consider the recurrence equation (1) where:

- x_n belongs to $\{0,1\}$ and represents the state of the automaton at time n,
- $a_i, i = 0, ..., k-1$, is a real number and will be called later the i-th weight, the set of a_i will be also called the memory coefficients,
- k is an integer and represents the memory size,
- θ is a real number called the threshold,
- **1** is the Heavyside function: **1** [u] is 0 if $u < 0$ and 1 else.

Main Question: Given k, a_i, θ, what are the behaviours of the sequence x_n.

No general answer exists for this problem in spite of several works during more than 30 years as can be seen in the list of references.

Consider the following k-dimensional recurrence relation:

$$Y(n+1) = A[Y(n)] \qquad (2)$$

where:

- $Y(n) \in \{0,1\}^k$,
- $A : \{0,1\}^k \longrightarrow \{0,1\}^k$ is defined by:

$$A \begin{bmatrix} y_1 \\ y_2 \\ \cdot \\ \cdot \\ y_{k-1} \\ y_k \end{bmatrix} = \begin{bmatrix} y_2 \\ y_3 \\ \cdot \\ \cdot \\ y_k \\ \mathbf{1}[\sum_{i=0}^{k-1} a_i y_{k-i} - \theta] \end{bmatrix}.$$

Hence (2) is a network of threshold automata of a particular form. We shall call it $A[k, a, \theta]$.

Consider the following k-dimensional recurrence relation:

$$Y(n+1) = \mathbf{1}[MY(n) - \theta] \qquad (3)$$

where:

- $Y(n) \in \{0,1\}^k$,
- $M : \{0,1\}^k \longrightarrow \{0,1\}^k$ is defined by:

$$M = \begin{bmatrix} 0 & 1 & 0 & . & . & 0 \\ 0 & 0 & 1 & 0 & . & 0 \\ . & . & . & . & . & . \\ a_{k-1} & a_{k-2} & . & . & a_1 & a_0 \end{bmatrix}.$$

Property 1 *(Network representation using threshold functions)*

$$(1) \iff (2) \iff (3).$$

This fundamental property, although easy to prove, provides an important relation between equation (1) and the automata network domain. First of all, remark that the number of states is 2^k and hence that the sequence converges to a cycle of length at most 2^k after a transient of length at most 2^k.

The list of questions that are generally handled are the following:

- given k, a_i, θ
  1. given $Y(0)$, what is the limiting set of $Y(n)$ when n goes to infinity?
  2. for any $Y(0)$, what are the limiting sets of $Y(n)$ when n goes to infinity?
  3. what is the length of the longest cycle?
  4. what is the number of cycles?
  5. what is the length of the maximal transient?
  6. are there conditions for the existence of fixed points?
- given k and a_i, what is the bifurcation structure of the system when θ varies?
- given k
  1. what is the length of the longest cycle?
  2. what is the length of the maximal transient?

As an illustration consider the following example:

$$x_{n+1} = \mathbf{1}[-4x_n - 2x_{n-1} - x_{n-2} + 2] \qquad (4).$$

Using the main property we can rewrite this recurrence relation under the following form:

x_n	x_{n-1}	x_{n-2}		x_{n+1}	x_n	x_{n-1}
0	0	0		1	0	0
0	0	1		1	0	0
0	1	0		1	0	1
0	1	1	$\longrightarrow$	0	0	1
1	0	0		0	1	0
1	0	1		0	1	0
1	1	0		0	1	1
1	1	1		0	1	1

By enumeration, it is easy to show that the system has a unique cycle of length 2 and that after a transient of length at most 4, the sequence converges towards this cycle. In other word, after at most 4 iterations the sequence generated by (4) oscillates (0101010101....).

The most general results concerning threshold automata networks are due to (Goles and Martinez, 1990):

Property 2 *(General property of threshold automata networks)*
Consider the network given by the recurrence

$$Y(n+1) = \mathbf{1}[QY(n) - \theta] \qquad \text{(4a).}$$

where Q is a real matrix.

- *If Q is symmetric, then $Y(n)$ converges, for any $Y(0)$, to either a fixed point or to a cycle of length 2.*
- *If Q is non-symmetric, then there could exist cycles of exponential length.*

3. Positive coefficients

Assume that $\forall a_i, a_i \geq 0$.

Property 3 *(Positive coefficients)*

- *If $0 \leq a_{k-1} \leq a_{k-2} \leq .... \leq a_1 \leq a_0$, then $Y(n)$ converges, for any $Y(0)$, to a fixed point: either $\{0^k\}$ or $\{1^k\}$. Hence the sequence x_n stabilizes either to $\{0\}$ or to $\{1\}$.*
- *If $\exists i, a_i \geq \theta$, then the length of the cycles divides $i+1$. This implies that the cycles are at most of length linear in k.*

4. Reversible systems

We say that $A[k, a, \theta]$ is reversible if A is a bijection of $\{0,1\}^k$. Recall that an application of $\{0,1\}^k$ is a shift if

$$A[z_0, z_1, ..., z_{k-1}] = (z_1, ..., z_{k-1}, z_0).$$

It is an antishisft if

$$A[z_0, z_1, ..., z_{k-1}] = (z_1, ..., z_{k-1}, 1 - z_0).$$

Property 4 *(Reversible systems)*

- *$A[k, a, \theta]$ is reversible if and only if*
  - *either $A[k, a, \theta]$ is a shift,*
  - *either $A[k, a, \theta]$ is an antishift*
- *The length of the cycles of a reversible system*
  - *divides k if $A[k, a, \theta]$ is a shift,*
  - *divides $2k$ if $A[k, a, \theta]$ is an antishift.*

Since a reversible automata is bijective, the trajectories are only cycles. Hence, in all cases the set of trajectories is composed of an exponential number of cycles of at most linear (in k) length.

5. Symmetric memory

We shall say that the memory is symmetric if the a_i form a symmetric word:

$$\forall i \in \{0, ..., k-1\}, a_i = a_{k-1-i}.$$

In order to study such systems, we introduce the following energy operator:

$$E(x_0, ..., x_k) = -\sum_{j=0}^{k-1} x_{k-j} \sum_{s=j+1}^{k} a_{s-j-1} + \theta \sum_{j=0}^{k} x_j.$$

Consider a sequence $\{x_n\}$ generated by $A(k, a, \theta)$. We can prove that:

$$\delta_n = E(x_{n-k}, ..., x_n) - E(x_{n-k-1}, ..., x_{n-1}) = -(x_n - x_{n-k-1})(\sum_{s=1}^{k} a_s x_{n-s} - \theta).$$

We deduce from this that:

- if $x_n \neq x_{n-k-1}$ then $\delta_n < 0$,
- if $x_n = x_{n-k-1}$ then $\delta_n = 0$.

We deduce that δ is decreasing on any trajectory generated by $A(k, a, \theta)$. Hence, δ is a Lyapunov function for $A(k, a, \theta)$.

Property 5 *(Symmetric memory)*
If $A(k, a, \theta)$ has a symmetric memory, then

- *the length of a cycle of $A[k, a, \theta]$ divides $k+1$.*
- *the length of any transient of $A[k, a, \theta]$ bounded by $(k+1)^2[2|\theta| + \sum_{s=0}^{k-1} |a_s|]$, and hence if the memory coefficients are polynomials in k then the length of any transient is also a polynomial in k.*

6. Geometric coefficients

In this section, we consider the case where the coefficients form a geometric sequence. Assume that the automaton is defined by :

$$x_{n+1} = \mathbf{1}[\alpha - \sum_{i=0}^{k-1} b^i x_{n-i}] \qquad (4b).$$

which corresponds to $A(k, a, \theta)$ with $a_i = -b^i$ and $\theta = -\alpha$. Let us specify the automaton for $b = 2$.

$$x_{n+1} = \mathbf{1}[\lambda - \sum_{i=0}^{k-1} 2^i x_{n-i} -] \qquad (5).$$

Consider now the following arithmetic game where $z_n \in [0, ..., 2^k - 1]$ for a given integer k:

$$z_{n+1} = \begin{cases} \lfloor \frac{z_n}{2} \rfloor & \text{if } z_n \geq h \\ \lfloor \frac{z_n}{2} \rfloor + 2^{k-1} & \text{if } z_n < h \end{cases} \qquad (6).$$

All these recurrence relations are in fact equivalent for $0 < b \leq \frac{1}{2}$.

Property 6
If $0 < b \leq \frac{1}{2}$ then (4) $\Longleftrightarrow$ (5) $\Longleftrightarrow$ (6).

The following result explicits the dynamic properties of these recurrences in details.

Property 7 *(Geometric coefficients)*
If $0 < b \leq \frac{1}{2}$ then :

- *$\forall \alpha$ and k, (4) contains a unique cycle of length not greater than $k+1$,*
- *if $p \leq k+1$ then there exists α such that (4) has a cycle of length p.*

In order to better understand the dynamic structure of (4), we can fix k, let α vary and study the corresponding cycle. Hence we compute the bifucation structure of the automaton. To this end consider a given cycle C as a word on the alphabet $\{0,1\}$ and define the rotation number associated to C as :

$$\rho(C) = \frac{\text{number of 1 in } C}{\text{length of } C}.$$

For example, if $C = (100001000)$ which is written as $C = (10^4 10^3)$ then $\rho(C) = \frac{2}{9}$. Since to a given α there is a unique corresponding cycle, we use also the notation $\rho(\alpha)$ instead of $\rho(C)$. The following property gives a complete description of the bifurcation structure.

Property 8 *(Bifurcation structure)*
ρ is a monotone increasing surjection from R onto the set of irreducible fractions with denominators not greater than $k+1$.

Hence ρ is a discrete approximation of a devil staircase. The preceding result gives a way to obtain the bifurcation structure of (4). Consider the set of irreducible fractions with denominators not greater than $k+1$ and list them in increasing order. We obtain the so called Farey sequence of order $k+1$. The list of their denominators is the ordered list of the lengths of the cycles of (4). Using the numerators we can compute the corresponding values of α. The following table illustrates this construction for $k = 8$.

α	C	Length(C)	$\rho(\alpha)$
0	0	1	0
b^7	10^8	9	$\frac{1}{9}$
b^6	10^7	8	$\frac{1}{8}$
b^5	10^6	7	$\frac{1}{7}$
b^4	10^5	6	$\frac{1}{6}$
$b^3 + b^7$	$10^4 10^3$	9	$\frac{2}{9}$
$b^2 + b^6$	10^3	4	$\frac{1}{4}$
$b^2 + b^5$	$10^3 10^2$	7	$\frac{2}{7}$
$b + b^4 + b^7$	10^2	3	$\frac{1}{3}$
$b + b^4 + b^6$	$10^2 10^2 10$	8	$\frac{3}{8}$
$b + b^3 + b^6$	$10^2 10$	5	$\frac{2}{5}$
$b + b^3 + b^5$	$10^2 1010$	7	$\frac{3}{7}$
$b + b^3 + b^5 + b^7$	$10^2 101010$	9	$\frac{4}{9}$
$1 + b^2 + b^4 + b^6$	10	2	$\frac{1}{2}$

7. Unbounded memory

$A[k, a, \theta]$ can be extended to take into account all the previously generated states. Such an automaton is said to have an infinite memory:

$$x_{n+1} = 1[\sum_{i=0}^{n-1} a_i x_{n-i} - \theta] \qquad (7).$$

No general result is known on this family of automata. We shall present a special case corresponding to the extension of (4):

$$x_{n+1} = 1[\alpha - \sum_{i=0}^{n-1} b^i x_{n-i}] \qquad (8)$$

where $0 < b \leq \frac{1}{2}$ and $\alpha \geq 0$.

Let us define :

$$y_n = \alpha - \sum_{i=0}^{n-1} b^i x_{n-i}.$$

We can rewrite (8) as $x_{n+1} = 1[y_n]$ and deduce that

$$y_{n+1} = by_n + (1 - b)\alpha - 1[y_n] \qquad (9).$$

Relation (9) is a one-dimensional real recurrence relation which is equivalent to (8). This topic has been intensively studied during the last twenty years. Summing-up the known results we obtain:

Property 9 *(Unbounded memory)*

If $0 < b \leq \frac{1}{2}$ then:

- *$\forall\alpha$, (8) contains a unique global attractor which is either a cycle or a Cantor set,*
- *for any given p there exists α such that (8) has a cycle of length p,*
- *the set of α values such that (8) has a Cantor attractor is a Cantor set of Lebesgue measure 0.*

As in the previous section we can define the rotation number associated to (8):

$$\rho(\alpha) = \lim \frac{1}{q} \sum_{n=0}^{q-1} 1[y_n] = \lim \frac{1}{q} \sum_{n=0}^{q-1} x_{n+1}.$$

Property 10 *(Bifurcation structure)*

- *$\forall b$ and α, $\rho(\alpha)$ is independent of y_0:*
  - *if $\rho(\alpha)$ is a rational number $\frac{p}{q}$, then (8) has a globally attracting cycle of length q,*

- *if $\rho(\alpha)$ is an irrational number, then (8) has a globally attracting Cantor set.*

– *Let b be fixed; ρ is a continuous monotone increasing function from R onto $[0, 1]$ whose derivative is zero almost everywhere (devil's staircase).*

We can remark the close analogy between the two last sections. (4) can be viewed as a truncation of (8): if we consider an unbounded memory, we can obtain a cycle of infinite length (Cantor set), but if we restrict to a k-size memory, the length of the cycle is bounded by $k+1$.

8. General bounds

Let us come back to the general automaton (1). The following results have been obtained in collaboration with M. Tchuente, D. Moumida and G. Tindo (Cosnard *et al.*, 1998).

From the previous sections one can conjecture that:

1. If $a_i \geq 0$ then the length of the cycles of (1) are not greater than k.
2. For general a_i then the length of the cycles of (1) are not greater than $2k$.
3. Transients cannot be exponential.

In order to prove or disprove the previous conjectures we used a clever search among possible solutions by reducing, and then solving, the problem to a linear programming system.

8.1. POSITIVE COEFFICIENTS

It can be proved by this method that conjecture 1 is true for $k = 2, 3, 4, 5, 6$ and that it is false for $k \geq 7$. In this case the following automaton:

$$a = (0, 0, 0, 2, 1, 1, 2), \theta = 2.5$$

has a cycle of length 11. Remark that there are almost 10^{10} threshold functions.

8.2. GENERAL COEFFICIENTS

Using a similar method we can prove that if

$$a_i = -1 \forall i \neq 2, k-1; a_2 = 0, a_{k-1} = -2; \theta = -1$$

then $A[k, a, \theta]$ has the following cycle of length $2k+6$:

$$C = (0^k 1^2 0 1 0^{k-2} 1 0 1^2).$$

However we shall prove a better result by using the composition property.

Property 11 *(Composition of automata)*
If $A[k, a_0a_1...a_{k-1}, \theta]$ has 2 cycles C_1 and C_2 of respective lengths $L(C_1)$ and $L(C_2)$ then $A[k, a_00a_10...0a_{k-1}, \theta]$ has a cycle C of length $L(C) = lcm(L(C_1)L(C_2))$ where lcm denotes the least common multiple.

Using this composition property, one can prove that:

Property 12 *(Polynomial bound)*
There exists $A[k, a, \theta]$ with a cycle of length $O(k^3)$.

9. Conclusion

We have presented a survey of the known properties of a threshold automaton with memory. The dynamical behaviour has been proven to be rich and complicated. Some mathematical properties are still very intriguing and many problems still open. As an example, it is open whether there exists $A[k, a, \theta]$ with cycles or tansients of exponential length. This shows that one should be very careful in using networks of automata with memory.

References

Caianiello E.R. and de Luca A. Decision equation for binary systems: application to neuronal behavior. *Kybernetic* **Vol. no. 3**: 33-40, 1966.

Collet P. and Eckmann J.P. *Iterated Maps of the Interval as Dynamical Systems.* Birkhauser, Basel, 1980.

Cosnard M., Goles E. and Moumida D. Bifurcation structure of a discrete neuronal equation. *Discrete Applied Mathematics* **Vol. no. 21**: 21-34, 1988.

Cosnard M., Goles E., Moumida D. and de St. Pierre T. Dynamical behavior of a neural automaton with memory. *Complex Systems* **Vol. no. 2**: 161-176, 1988.

Cosnard M., Tchuente M. and Tindo G. Sequences Generated by Neuronal Automata with Memory. To appear, 1998.

Goles E. and Martinez S. *Dynamical Behaviour and Applications.* Kluwer Academic Publishers, 1990.

Kitagawa T. Dynamical Systems and Operators Associated with a Single Neuronic Equation. *Mathematical Biosciences* **Vol. no. 18**, 1973.

Nagami H., Kitahashi and Tanaka Characterization of Dynamical Behaviour Associated with a Single Neuronic Equation. *Mathematical Biosciences* **Vol. no. 32**, 1976.

Yamaguti M. and Hata M. On the Mathematical Neuron Model. *Lecture Notes in Biomathematics* **Vol. no. 45**: 171-177, 1982.

Yoshizawa S. Periodic Pulse Generated by an Analog Neuron Model. *Lecture Notes in Biomathematics* **Vol. no. 45**: 155-170, 1982.

LINEAR CELLULAR AUTOMATA AND DE BRUIJN AUTOMATA

K. SUTNER
Carnegie Mellon University
Pittsburgh, PA 15213

Abstract. Linear cellular automata have a canonical representation in terms of labeled de Bruijn graphs. We will show that these graphs, construed as semiautomata, provide a natural setting for the study of cellular automata. For example, we give a simple algorithm to determine reversibility and surjectivity of the global maps. We also comment on Wolfram's question about the growth rates of the minimal finite state machines associated with iterates of a cellular automaton.

1. Introduction

Historically, there are two main sources of interest in cellular automata. One the one hand, in symbolic dynamics they appear as morphisms of shift spaces, see Hedlund's (Hedlund, 1969) for a detailed discussion. On the other hand, von Neumann and Ulam used two-dimensional cellular automata to address the problem of self-reproduction, see (Burks, 1970). More recent work by Wolfram focuses on computational and complexity aspects (Wolfram, 1986) of the evolution of configurations on cellular automata, and is somewhat more in the spirit of the second approach. The arguments used in symbolic dynamics typically involve combinatorics, topology, measure theory and matrix theory, whereas computational studies often focus on automata and formal languages. Notably for one-dimensional cellular automata, automata theory and in particular the theory of finite state machines provides a very natural setting, and many of the basic results in Hedlund's paper can be expressed conveniently in terms of standard notions of automata theory. For example, Hedlund's result that all finite blocks have the same number of predecessors under a surjective map translates into the assertion that the de Bruijn automaton associated to the cellular automaton is balanced: all words have the same multiplicity.

More importantly, automata theory motivates many algorithms that are of use in the study of cellular automata. Below we will present a simple algorithm that tests whether a cellular automata is reversible, m-to-one or surjective. For reversible cellular automata, the algorithm also provides the necessary information for the actual construction of the inverse automaton. Lastly, there are well-known normal forms for finite state machines that can be used as complexity measures for sofic shifts, and in particular the ranges of the global maps of cellular automata. We will show that there are cellular automata of any width whose associated minimal automata have maximum size. However, for iterates ρ^t there is an exponential gap to the maximum size.

We begin with a brief review of some of the basic concepts in symbolic dynamics, with a view towards cellular automata.

1.1. SHIFT SPACES AND THEIR MORPHISMS

Let Σ be a finite alphabet, and denote by ${}^\omega\Sigma^\omega$ the collection of all biinfinite words over Σ, usually referred to as *configurations* in the context of symbolic dynamics. We can associate every configuration X with its *cover*, $\mathrm{cov}(X) \subseteq \Sigma^*$, the set of all finite factors of X. For a set $\mathfrak{X} \subseteq {}^\omega\Sigma^\omega$ of configurations define its cover $\mathrm{cov}(\mathfrak{X})$ to be the union of all the covers $\mathrm{cov}(X)$ where $X \in \mathfrak{X}$. A *shift space* or *subshift* is a subset $\mathfrak{X}$ of ${}^\omega\Sigma^\omega$ that is topologically closed in the sense of the natural product topology on ${}^\omega\Sigma^\omega$ and shift-invariant: $\sigma(\mathfrak{X}) = \mathfrak{X}$ where $\sigma(X)_i = X_{i+1}$. Since ${}^\omega\Sigma^\omega$ is compact, a shift space can be reconstructed from its cover: a configuration X is in $\mathfrak{X}$ iff it is the limit of a sequence of words in the cover.

A shift space is *sofic* or a *sofic system* if its cover is a regular language. Any shift space $\mathfrak{X}$ can be described in terms of *forbidden blocks*, i.e., a collection of words $W \subseteq \Sigma^*$ such that $\mathrm{cov}(\mathfrak{X}) = \Sigma^* - \Sigma^* W \Sigma^*$. If W can be chosen as a finite set, the space is a *subshift of finite type*. The appropriate morphisms between shift spaces are continuous maps $f : \mathfrak{X} \to \mathfrak{Y}$ that commute with the shift. Weiss has shown that sofic systems are the smallest class of subshifts that contains all shifts of finite type and is closed under morphisms, see (Weiss, 1973). The morphism can always be chosen to be finite-to-one according to (Coven and Paul, 1975).

By the Lyndon-Curtis-Hedlund theorem (Hedlund, 1969), in the special case $\mathfrak{X} = \mathfrak{Y} = {}^\omega\Sigma^\omega$, these maps are precisely the *global maps* of one-dimensional *cellular automata*. For our purposes, a cellular automaton is simply a *local map* $\rho : \Sigma^w \to \Sigma$ that extends naturally to configurations via $\rho(X)(i) = \rho(X(i - w + 1), \ldots, X(i))$. The extended function ρ is the *global map* of the cellular automaton. In coding theory (Lind and Marcus, 1995) global maps are referred to as sliding block codes and it is customary to divide w into memory and anticipation, but for our purposes the last definition suffices.

The language of a semiautomaton $\mathcal{A} = \langle Q, \Sigma, \tau \rangle$ will be denoted by $\mathcal{L}(\mathcal{A})$, and we write $\mathcal{L}(\rho)$ for the cover of $\rho({}^\omega\Sigma^\omega)$. As usual, when τ is clear from context we will think of the powerset of Q as a semimodule over Σ^* and write $P \cdot x$ rather than $\tau(P, x)$. The growth function $\gamma(L, n)$ denotes the number of words in $L \subseteq \Sigma^*$ of length n.

1.2. DE BRUIJN AUTOMATA AND CELLULAR AUTOMATA

A one-dimensional cellular automaton ρ of width $w \geq 2$ is naturally associated with a semiautomaton $B(\rho) = \langle \Sigma^{w-1}, \Sigma, \tau \rangle$ where

$$\tau = \{ (ax, \rho(axb), xb) \mid a, b \in \Sigma, x \in \Sigma^{w-2} \}.$$

Hence, the underlying transition diagram of $B(\rho)$ is a de Bruijn graph. From now on we will assume that $k = |\Sigma| \geq 2$ is fixed, so that the de Bruijn automaton has size k^{w-1}. It is clear that $B(\rho)$ accepts the cover of $\rho({}^\omega\Sigma^\omega)$.

One may construe $B(\rho)$ as an $\omega\omega$-transducer (Culik *et al.*, 1989), but we will confine ourselves to the interpretation as an acceptor and emphasize computations in these machines. By a *computation* of a semiautomaton $\mathcal{A} = \langle Q, \Sigma, \tau \rangle$ we mean an alternating sequence $\pi = p_0, x_1, p_1, \ldots, x_r, p_r$ where $p_i \in Q$, $x_i \in \Sigma$ and $\tau(p_{i-1}, x_i, p_i)$ for all $i = 1, \ldots, r$. State p_0 is the *source state* of the computation and p_r the *target state.* We write $\pi(i)$ for p_i. The word $x = x_0 \ldots x_r$ is the *label* of the computation. We also say that π *carries* x. $r = |x|$ is the length of the computation, but we will use similar notation for biinfinite computations. Thus, $\rho({}^\omega\Sigma^\omega)$ is the set of all labels of biinfinite computations in $B(\rho)$.

Note that for biinfinite computations we may safely assume that the underlying graph of the semiautomaton is *non-transient* (i.e., every node has indegree and outdegree at least 1). In particular a semiautomaton is *transitive* if it contains only one strongly connected component. A *diamond* in a graph is a pair of paths (ignoring labels) of the same length, with the same source, and the same target. The following proposition is easy to establish.

Lemma 1 *Diamond Lemma*

Let $G = \langle V, E \rangle$ be a finite non-transient graph. Then G contains a diamond iff $|V| < |E|$.

We will often consider sets of computations, finite or biinfinite, with the same label. In particular $\Pi(x)$ denotes the set of all computations on

x. Two computations π and π' are *separated* if $\pi(t) \neq \pi'(t)$ for all t, and Π is separated if all its elements are. Lastly, a semiautomaton $\mathcal{A}$ is separated if any two biinfinite computations in $\mathcal{A}$ that carry the same label are separated. Two computations that fail to be separated pass through the same *bottleneck* $p \in Q$: for some t, $\pi(t) = p = \pi'(t)$. Since the fusion of the trace (p_i) of π is a predecessor of x, separated computations produce in particular distinct predecessors.

2. Surjective Cellular Automata

In symbolic dynamics, epimorphisms and isomorphisms, or conjugacies, of shift spaces are of particular interest. In fact, Hedlund's paper (Hedlund, 1969) essentially undertakes a detailed study of these maps. We will rephrase the basic results with a view towards exploiting them in the classification algorithm in section 2.4.

2.1. CHARACTERIZATIONS OF SURJECTIVITY

There are several characterization of surjectivity of a global map that can be expressed naturally in terms of the associated automata. First, consider the *multiplicity* of finite words with respect to $B(\rho)$. Let

$$\mu(x) := \text{number of computations in } B(\rho) \text{ with label } x.$$

A semiautomaton is *balanced* if all words in its language have the same multiplicity. Note that in a de Bruijn automaton the number of computations of length $r \geq w - 1$ with fixed source and target is k^{r-w+1}. It follows from path-counting, that balance in a de Bruijn automaton is equivalent to the condition $\mu(x) = k^{w-1}$ for all $x \in \Sigma^*$.

Second, there is the notion of *local injectivity*, an idea that goes back to the Garden-of-Eden theorem by Moore and Myhill, (Moore, 1962; Myhill, 1963). Let us say that two finite computations are *coterminal* if they have the same source and target state. A semiautomaton is *unambiguous* if it has no coterminal computations with the same label.

Theorem 1

The following conditions are equivalent.

1. *The cellular automaton ρ is surjective.*
2. *The de Bruijn automaton $B(\rho)$ is balanced.*
3. *The de Bruijn automaton $B(\rho)$ is unambiguous.*

Proof Balance obviously implies surjectivity, so suppose ρ is surjective. Let u be a word of minimal multiplicity $m > 0$. It follows that every extension ux also has multiplicity m. In particular, all the words uxu where

$x \in \Sigma^{w-1}$ have multiplicity m. By path-counting, the computations with label uxu, $x \in \Sigma^{w-1}$, are already uniquely determined by their initial and final segment of length $|u|$. Since u has multiplicity m, there must be m^2 such computations. But each fixed uxu has multiplicity m, so we must have $m = k^{w-1}$. It follows that any word has multiplicity at least k^{w-1}. Again by path-counting it follows that in fact $\mu(z)$ is equal to this bound for all words z.

Now suppose that ρ is surjective but $B(\rho)$ is ambiguous. Since the automaton is transitive, we can choose a state p such that there are two computations with the same label x and source and target p. But then the multiplicity of x^r is at least 2^r, contradicting the previous result.

On the other hand, if $B(\rho)$ is unambiguous, pick some state p and consider the language obtained from $B(\rho)$ by fixing p as the sole initial and final state. There are k^{n-w+1} computations from p to p in $B(\rho)$ of length $n \geq w-1$, and since the automaton is unambiguous they must all have distinct labels. Hence $\gamma(L,n) = k^{n-w+1} = c{\cdot}k^n$ for some constant $0 < c < 1$. Now suppose for the sake of a contradiction that ρ is not surjective and pick a word x not accepted by $B(\rho)$. A counting argument shows that the limit of $\gamma(\Sigma^* x \Sigma^*, n)/k^n$ is 1 as n tends to infinity. But $\Sigma^* x \Sigma^*$ is contained in the complement of $\mathcal{L}(\rho)$, and we have a contradiction. □

Needless to say, these arguments generalize to other graphs with appropriate regularity properties, see (Nasu, 1981). Also note that the last argument actually shows that ρ is surjective iff the cardinality of the set of preimages of any configuration X is uniformly bounded by some finite number. In fact, we must have $|\rho^{-1}(X)| \leq k^{w-1}$ for any surjective cellular automaton and any configuration X. On the other hand, if ρ fails to be surjective there is a configuration with uncountably many predecessors.

The multiplicity map can also be construed as a monoid homomorphism

$$\mu : \Sigma^* \to Q \times Q \to \mathbf{N}$$

that maps finite words into Q by Q matrices of nonnegative integers: $\mu(x)(p,q)$ is the number of computations carrying x with source p and target q. Our definition of multiplicity is then simply the 1-norm of these matrices. However, dealing directly with the matrices affords an alternative way to develop many of the basic results about shift endomorphisms, see (Gleason, 1992) for this approach. Most importantly, it follows from the Frobenius-Perron theorem that the matrix semigroup generated by $\mu(a)$, $a \in \Sigma$, is finite iff it fails to contain the null matrix. The matrices of minimal rank form an ideal in the semigroup, that provides information about the number of predecessors of configurations, see section 2.3.

2.2. WELCH AUTOMATA

Suppose that ρ is surjective. As we have seen, the number of computations with a given finite word x as label is simply k^{w-1}. The number of biinfinite computations with a given label $X \in {}^\omega\Sigma^\omega$ is thus bounded by k^{w-1} but it is more difficult to determine its value in general. Let

$$w_\rho = \min(\,|\rho^{-1}(X)| \mid X \in {}^\omega\Sigma^\omega)$$

be the least number of predecessors of any infinite configuration. For certain configurations that are of special interest in symbolic dynamics, the number of predecessors is exactly w_ρ. For example, bilaterally transitive configurations have this property. Here, Y is bilaterally transitive if every finite word occurs as a factor of Y infinitely often in both directions. Since ${}^\omega\Sigma^\omega$ is compact, for any configuration X we can choose a strictly ascending (or descending) sequence $(t_i)_{i \geq 0}$ such that $\lim \sigma^{t_i}(Y) = X$ where σ denotes the shift.

Suppose Π is a set of biinfinite computations with the same label. For any integer t let $\Pi(t) = \{\,\pi(t) \mid \pi \in \Pi\,\}$. It follows from lemma 1 that whenever $s < t$ and $\Pi(s) = \Pi(t)$, then for all i, $s \leq i \leq t$, we have $|\Pi(i)| = |\Pi(s)|$. For otherwise consider the graph $G = \langle P(s), E\rangle$ where $E = \{\,(\pi(s), \pi(t)) \mid \pi \in \Pi\,\}$. If $|\Pi(i) \neq |\Pi(s)|$ for some i then G has a diamond by the lemma. Let $x \in \Sigma^*$ be the label of the computation from s to t. Then ${}^\omega x^\omega$ has infinite multiplicity, contradicting surjectivity.

We say that a set of states P is *persistent* with respect to Π if there exists a strictly ascending sequence $(t_i)_{i \geq 0}$ of integers such that for all i: $P = \Pi(t_i)$. Likewise, P is *copersistent* if there is a strictly descending sequence $(s_i)_{i \geq 0}$ such that for all i: $P = \Pi(s_i)$. Lastly, P is *bipersistent* if P is both persistent and copersistent. For cardinality reasons, there always is a persistent as well as a copersistent set of states. This fact was used in (Sutner, 1991) to answer a question by Golze (Golze, 1976).

Theorem 2

Every recursive configuration that has a predecessor already must have a recursive predecessor.

Of course, the last result fails in dimensions 2 and higher. In the presence of a bipersistent state set all biinfinite computations must be separated. In conjunction with a limit argument, persistence can be used to show that transitive configurations have only separated predecessors.

Theorem 3

Let $Y \in {}^\omega\Sigma^\omega$ be bilaterally transitive and ρ surjective. Then the computations with label Y in $B(\rho)$ are separated.

Consider an arbitrary configuration X and a, say, strictly ascending sequence $(t_i)_{i \geq 0}$ such that $\lim \sigma^{t_i}(Y) = X$. By selecting appropriate subsequences, the limit will transport the $|\rho^{-1}(Y)|$-many separated computations with label Y into $|\rho^{-1}(Y)|$-many separated computations with label X. Hence we actually have $w_\rho = |\rho^{-1}(Y)|$ for any bilaterally transitive Y.

Now suppose we truncate the computations in $\Pi(Y)$ to a finite segment with label y, yielding computations $\pi_1, \ldots, \pi_r$. $\Pi(y)$ may contain additional finite computations since $|\Pi(y)| = k^{w-1}$. However, by compactness, if y is of sufficient length, all the computations in $\Pi(y)$ will share a bottleneck with one of the π_i. The question arises how many computations attach themselves to a given π_i.

To answer this question, recall that the *kernel automaton* of a semiautomaton $\langle Q, \Sigma, \tau \rangle$ is the part of the full power automaton that is generated by the singletons $\{p\}$, $p \in Q$. Now define the *forward Welch automaton* W_ρ^+ of ρ to be the subautomaton of the kernel automaton of $B(\rho)$ induced by the states of maximal cardinality. To see that W_ρ^+ is well-defined, note that for any $P \subseteq Q$ of maximal cardinality we must have $|P \cdot a| = |P|$. For consider P in the kernel automaton. There are $k|P|$ edges leaving P in $B(\rho)$, and, since the de Bruijn automaton is unambiguous, for any two states p and q in P we must have that $p \cdot a$ and $q \cdot a$ are disjoint: $|P \cdot a| = \sum_{p \in P} |p \cdot a|$.

The *forward Welch coefficient* w_ρ^+ is the cardinality of the state sets in W_ρ^+. The backward automaton W_ρ^- and coefficient w_ρ^- are defined analogously where $B(\rho)$ is replaced by the corresponding reverse automaton. The following properties of the Welch automata are immediate from the definitions.

Lemma 2

Let ρ be surjective. The Welch automata for ρ are deterministic, complete and transitive and their state sets cover Σ^{w-1}. For all $P' \in W_\rho^-$, $P \in W_\rho^+$, $|P' \cap P| \leq 1$.

From the definitions of the Welch automata, we can choose a word x such that there is a set Π_1 of size $w_\rho^- w_\rho^+$ of computations with label x that all pass through some common bottleneck, and this number is maximal. Any extension uxv will preserve these computations. If there is a computation with label x not in Π_1, we can thus construct an extension uxv of x such that there is another collection of computations Π_2 with label uxv of size $w_\rho^- w_\rho^+$, passing through a common bottleneck, and all the computations in Π_1 are separated from all the computations in Π_2. Proceeding by induction, we obtain a word z and a partition $\Pi(z) = \Pi_1 \cup \ldots \cup \Pi_m$ where all Π_i have size $w_\rho^- w_\rho^+$, and the collections are pairwise separated. It follows that $k^{w-1} = \mu(z) = w_\rho^- \cdot m \cdot w_\rho^+$. Any configuration $Z \in {}^\omega\Sigma^\omega$ containing z as a factor must have m separated computations. If we choose a bilaterally transitive extension, we see that $m = w_\rho$. Hence we have the following lemma.

Lemma 3
For any surjective cellular automaton ρ: $w_\rho^- \cdot w_\rho \cdot w_\rho^+ = k^{w-1}$.

In terms of the Gleason-Rothaus multiplicity semigroup, w_ρ is the minimal rank of these matrices, see (Gleason, 1992). By counting computations it is not hard to show that the Welch coefficients are morphisms from the monoid of epimorphisms of ${}^\omega\Sigma^\omega$ (Hedlund, 1969) to the monoid of natural numbers under multiplication. Thus, they are helpful in answering questions about the decomposition of epimorphisms.

Clearly, $w_\rho^+ = 1$ iff $B(\rho)$ is deterministic, and $w_\rho^- = 1$ iff $B(\rho)$ is codeterministic. Hence, for a permutation automaton $B(\rho)$, we have $w_\rho = k^{w-1}$ and all configurations have precisely this number of predecessors, and all the computations are separated. In general, computations fail to be separated, but there is another special case where separation occurs: when ρ is a m-to-one map. This situation is discussed in the next section.

Example: Consider the binary cellular automaton ρ of width 4 with rule number 7230. The local rule is given by the binary expansion of the rule number, padded to 16 digits:

$$\begin{array}{llll} 0000 \to 0 & 0100 \to 1 & 1000 \to 0 & 1100 \to 1 \\ 0001 \to 1 & 0101 \to 1 & 1001 \to 0 & 1101 \to 0 \\ 0010 \to 1 & 0110 \to 0 & 1010 \to 1 & 1110 \to 0 \\ 0011 \to 1 & 0111 \to 0 & 1011 \to 1 & 1111 \to 0 \end{array}$$

The global map is surjective, and the Welch coefficients are $(2,1,4)$. The kernel automata have size 15 and 20, and the Welch automata themselves have size 6 and 4, respectively. The transfer sequences in the kernel automata show that any configuration containing the factor 1011 has a unique predecessor. On the other hand, ${}^\omega 01^\omega$ is an example of a configuration with two predecessors. Thus, the map is not m-to-one.

2.3. *M*-TO-ONE GLOBAL MAPS

Cellular automata whose global maps are m-to-one can be characterized in various ways. Topologically, these are precisely the global maps that are open in the sense of the usual product topology on ${}^\omega\Sigma^\omega$, see (Hedlund, 1969) for a detailed discussion. We will here focus on a characterization that relates directly to computations in de Bruijn automata.

A semiautomaton is *d-deterministic* for some integer $d \geq 0$ if for any two computations π and π' of length $d+1$ with the same label and the same source we have $\pi(1) = \pi'(1)$. Thus, a 0-deterministic automaton is simply a deterministic automaton. The definition for d-codeterministic automata is analogous.

Theorem 4

Let ρ be a cellular automaton and $B(\rho)$ its de Bruijn automaton. Then the following are equivalent:

1. *The global map of ρ is m-to-one.*
2. *The biinfinite computations in $B(\rho)$ are separated.*
3. *$B(\rho)$ is d^+-deterministic and d^--codeterministic for some $d^+, d^- \geq 0$.*
4. *The Welch automata are the only non-trivial strongly connected components in the kernel automata of $B(\rho)$ and its reverse.*

Proof Suppose ρ is m-to-one. Clearly, ρ is surjective in this case, so consider a configuration Y and its predecessors $X_1, \ldots, X_m$. If Y is bilaterally transitive, then all the corresponding computations π_i must be separated by theorem 3. But since Y is bilaterally transitive we can use a limit argument to show that for any configuration Y' there are already m separate computations with label Y'. Thus, all computations are separated.

Now assume that computations are separated in $B(\rho)$ but that $B(\rho)$ fails to be d-deterministic for all $d \geq 0$. Then we can produce pairs of finite computations of arbitrary length with the same source and the same label that disagree everywhere except for the source. By a limit argument, there is as pair of biinfinite computations that pass through a common state, but fork thereafter, and we have the desired contradiction. The argument for d-codeterminism is entirely similar.

Next, suppose $B(\rho)$ is d^+-deterministic. Pick a computation of length at least d^+ and label x, say, from state p to state q. Let q_0 be any state in $B(\rho)$ and choose a word y such that $q_0 \cdot y$ lies in the forward Welch automaton and contains p. Then $q_0 \cdot yx$ also lies in the Welch automaton. Moreover, all the computations associated with yx must pass through p since $B(\rho)$ is d^+-deterministic. Thus, there cannot be any non-trivial strongly connected components in the kernel automaton of $B(\rho)$ other than the Welch automaton itself. The argument for the backward Welch automaton is entirely similar.

Lastly, suppose the Welch automata are the only non-trivial components in the kernel automata. Let Π be the set of all computations on label $x \in \Sigma^*$, $|x| = 2n$, that pass through bottleneck p at time n. By our assumption, for sufficiently large n, we must have $\Pi(0) \in W_\rho^-$ and $\Pi(2n) \in W_\rho^-$. It is easy to see that for any other such collection Π', any computation in Π is separate from all computations in Π'. Hence there are precisely $m = k^{w-1}/w_\rho^- \cdot w_\rho^+$ such collections, giving rise to precisely m predecessors for any configuration. $\square$

Example: The binary cellular automaton ρ of width 4 with rule number 23130 has Welch coefficients $(2, 4, 1)$. Thus, the de Bruijn automaton is deterministic. The global map is 4-to-one, as can be seen from the observation $\rho(v, x, y, z) = x + z \pmod 2$: ρ is in essence the elementary rule 90, and the latter is a permutation rule.

2.4. THE CLASSIFICATION AND INVERSION ALGORITHM

We now turn to question of deciding whether the global map of a cellular automaton is an epimorphism or isomorphism. We present a uniform algorithm that additionally also tests whether the map is m-to-one, see (Amoroso and Patt, 1972) and (Head, 1989) for alternative methods.

Of course, surjectivity is by compactness equivalent to the cover being all of Σ^*, but it is PSPACE-complete to test this property for regular languages in general, even when they are given by transitive semiautomata, see (Yu, 1997). Moreover, we will see in section 3 that specifically for de Bruijn automata exponential blow-up during deterministic simulation occurs with some frequency. However, we can test ambiguity of $B(\rho)$ as follows. Let $B^2(\rho)$ be the product automaton of $B(\rho)$ with itself and denote by D the diagonal in $B^2(\rho)$, i.e., the points of the form (x, x). Since D is strongly connected, it must be contained in one strongly connected component C. But then $B(\rho)$ is unambiguous iff $D = C$. We can construct the product automaton in quadratic time, and compute the components in time linear in the size of $B^2(\rho)$ using, say, Tarjan's algorithm, see (Mehlhorn, 1984). Clearly, the global map is injective iff in addition there are no other non-trivial strongly connected components, so this property can be tested at the same time.

In fact, we can also check if the global map is m-to-one: this is equivalent to having no paths between $D = C$ and any other non-trivial strongly connected component. The latter property can be tested by computing the collapse G of the graph of $B^2(\rho)$ and testing reachability in G, all in linear time.

Summarizing, we have

- ρ is surjective iff $C = D$,
- ρ is m-to-one iff $C = D$, and C is isolated in the collapse of $B^2(\rho)$,
- ρ is injective iff the only non-trivial strongly connected component in $B^2(\rho)$ is $C = D$.

Hence we have the following result.

Theorem 5

There is a quadratic time algorithm to test whether the global map of a cellular automaton is injective, m-to-one or injective.

As a practical matter, since the computation of the product graph dominates the running time of the algorithm, one can eliminate obviously non-surjective rules by first testing whether $B(\rho)$ is, say, 2-balanced (i.e., whether all words of length 2 have the proper multiplicity). This precomputation takes only $O(k^{w+1})$ steps. Secondly, we can use the reduced product automaton whose states are subsets of Q of cardinality 1 and 2, rather than the ordinary Cartesian product. The size of this automaton is $n(n+1)/2$ rather than n^2.

A continuous map from any compact Hausdorff space to another is automatically closed, and it follows that injective global maps are homeomorphisms. In particular, their inverses are again given by a cellular automaton. Hence, for ρ injective the question arises how to compute the inverse cellular automaton τ. A slight modification of the classification algorithm also produces the inverse. Recall that D is the only non-trivial strongly connected component in the product automaton. Hence the points in the weak component (ignoring the direction of the edges) of D other than D itself must form an acyclic graph. Indeed, there are two acyclic graphs: one that admits paths to the diagonal, and one that admits paths from the diagonal. Let d^- and d^+ be the corresponding maximum path lengths of these graphs (we include the points where the graphs are anchored to the diagonal). Thus, $B(\rho)$ is d^+-deterministic and d^--codeterministic.

It follows that any pair of computations with label $x \in \Sigma^*$ of length at least $d^- + d^+$ must have a common bottleneck. Hence, we can effectively determine an inverse cellular automaton τ for ρ of width $w' = d^- + d^+$ by setting $\tau(u) = p_1$ where $p \in \Sigma^{w-1}$ is the bottleneck of a computation carrying $u \in \Sigma^{w'}$ and p_1 the first character in p.

Plainly, $\tau(u)$ is independent of the last $n-2$ arguments, so we can get a smaller local map of width $w' = d^- + d^+ - n + 2$. Also note that this method only produces an inverse map up to a shift. A slightly different method of inversion, based on the fact that injectivity is characterized by both Welch automata being definite is given by Nasu (Nasu, 1981).

Example: The binary cellular automaton ρ of width 4 and rule number 13155 has Welch coefficients $(4, 1, 2)$. The automaton is reversible, and the inverse automaton is also of width 4 and has rule number 14643.

3. Irreducible Sofic Shifts and Fischer Automata

Since shift spaces of the form $\mathcal{X} = \rho({}^{\omega}\Sigma^{\omega})$ are sofic, one natural measure for their complexity is the size of the minimal automaton that recognizes the cover $\mathcal{L}(\rho) = \text{cov}(\mathcal{X})$, a regular language. In the mid-eighties Wolfram and his coworkers performed extensive calculations for elementary cellular automata and their iterates to determine these sizes, see (Wolfram, 1986).

In particular in (Wolfram, 1985), see items 12 and 13, Wolfram raises the question whether the sizes of the minimal automata for $\mathcal{L}(\rho^t)$ typically grow as the iteration number t increases. The limit $\mathcal{L}_\infty(\rho) = \bigcap \mathcal{L}(\rho^t)$ is trivially co-recursively enumerable, but may fail to be regular. Indeed, Hurd (Hurd, 1987) has shown that the limit language can be context-free and even non-recursive. Even if the limit language is regular, it may well happen that $\mu(\rho^t)$ approaches infinity. Intriguingly, Goles et. al. (Goles *et al.*, 1993) have shown that ρ may be computationally universal, but still have regular limit language. Thus, one should expect certain difficulties in answering Wolfram's question.

The (accessible part of the) power automaton of a semiautomaton of size n can be constructed in time $O(n \cdot s)$ where s is actual size of the power automaton, and minimization is then $O(s \log s)$, see (Hopcroft, 1971). Thus, the principal difficulty in obtaining computational data is the fact that for a cellular automaton of width w the size of the de Bruijn automaton is k^{w-1}, so that the only obvious upper bound on the size of the power and minimal automaton is $2^{k^{w-1}}$. We will see that there are binary cellular automata that realize this bound. Note that in particular the study of iterates becomes quickly infeasible, since the width of ρ^t is $t(w-1)+1$.

3.1. THE MINIMAL FISCHER AUTOMATON

Minimal automata are actually only one possible choice of normal form. The cover of a sofic shift of the form $\rho({}^\omega\Sigma^\omega)$ is a factorial, extensible and transitive regular language: $uv \in L \Longrightarrow u, v \in L$, $u \in L \Longrightarrow \exists a, b \in \Sigma(aub \in L)$, and $u, v \in L \Longrightarrow \exists x(uxv \in L)$. For languages of this type, there is an alternative notion of minimal automaton, first introduced by Fischer (Fischer, 1975) and discovered independently by Beauquier (Beauquier, 1989) in the guise of the 0-minimal ideal in the syntactic semigroup of $\mathcal{L}(\rho)$. The minimal Fischer automaton is unique up to isomorphism, and is the smallest deterministic semiautomaton accepting a given factorial, extensible, transitive regular language. Behavioral equivalence provides a natural epimorphism from any deterministic semiautomaton to the minimal Fischer automaton. The following argument also establishes a connection between ordinary minimal automata and the minimal Fischer automaton, see (Sutner, 1997a) for details.

Theorem 6

The minimal Fischer automaton for an irreducible sofic shift is a subautomaton of the partial minimal automaton induced by a strongly connected component of out-degree 0. *The component is uniquely determined by this condition.*

Proof Let F be the minimal Fischer automaton accepting a language L, M' the power automaton of F, and M the minimal automaton, obtained by factoring M' with respect to behavioral equivalence. We assume that the sink is removed from M. Since F is reduced it has a synchronizing word, and therefore it embeds into M. But, again since F is reduced, it also embeds into M. Call the image F'. By transitivity, F' forms one strongly connected component of M, and has no out-edges.

Suppose C is another such component. Since the semiautomaton C is also reduced, we can synchronize C by u, and F' by v. L is transitive, so it contains a word uxv, which synchronizes both C and F', say, to states p and q respectively. Since M is reduced, there is a word in the behavior of p that is not in the behavior of q, or conversely. Thus, the behavior of C is different from L. But C has out-degree 0, and therefore must have behavior L, a contradiction. □

The decomposition into strongly connected components is quite familiar in symbolic dynamics in calculations of topological entropy. It is easy to see that for any factorial language L the sequence $(\log \gamma(L,n))_{n\geq 0}$ is subadditive, and therefore the limit $\lim_{n\to\infty} 1/n \log \gamma(L,n)$ exists. If L is in particular the cover of a shift space $\mathfrak{X}$, this limit is the *entropy* of the space. So suppose we have some non-transient semiautomaton $\mathcal{A}$ describing a sofic shift. Associate each strongly connected component of $\mathcal{A}$ with the adjacency matrix of the corresponding subgraph, construed as a nonnegative integer matrix. By transitivity, these matrices are irreducible. One can show that the entropy of the shift is the logarithm of the maximum of the Perron eigenvalues of these matrices. Moreover, in the case where $\mathcal{A}$ is (the non-transient part of) the minimal DFA the subshifts defined by the components other than the minimal Fischer automaton all have strictly smaller entropy than the whole shift, see (Lind and Marcus, 1995). Incidentally, the main result in Fischer's original paper (Fischer, 1975) also relates to entropy: every transitive sofic system is intrinsically ergodic.

As mentioned previously, all sofic shifts are factors of shifts of finite type. One can use the minimal Fischer automaton to test whether $\rho({}^\omega\Sigma^\omega)$ is of finite type.

Lemma 4

The minimal Fischer automaton of ρ is definite iff $\rho({}^\omega\Sigma^\omega)$ is of finite type.

Proof Let $\mathcal{A}$ be the minimal Fischer automaton for $\mathcal{L}(\rho)$. Suppose that for $\mathcal{A}$ is definite, so that for all $x \in \mathcal{L}(\rho)$ of length at least n all computations in $\Pi(x)$ terminate in the same state. Consider the de Bruijn automaton $\mathcal{B} = \langle \Sigma^n, \Sigma, \delta \rangle$ where $(az, b, zb) \in \delta$ iff $azb \in \mathcal{L}(\rho)$. It is not hard to see that $\mathcal{L}(\mathcal{B}) = \mathcal{L}(\rho)$, so that $\Sigma^n - \mathcal{L}(\rho)$ is the finite set of forbidden factors in $\mathcal{L}(\rho)$.

For the opposite direction we may assume that all forbidden blocks have length $n+1$ and construct the same de Bruijn automaton $\mathcal{B}$ as before. The automaton is clearly deterministic and definite, and since the minimal Fischer automaton is just a factor automaton of $\mathcal{B}$ it follows that it too must be definite. □

Example: Consider the elementary cellular automaton number 160, whose local map is given by $\rho(x, y, z) = x \wedge z$. The minimal Fischer automaton of ρ^t has size $(t+1)^2$. The complement of the cover of $\rho^t({}^\omega\Sigma^\omega)$ consists of all strings that are of the form $x = x_1 x_2 \ldots x_{2s+1}$ where $x_1 = 1 = x_{2s+1}$ but $x_3 = 0$, for some $s \leq t+1$. The limit language fails to be transitive, but is regular and has a partial minimal DFA of size 9.

How large can the minimal Fischer automaton for ρ be? The only obvious upper bound on the size of the power and minimal automaton associated with $B(\rho)$ is $2^{k^{w-1}}$. Here is one way to construct cellular automata that realize this bound, and also have Fischer automata of maximal size. Consider a transitive permutation automaton $\mathcal{A} = \langle Q, \Sigma, \tau \rangle$ with n states and an alphabet of size at least 2. let $\mathcal{A}'$ be an automaton obtained by changing the label of one transition, a so-called 1-*permutation* automaton. It is shown in (Sutner, 1997a) that in the special case when the switched label belongs to a self-loop the power automaton of $\mathcal{A}'$ has size 2^n, and that the minimal Fischer automaton is obtained by deleting the sink from the power automaton. Since de Bruijn automata have loops and admit permutation labelings, we have the following result.

Theorem 7

There are cellular automata of arbitrary width w such that the minimal Fischer automaton of $\mathcal{L}(\rho)$ has size $2^{k^{w-1}} - 1$.

Switching the labels of transitions other than the self-loops in de Bruijn permutation automata also frequently produces exponential blow-up. The following table shows the sizes of all binary 1-permutation automata of width $w = 5$. Δ indicates the difference of the size of the power automaton of $B(\rho)$ to the maximum 2^{16}.

Δ	0	1	2	4	8	16	32	64
freq.	4096	512	896	480	240	208	328	352
Δ	124	128	170	256	512	1024	1052	2048
freq.	8	296	8	224	160	120	8	256

There are $2^{2^{w-2}}$ binary permutation automata of width w. According to the table, there are exactly $2^{w-1} 2^{2^{w-2}}$ 1-permutation automata of width

5 that exhibit full exponential blow-up, accounting for exactly half of all possible combinations of permutation labelings and edges. We currently have no proof that this holds in general.

For binary cellular automata full blow-up already implies that the automaton is a 1-permutation automaton. On the other hand, it is easy to see that the Hamming distance to the nearest permutation automaton in $B(\rho^t)$ grows exponentially with t. Hence we have the following result.

Lemma 5

None of the iterates ρ^t, $t > 1$, of a binary cellular automaton has a minimal Fischer automaton of maximal size.

In general, there appears to be a connection between the Hamming distance of a given de Bruijn automaton $B(\rho)$ to the nearest permutation automaton and the size of the minimal Fischer automata, but the precise nature of this connection is not clear at this time. Note that since the expected value of the Hamming distance of a binary cellular automaton of width w to the nearest permutation automaton is $5/16 \cdot 2^w$, one consequence of this correlation would be that with increasing width random cellular automata would have relatively smaller minimal Fischer automata.

3.2. HARDNESS CONSIDERATIONS

With a view towards Wolfram's questions it would be interesting to be able to obtain more computational data, see the table in (Wolfram, 1986). Calculations for minimal Fischer automata even for binary cellular automata become problematic for width 5 and prohibitive for higher widths. It is therefore natural to attempt to avoid the actual construction of the power automaton of $B(\rho)$. For example, one might try to compute the size of the power automaton without actually building it.

It is well-known that it is PSPACE-complete to determine whether a regular language is equal to Σ^* if the language is represented by a nondeterministic machine, see (Yu, 1997) for an overview of this and related results. A straightforward modification of the argument shows that the problem remains hard for semiautomata, at least for alphabets of size at least 3. Thus there is little hope to determine the size of the minimal automaton or the minimal Fischer automaton.

Since minimization can be accomplished in time polynomial in the size of the power automaton, it would be of interest to be able to predict the size of the power automaton of a semiautomaton $\mathcal{A}$ in time polynomial in the size of $\mathcal{A}$. More precisely, we would like to compute the function $\pi(\mathcal{A})$ defined to be the size of the accessible part of the full power automaton. It is clear that is a special case of the problem of determining the size of a weakly connected component in a digraph. The machinery of the Immerman/Szelepsényi theorem shows that π can be computed in nondeterministic linear space, see (Papadimitriou, 1994) for necessary background. Unfortunately, this bound is tight as expressed in the following result.

Theorem 8

The problem of testing whether the power automaton of a given semiautomaton has size at most B is PSPACE-complete.

An analogous result holds for kernel automata. The problem in both cases lies in the fact that the nondeterministic machine provides a succinct representation of the corresponding deterministic machine, and the latter may be of size exponential in the size of the original machine. The proof given in (Sutner, 1997b) does not directly apply to de Bruijn automata, but it is doubtful that the problem can be solved in polynomial for this restricted class of instances.

References

Amoroso S. and Patt Y.N. Decision procedures for surjectivity and injectivity of parallel maps for tesselation structures. *Journal of Computer and Systems Sciences* **Vol. no. 6**: 448–464, 1972.

Beauquier D. Minimal automaton for a factorial, transitive, rational language. *Theoretical Computer Science* **Vol. no. 67**: 65–73, 1989.

Burks E.W. *Essays on Cellular Automata* University of Illinois Press, 1972.

Coven E.M. and Paul M.E. Sofic systems. *Israel J. of Mathematics* **Vol. 20 no. 2**: 165–177, 1975.

Čulik II K., Pachl J. and Yu S. On the limit sets of cellular automata. *SIAM Journal on Computing* **Vol. 18 no. 4**: 831–842, 1989.

Fischer R. Sofic systems and graphs. *Monatshefte für Mathematik* **Vol. no. 80**: 179–186, 1975.

Gleason A. Semigroups of shift register counting matrices. *Math. Systems Theory* **Vol. no. 25**: 253–267, 1992, Revised by Kochman F., Neuwirth L.

Goles E., Maass A. and Martinez S. On the limit set of some universal cellular automata. *Theoretical Computer Science* **Vol. no. 110**: 53–78, 1993.

Golze U. Differences between 1- and 2-dimensional cell spaces. in Lindenmayer A. and Rozenberg G., editors, *Automata, Languages and Development* North-Holland, 369–384, 1976.

Head T. Linear CA: injectivity from ambiguity. *Complex Systems* **Vol. 3 no. 4**: 343–348, 1989.

Hedlund G. A. Endomorphisms and automorphisms of the shift dynamical system. in Z. Kohavi, editor, *The theory of machines and computation* Academic Press, New York, 320–375, 1971.

Hopcroft J.E. An n log n algorithm for minimizing states in a finite automaton. *Math. Systems Theory* **Vol. no. 3**: 189–196, 1971.

Hurd L. Formal language characterizations of cellular automata limit sets. *Complex Systems* **Vol. 1 no. 1**: 69–80, 1987.

Lind D. and Marcus B. *Introduction to Symbolic Dynamics and Coding.* Cambridge University Press, 1995.

Mehlhorn K. *Data Structures and Algorithms, vol. II: Graph Algorithms and NP-Completeness.* Springer-Verlag, 1984.

Moore E. F. Machine models of self-reproduction. *Proc. Symp. Appl. Math.* **Vol. no. 14**: 17–33, 1962.
Myhill J. The converse of moore's garden-of-eden theorem. *Proc. AMS* **Vol. no. 14**: 685–686, 1963.
Nasu M. Homomorphisms of graphs and their global maps. in Saito N. and Nishiziki T., editors, *Graph theory and algorithms*, Springer Verlag, **LNCS 108**:159–170, 1981.
Papadimitriou C.H. *Computational Complexity*. Addison-Wesley, 1994 .
Sutner K. Linear cellular automata and Fischer automata. To appear in *Parallel Computing.*
Sutner K. De Bruijn graphs and linear cellular automata. *Complex Systems* **Vol.5 no. 1**: 19–30, 1991.
Sutner K. The size of power automata. Submitted, 1997.
Weiss B. Subshifts of finite type and sofic systems. *Monatshefte für Mathematik* **Vol. no. 77**: 462–474, 1973.
Wolfram S. Twenty problems in the theory of cellular automata. *Physica Scripta* **Vol. no. 79**: 170–183, 1985.
Wolfram S. *Theory and Applications of Cellular Automata*, World Scientific, 1986.
Yu S. Regular languages. in Rozenberg G. and Salomaa A., editors, *Handbook of Formal Languages*, Springer Verlag, **volume 1, chapter 2**, 1997.

CELLULAR AUTOMATA, FINITE AUTOMATA, AND NUMBER THEORY

J.-P. ALLOUCHE
CNRS, LRI
Bâtiment 490
F-91405 Orsay Cedex (France)

1. Introduction

Although general cellular automata have a long story, a large number of papers devoted to them only study one-dimensional cellular automata. (For but one example of a two-dimensional cellular automaton, that is related to the Belouzov-Zhabotinski-Zaikin oscillating chemical reaction, see (Greenberg *et al.*, 1980; Greenberg *et al.*, 1978a; Greenberg *et al.*, 1978b; Allouche and Reder, 1984).) Furthermore, the authors usually restrict the study to *additive* (sometimes called *linear*) cellular automata: first they are easier to study, second they already have complex behaviors.

Since general cellular automata are equivalent to Turing machines, i.e., the "most general machines" we can think of, it is interesting to find out whether some restricted classes of cellular automata (for example the additive $1-D$ cellular automata) can be proven equivalent to simpler machines, such as finite automata. On the other hand, patterns generated by cellular automata are often regular but intriguing, and one might try to find out arithmetic properties underlying these patterns.

We will survey recent results concerning $1-D$ additive cellular automata over the ring $\mathbb{Z}/d\mathbb{Z}$, their relationship to $2-D$ finite automata, and arithmetic properties of the patterns produced by both kinds of automata.

2. Additive cellular automata

Since this is discussed elsewhere in this volume, we just recall informally what a cellular automaton is. A general cellular automaton consists of a lattice considered as a set of nodes, a neighborhood relation for the nodes, a map assigning to each node a value in a commutative ring, and an evolution map, that maps the value of each node to a value depending only on the previous value of the node and the previous values of its neighbors. We also make the hypothesis that a node with value zero, with neighbors with values zero, has still value zero after applying the evolution map. At the beginning, a value is assigned to each node: the set of these values and the corresponding nodes are called the *initial condition* or *initial configuration*. Starting from this configuration and iterating the evolution map gives the *time evolution* of the automaton, the number of iterations being the number of time units. The set of values and the corresponding nodes after t iterations of the evolution map are called the *configuration at time t*.

In what follows, the cellular automata we consider are one-dimensional, and the ring is $\mathbb{Z}/d\mathbb{Z}$, for a fixed integer $d \geq 2$. Each node has a finite number of neighbors, and the neighborhood relation is homogeneous. Furthermore, the automaton is *additive* (or linear), in the sense that the evolution map is additive. It is easy to see that, if the initial condition contains only a finite number of nodes with non-zero values, the behavior of the automaton is entirely given by two polynomials R and G (possibly Laurent polynomials, i.e., polynomials in X and $1/X$), such that the configuration at time t corresponds to the polynomial GR^t.

The configurations at $t = 0, 1, \cdots$ are drawn on a two-dimensional lattice, giving a two-dimensional pattern.

3. Finite automata

In this section, we first describe one-dimensional finite automata. To make the discussion simpler, we will actually speak of morphisms of constant length. These morphisms generate the same sequences as finite automata with output (or *tag-systems*), see (Cobham, 1972). Let us take an example. We start with the finite *alphabet* $\mathcal{A} = \{a, b, c, d\}$, and we denote by $\mathcal{A}^*$ the set of finite words on $\mathcal{A}$ equipped with the concatenation law ($\mathcal{A}^*$ is the free monoid generated by $\mathcal{A}$). We define a map σ that associates to each letter in $\mathcal{A}$ a two-letter word on $\mathcal{A}$ by: $a \longrightarrow ab$, $b \longrightarrow cb$, $c \longrightarrow ad$, and $d \longrightarrow cd$. This map is extended to a morphism on $\mathcal{A}^*$ by defining the image of a word as the concatenation of the images of its letters, e.g., $\sigma(abc) = abcbad$. Now, we can iterate σ, starting from the letter a:

$$\begin{aligned}
\sigma^0(a) &= a \\
\sigma^1(a) &= ab \\
\sigma^2(a) &= \sigma(ab) = abcb \\
\sigma^3(a) &= \sigma(abcb) = abcbadcb \\
&\cdots
\end{aligned}$$

We immediately see that each new word $\sigma^{i+1}(a)$ begins with the previous one $\sigma^i(a)$, and that this easily follows from the fact that $\sigma(a)$ begins with a. As a consequence, the sequence of words $(\sigma^i(a))_{i\geq 0}$ converges to an infinite word (for the topology of simple convergence), and this infinite word is a fixed point of the morphism σ (actually σ is extended to infinite words by *continuity*, i.e., the image of an infinite word by σ is obtained as the limit of the images of longer and longer finite prefixes).

Now, suppose we take the pointwise image of this fixed point by the map φ defined by $\varphi(a) = \varphi(b) = 0$ and $\varphi(c) = \varphi(d) = 1$. We thus obtain the infinite word $00100110\cdots$. Such a word is called a 2-*automatic* word. More generally, a k-*automatic* word on a finite alphabet is an infinite word that can be obtained as a pointwise image of a fixed point of a morphism of length k (i.e., the image of each letter is a k-letter word) on a possibly larger alphabet.

Note that the particular infinite word we give above is the so-called paperfolding sequence. To know more about this sequence, the reader can read (Dekking *et al.*, 1982; Allouche, 1987).

The following characterization is useful and not hard to obtain from the definition above: *a sequence* $(u_n)_{n\geq 0}$ *is* k-*automatic if and only if its* k-*kernel, i.e., the set of subsequences* $\{(u_{k^j n+r})_{n\geq 0},\ j \geq 0,\ 0 \leq r \leq k^j - 1\}$, *is finite.*

This notion of morphism of constant length can be generalized to multidimensional words. Let us explain, on an example, the two-dimensional case. Let $\mathcal{A} = \{a, b, c\}$ be a finite alphabet. We are interested in two-dimensional square words, such as

$$\begin{matrix} a & b & b & a & c \\ c & a & b & b & a \\ b & c & a & c & b \\ c & b & c & a & c \\ a & b & c & c & b \end{matrix} .$$

We define the map τ on $\mathcal{A}$ by

$$\tau(a) = \begin{matrix} a & b \\ c & b \end{matrix} \qquad \tau(b) = \begin{matrix} c & a \\ b & c \end{matrix} \qquad \tau(c) = \begin{matrix} b & c \\ a & a \end{matrix} .$$

The map can be extended to square words, and then can be iterated, starting, say from a, and replacing at each step each letter by the "square" given by the image by τ of this letter:

$$a \to \begin{matrix} a & b \\ c & b \end{matrix} \to \begin{matrix} a & b & c & a \\ c & b & b & c \\ b & c & c & a \\ a & a & b & c \end{matrix} \to \begin{matrix} a & b & c & a & b & c & a & b \\ c & b & b & c & a & a & c & b \\ b & c & c & a & c & a & b & c \\ a & a & b & c & b & c & a & a \\ c & a & b & c & b & c & a & b \\ b & c & a & a & a & a & c & b \\ a & b & a & b & c & a & b & c \\ c & b & c & b & b & c & a & a \end{matrix} \to \cdots$$

As previously, the image of a by τ "begins" with a, and hence each new square "begins" with the previous one. Thus, the sequence $(\tau^i(a))_{i\geq 0}$ converges. Its limit is a two-dimensional infinite word, that is a fixed point of (the extension) of τ. As previously we can take a pointwise image of this infinite word to obtain a 2-*automatic* two-dimensional sequence. We define k-*automaticity* in a similar way. We also have that *a double sequence* $(u_{m,n})_{m,n\geq 0}$ *is* k-*automatic if and only if its* k-*kernel, i.e., the set of subsequences* $\{(u_{k^jm+r,k^jn+s})_{m,n\geq 0},\ j \geq 0,\ 0 \leq r, s \leq k^j - 1\}$, *is finite.*

4. Arithmetic properties of finite automata

In 1979 Christol obtained in (Christol, 1979) (see also (Christol *et al.*, 1980)) an important result relating one-dimensional automatic sequences and algebraic formal power series.

Theorem 1 (Christol)

A sequence $(u_n)_{n\geq 0}$ *with values in the finite field* $\mathbf{F}_q$ *is* q-*automatic if and only if the formal power series* $\sum_{n\geq 0} u_n X^n$ *is algebraic over the field of rational functions* $\mathbf{F}_q(X)$.

Remarks

- This theorem has nothing to do with the Chomsky-Schützenberger theorem (Chomsky and Schützenberger, 1963).

- To give an example, let us take again the paperfolding sequence we considered above. It is not difficult to see that this sequence $(u_n)_{n\geq 0}$ has the property

$$\forall n \geq 0,\ u_{4n} = 0,\ u_{4n+2} = 1,\ u_{2n+1} = u_n.$$

Hence, we have over $\mathbf{F}_2$, i.e., modulo 2,

$$\begin{aligned} F(X) := \sum_{n\geq 0} u_n X^n &= \sum_{n\geq 0} u_{2n} X^{2n} + \sum_{n\geq 0} u_{2n+1} X^{2n+1} \\ &= \sum_{n\geq 0} X^{4n+2} + X \sum_{n\geq 0} u_n X^{2n} \\ &= \frac{X^2}{1-X^4} + XF^2, \text{ (remember this is modulo 2).} \end{aligned}$$

Hence the formal power series $F = \sum_{n\geq 0} u_n X^n$ satisfies, over the field $\mathbf{F}_2(X)$, the algebraic (quadratic) equation $X(1+X)^4F^2 + (1+X)^4F + X^2 = 0$.

In 1987, Salon (Salon, 1986; Salon, 1987) proved that Christol's theorem holds in higher dimension.

Theorem 2 (Salon)
A sequence $(u_{m,n})_{m,n\geq 0}$ with values in the finite field $\mathbf{F}_q$ is q-automatic if and only if the formal power series $\sum_{m,n\geq 0} u_{m,n}X^mY^n$ is algebraic over the field of rational functions $\mathbf{F}_q(X,Y)$.

In other words, a combinatorial property of sequences (one or more dimensions) translates into an algebraic property of formal power series. This was one of the motivations for the results in the next section. Note that these two theorems permit to give "elementary" proofs of algebraicity or transcendence of formal power series: Christol's theorem has been used for proving the transcendence of values of Carlitz functions, for example in (Berthé, 1994; Allouche, 1996; Mendès France and Yao, 1997). Salon used his theorem to give a simple proof, in the case where the ground field is finite, of a theorem of Deligne (Deligne, 1984): *let K be a commutative field of positive characteristic. If the formal power series $\sum_{m,n} u_{m,n}X^mY^n \in K[[X,Y]]$ is algebraic over $K(X,Y)$, then its diagonal $\sum_n u_{n,n}X^n$ is algebraic over $K(X)$.*

5. Arithmetic properties of patterns generated by cellular automata

We start by considering the Pascal triangle modulo 2,

$$\begin{array}{ccccccccccccccccc}
1 & 0 & 0 & 0 & 0 & 0 & 0 & 0 & 0 & 0 & 0 & 0 & 0 & 0 & 0 & 0 & \cdots \\
1 & 1 & 0 & 0 & 0 & 0 & 0 & 0 & 0 & 0 & 0 & 0 & 0 & 0 & 0 & 0 & \cdots \\
1 & 0 & 1 & 0 & 0 & 0 & 0 & 0 & 0 & 0 & 0 & 0 & 0 & 0 & 0 & 0 & \cdots \\
1 & 1 & 1 & 1 & 0 & 0 & 0 & 0 & 0 & 0 & 0 & 0 & 0 & 0 & 0 & 0 & \cdots \\
1 & 0 & 0 & 0 & 1 & 0 & 0 & 0 & 0 & 0 & 0 & 0 & 0 & 0 & 0 & 0 & \cdots \\
1 & 1 & 0 & 0 & 1 & 1 & 0 & 0 & 0 & 0 & 0 & 0 & 0 & 0 & 0 & 0 & \cdots \\
1 & 0 & 1 & 0 & 1 & 0 & 1 & 0 & 0 & 0 & 0 & 0 & 0 & 0 & 0 & 0 & \cdots \\
1 & 1 & 1 & 1 & 1 & 1 & 1 & 1 & 0 & 0 & 0 & 0 & 0 & 0 & 0 & 0 & \cdots \\
1 & 0 & 0 & 0 & 0 & 0 & 0 & 0 & 1 & 0 & 0 & 0 & 0 & 0 & 0 & 0 & \cdots \\
1 & 1 & 0 & 0 & 0 & 0 & 0 & 0 & 1 & 1 & 0 & 0 & 0 & 0 & 0 & 0 & \cdots \\
1 & 0 & 1 & 0 & 0 & 0 & 0 & 0 & 1 & 0 & 1 & 0 & 0 & 0 & 0 & 0 & \cdots \\
1 & 1 & 1 & 1 & 0 & 0 & 0 & 0 & 1 & 1 & 1 & 1 & 0 & 0 & 0 & 0 & \cdots \\
1 & 0 & 0 & 0 & 1 & 0 & 0 & 0 & 1 & 0 & 0 & 0 & 1 & 0 & 0 & 0 & \cdots \\
1 & 1 & 0 & 0 & 1 & 1 & 0 & 0 & 1 & 1 & 0 & 0 & 1 & 1 & 0 & 0 & \cdots \\
1 & 0 & 1 & 0 & 1 & 0 & 1 & 0 & 1 & 0 & 1 & 0 & 1 & 0 & 1 & 0 & \cdots \\
1 & 1 & 1 & 1 & 1 & 1 & 1 & 1 & 1 & 1 & 1 & 1 & 1 & 1 & 1 & 1 & \cdots \\
\vdots & \vdots & \vdots & \vdots & \vdots & \vdots & \vdots & \vdots & \vdots & \vdots & \vdots & \vdots & \vdots & \vdots & \vdots & \vdots & \vdots
\end{array}$$

thus obtaining a picture analogous to the familiar Sierpinski triangle. The "regularity" of the patterns can be seen in many ways. One of them is the following proposition: *the double sequence* $\left(\binom{m}{n}\right)_{m,n\geq 0}$ *is 2-automatic.*

This can be seen by studying the 2-kernel of this sequence, and using a well-known result of Lucas in the case $p = 2$: *if p is a prime number, and* $0 \leq i, j \leq p-1$, *then* $\binom{pm+i}{pn+j} \equiv \binom{m}{n}\binom{i}{j} \pmod{p}$.

More generally what can be said about the Pascal triangle reduced modulo any integer d? The following theorem answers this question.

Theorem 3 (see (Allouche *et al.*, 1996)**)**

Let $d \geq 2$ be an integer. The Pascal triangle modulo d is a k-automatic sequence if and only if d and k are powers of a same prime number p.

Proof.

We sketch the proof (see (Allouche *et al.*, 1996), see also (Korec, 1990)). If $d = p$ is a prime number, Lucas theorem shows as above that the Pascal triangle is p-automatic (hence p^i-automatic for any $i \geq 1$). This can be seen also by noticing that the double sequence of binomial coefficients can be obtained by expanding the polynomial $R(X) = (1+X)^m$. But this polynomial has the *p-Fermat property* modulo p: $R(X^p) \equiv R^p(X) \pmod{p}$.

If now $d = p^\alpha$ is a prime power, the result can be obtained, after noting that the polynomial $R^{p^{\alpha-1}}$ has the p-Fermat property.

To show that the Pascal triangle modulo d is not k-automatic, for any k, when d is not a prime power, we use a theorem due to Cobham (Cobham,

1969): *let k and ℓ be two integers that are multiplicatively independent, i.e., not powers of a same integer. Then, if a sequence is both k-automatic and ℓ-automatic, it must be ultimately periodic.* Now let us suppose for example that $d = 6$, to show on an example how the proof works. If the Pascal triangle modulo 6 is a k-automatic sequence, then, re-reducing modulo 2, we obtain that the sequence $(\binom{m}{n})_{m,n \geq 0}$ modulo 2 is k-automatic, but we also know it is 2-automatic. Similarly, the sequence $(\binom{m}{n})_{m,n \geq 0}$ modulo 3 is k-automatic, but we also know it is 3-automatic.

Now, if a sequence $(u_{m,n})_{m,n \geq 0}$ is k-automatic, then any sequence

$$(u_{am+bn+c,dm+en+f})_{m,n \geq 0}$$

is k-automatic. Hence, on one hand the sequence $(\binom{3n}{n})_{n \geq 0}$ modulo 2 is both k-automatic and 2-automatic. On the other hand, the sequence $(\binom{2n}{n})_{n \geq 0}$ modulo 3 is both k-automatic and 3-automatic. If we show that the sequence $(\binom{3n}{n})_{n \geq 0}$ modulo 2 is not ultimately periodic, this will imply from Cobham's theorem, that k must be a power of 2. Similarly, if we show that the sequence $(\binom{2n}{n})_{n \geq 0}$ modulo 3 is not ultimately periodic, this will imply that k is a power of 3. Comparing both results for k gives the desired contradiction.

Using Lucas theorem the reader can show that, if $T(X) = \sum_{n \geq 0} \binom{3n}{n} X^n$, then

$$XT^3 + T + 1 \equiv 0 \pmod 2.$$

It is easy to show that this equation cannot have rational solutions, i.e., solutions that are rational functions. Hence the sequence $(\binom{3n}{n})_{n \geq 0}$ modulo 2 is not ultimately periodic.

The reader can easily show, using Lucas theorem again, or a well-known expansion, that

$$\left(\sum_{n \geq 0} \binom{2n}{n} X^n \right)^2 = (1 - 4X)^{-1},$$

hence

$$(1 - 4X) \left(\sum_{n \geq 0} \binom{2n}{n} X^n \right)^2 = 1.$$

This relation holds in $\mathbb{Z}$ but also in $\mathbb{Z}/3\mathbb{Z}$. It easily implies that the formal power series $\left(\sum_{n \geq 0} \binom{2n}{n} X^n \right)$ modulo 3 is not a rational function. Hence the sequence $(\binom{2n}{n})_{n \geq 0}$ modulo 3 is not ultimately periodic. ∎

Now one can ask whether an analogous theorem holds for general additive one-dimensional cellular automata over the ring $\mathbb{Z}/d\mathbb{Z}$. The complete answer is given in the following theorem, see (Allouche *et al.*, 1998) for details.

Theorem 4 (see (Allouche *et al.*, 1998))
Let $R(X)$ and $G(X)$ be two polynomials in $\mathbb{Z}[X]$, and let d be an integer ≥ 2. Suppose that for every prime number p dividing d we have $G(X) \not\equiv 0$ modulo p. We consider the additive cellular automaton C defined on $\mathbb{Z}/d\mathbb{Z}$ by the polynomial $R(X)$ modulo d, and the initial condition G modulo d, i.e., the configuration at time t is given by GR^t modulo d. Then, we necessarily have one of the following three cases.

- *There exist two different prime numbers p and q dividing d, such that neither $R(X)$ modulo p nor $R(X)$ modulo q are monomials. Then, the double sequence generated by the additive cellular automaton C is not k-automatic, for any $k \geq 2$.*
- *There exists one prime number p dividing d for which $R(X)$ modulo p is not a monomial, and for every other prime divisor q of d (if any), the polynomial $R(X)$ modulo q is a monomial. Then, the double sequence generated by the additive cellular automaton C is p^a-automatic, for every $a \geq 1$, and this sequence is not k-automatic for any $k \notin \{p^a;\ a \geq 1\}$.*
- *For every prime number p dividing m, the polynomial $R(X)$ modulo p is a monomial. Then, the double sequence generated by the additive cellular automaton C is k-automatic for every $k \geq 2$.*

6. Miscellanea

One of the key properties that is used for both theorems in Section 5 is that every polynomial R over $\mathbb{Z}/p\mathbb{Z}$, where p is a prime number, has the *p-Fermat property*: $R(X^p) \equiv R^p(X) \pmod{p}$. This implies that, for every $i \in [0, p-1]$, we have

$$R^{pn+i}(X) \equiv R^n(X^p)R^i(X) \pmod{p}.$$

A more general property would also work: a *p-Carlitz sequence of polynomials* is a sequence of polynomials $(R_n(X))_{n\geq 0}$, with $R_0(X) = 1$, such that, for every $n \geq 1$ and for every $i \in [0, p-1]$,

$$R_{pn+i}(X) \equiv R_n(X^p)R_i(X) \pmod{p}.$$

If we replace the sequence $(R^n)_{n\geq 0}$ for a fixed polynomial R by a sequence of polynomials $(R_n)_{n\geq 0}$ having the p-Carlitz property, many of the previous results still hold. An interesting example of such a sequence of polynomials is the sequence of Legendre polynomials, that has the p-Carlitz property for each odd prime number p. The reader can see (Allouche and Skordev, 1997) for more details.

For other arithmetic properties of additive cellular automata over the field of complex numbers or a field of positive characteristic, the reader is referred to the interesting paper (Litow and Dumas, 1993): the authors prove *inter alia* the following result. *Let K be either the field of complex numbers or a field with positive characteristic. Let C be a bilateral additive cellular automaton (bilateral meaning that the polynomials that give the initial condition and the evolution map are Laurent polynomials). Then, the vertical generating series of the two-dimensional sequence generated by the time-evolution of C are algebraic over the field of rational functions $K(X)$.* They furthemore show that such cellular automata can generate classical sequences such as the Catalan numbers, and, up to a slight generalization of the model, the paperfolding sequence as well as the Motzkin numbers.

References

Allouche J.-P. Automates finis en théorie des nombres. *Exposition. Math.* **Vol. no. 5**: 239–266, 1987.

Allouche J.-P. Transcendence of the Carlitz-Goss Gamma function at rational arguments. *Number Theory* **Vol. no. 60**: 318–328, 1996.

Allouche J.-P., von Haeseler F., Peitgen H.-O., Petersen A. and Skordev G. Automaticity of double sequences generated by one-dimensional linear cellular automata. *Theoret. Comput. Sci.* (to appear)

Allouche J.-P., von Haeseler F., Peitgen H.-O. and Skordev G. Linear cellular automata, finite automata and Pascal's triangle. *Discrete Appl. Math.* **Vol. no. 66**: 1–22, 1996.

Allouche J.-P. and Reder C. Oscillations spatio-temporelles engendrées par un automate cellulaire. *Discr. Appl. Math.* **Vol. no. 8**: 215–254, 1984.

Allouche J.-P. and Skordev G. Schur congruences, Carlitz sequences of polynomials and automaticity. preprint, *Inst. Dyn. Syst., Univ. Bremen* **Vol. no. 399**, 1997.

Berthé V. Automates et valeurs de transcendance du logarithme de Carlitz. *Acta Arith.* **Vol. no. 66**: 369–390, 1994.

Chomsky N. and Schützenberger M. The algebraic theory of context-free languages. in: *Computer programming and formal languages*, North Holland, Amsterdam, 118–161, 1963 .

Christol G. Ensembles presque périodiques k-reconnaissables. *Theoret. Comput. Sci.* **Vol. no. 9**: 141–145, 1979.

Christol G., Kamae T., Mendès France M. and Rauzy G. Suites algébriques, automates et substitutions. *Bull. Soc. Math. France* **Vol. no. 108**: 401–419, 1980.

Cobham A. On the base-dependence of sets of numbers recognizable by finite automata. *Math. Systems Theory* **Vol. no. 3**: 186–192, 1969.

Cobham A. Uniform tag sequences. *Math. Systems Theory* **Vol. no. 6**: 164–192, 1972.

Dekking F. , Mendès France M. and van der Poorten A. Folds!. *Math. Intelligencer* **Vol. no. 4**: 130–138, 173–181, 190–195, 1982.

Deligne P. Intégration sur un cycle évanescent. *Invent. Math.* **Vol. no. 76**: 129–143, 1984.

Greenberg J., Greene C. and Hastings S. A combinatorial problem arising in the study of reaction-diffusion equations. *SIAM J. Algebraic Discrete Methods* **Vol. no. 1**: 34–42, 1980.

Greenberg J., Hassard B. and Hastings S. Pattern formation and periodic structures in systems modeled by reaction-diffusion equations. *Bull. Am. Math. Soc.* **Vol. no. 84**: 1296–1327, 1978.

Greenberg J. . and Hastings S. Spatial patterns for discrete models of diffusion in excitable media. *SIAM J. Appl. Math.* Vol. no. **34**: 515–523, 1978.
Korec I. Pascal triangles modulo *n* and modular trellises. *Computers and Artificial Intelligence* Vol. no. **9**: 105–113, 1990.
Litow B. and Dumas P. Additive cellular automata and algebraic series. *Theoret. Comput. Sci.* Vol. no. **119**: 345–354, 1993.
Mendès France M. and Yao J.-Y. Transcendence and the Carlitz-Goss gamma function. *J. Number Theory* Vol. no. **63**: 396–402, 1997.
Salon O. Suites automatiques à multi-indices. *Séminaire de Théorie des Nombres de Bordeaux* Exposé **4**: 4-01–4-27; followed by an Appendix by J. Shallit, 4-29A–4-36A, 1986-1987.
Salon O. Suites automatiques à multi-indices et algébricité. *C. R. Acad. Sci. Paris, Série I* Vol. no. **305**: 501–504, 1987.

DECISION PROBLEMS ON GLOBAL CELLULAR AUTOMATA

K. ČULIK II
Department of Computer Science,
University of South Carolina
Columbia, S.C. 29208, U.S.A.

Abstract. Global cellular automata (GCA) are introduced as a generalization of 1-dimensional cellular automata allowing the next state of a cell to depend on a "regular" global context rather than just a fixed size neighborhood. We show that the global function of every ordinary CA can be implemented by a GCA. Hence, every undecidable problem for CA is also undecidable for GCA. We show that also all well known decidable results for CA can be extended to GCA.

1. Introduction

Cellular automata (CA) are models of complex systems in which an infinite lattice of cells is updated in parallel according to a simple local rule. A dynamical system on the lattice of cells is a continuous and shift-invariant function iff it can be specified by a CA.

We will generalize 1-dimensional CA to provide for a "regular" global context, while still using simple transition rules specified by a simple finite transducer called an $\omega\omega$-sequential machine. Our global cellular automaton (GCA) will retain most of the properties of CA and at the same time allow us to define many non-continuous transition functions. An important special case is the possibility of using two or more "classical" CA-rules in one dynamical system, with one of them being selected to be applied for the whole or a part of the current configuration according some "regular" conditions. Any "negative" result valid for 1-dimensional CA, for example,

[0]This work was supported by the National Science Foundation under Grant No. CCR-9417384. This is an abbreviated version of "Global Cellular Automata", to appear in *Complex Systems.*

any undecidability result is, of course, also valid for GCA. We will not consider such problems. However, quite surprisingly, most "positive" results can be extended to GCA. Thanks to some techniques known for finite transducers these extended results are proved rather easily. This is not so surprising when we note that in the simplified proof of the decidability of the injectivity problem for 1-dimensional CA in (Čulik II, 1987) we actually have implicitly used GCA.

We assume that the reader is familiar with basic notions of automata and language theory, see e.g. (Hopcroft and Ullman, 1979). In the next section we review the other necessary prerequisites and introduce $\omega\omega$-sequential machines ($\omega\omega$-SM). They are a special case (length preserving) of $\omega\omega$-transducers from (Čulik II and Yu, 1991).

In section 3, we introduce Global Cellular Automata (GCA) as a generalization of 1-dimensional CA. The global GCA-function is defined by a single-valued and complete $\omega\omega$-SM. We show that the definition is effective, we can test whether a given $\omega\omega$-SM has the required properties. We also show that GCA are indeed a generalization of CA; that is, every 1-dimensional CA-rule can be implemented by a complete and single-valued $\omega\omega$-SM. We give examples of GCA that cannot be implemented as CA and show some general techniques for constructing GCA. For example we can combine several 1-dimensional CA working on disjoint domains of configurations.

In section 4 we study the well known problems that are decidable for 1-dimensional CA, in particular, injectivity and surjectivity. We extend these results to GCA. Any problem that is undecidable for (ordinary) CA is, of course, also undecidable for GCA.

2. $\omega\omega$-finite automata and $\omega\omega$-sequential machines

First we recall the definition of the classical 1-dimensional Cellular Automaton (CA) with the neighborhood of a cell consisting of the cell itself, and its r neighbors to each side.

A CA is a triple $A = (S, r, f)$ where S is a finite set of states, r specifies the size of the neighborhood, and $f : S^{2r+1} \to S$ is the local function called also the CA rule.

A *configuration* c of the CA is a function $c : \mathbb{Z} \to S$, which assigns a state S to each cell of the CA. The set of configurations is denoted $S^{\mathbb{Z}}$. The local function f is extended to the *global function*

$$G_A : S^{\mathbb{Z}} \to S^{\mathbb{Z}}.$$

By definition, for $c, d \in S^{\mathbb{Z}}$, $G_A(c) = d$ if and only if $d(i) = f(c(i-r), c(i-r+1), \ldots, c(i+r))$ for all $i \in \mathbb{Z}$.

The configuration space $S^{\mathbf{Z}}$ is a product of infinitely many finite sets S. When S is endowed with the discrete topology, the *product topology* on $S^{\mathbf{Z}}$ is compact by Tychonoff's theorem (Kelley , 1955, theorem 5.13). A subbasis of open sets for the product topology consists of all sets of the form

$$\{c \in S^{\mathbf{Z}} \mid c(i) = a\}, \tag{1}$$

where $i \in \mathbf{Z}$ and $a \in S$. A subset of $S^{\mathbf{Z}}$ is open if and only if it is a union of finite intersections of sets of the form (1).

The shift $\sigma : S^{\mathbf{Z}} \to S^{\mathbf{Z}}$ is defined by $\sigma(c) = c'$ where $c'(i+1) = c(i)$ for each $i \in \mathbf{Z}$. Hedlund (Hedlund , 1969) gave the following characterization of CA as dynamical systems:

Theorem 1
$G : S^{\mathbf{Z}} \to S^{\mathbf{Z}}$ is the global function of a CA *if and only of it is shift-invariant and continuous.*

The set of bi-infinite words over S is denoted by $S^{\omega\omega}$. For $c \in S^{\mathbf{Z}}$ we denote by $c^{\omega\omega}$ the corresponding bi-infinite word in $S^{\omega\omega}$ similarly for a set $C \subseteq S^{\mathbf{Z}}$. If $d = \sigma^n(c), n \geq 1$ then $d^{\omega\omega} = c^{\omega\omega}$. Hence any set $R \subseteq S^{\omega\omega}$ of bi-infinite words corresponds to a shift-invariant subset of $S^{\mathbf{Z}}$.

Now we define our main tools namely $\omega\omega$-finite automaton (FA) (Čulik II and Yu, 1991) and $\omega\omega$-sequential machine ($\omega\omega$-SM). The latter is a special case of an $\omega\omega$-transducer of (Čulik II and Yu, 1991), and a generalization of a sequential machine of (Čulik II, 1998). Inputs of an $\omega\omega$-FA are bi-infinite words over an alphabet S which can be viewed as shift-invariant classes of configurations in $S^{\mathbf{Z}}$. A set $X \subseteq S^{\mathbf{Z}}$ is said to be shift-invariant if $\sigma(X) = X$.

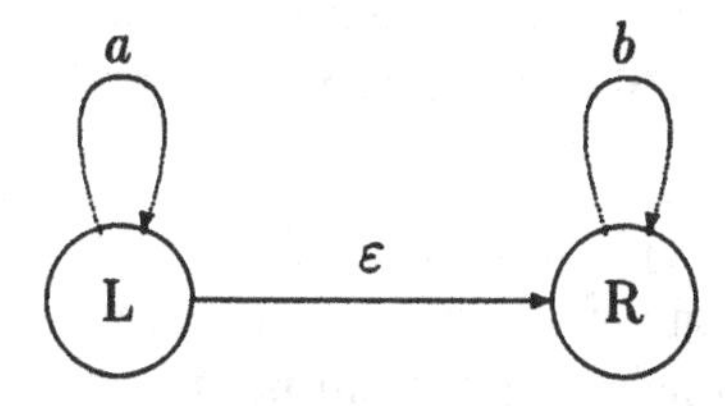

Figure 1. An $\omega\omega$-FA A

Finite automata that recognize sets of bi-infinite words have been defined by Nivat and Perrin (Nivat and Perrin, 1986), studied in (Beauquier, 1984; Čulik II *et al.*, 1989; Čulik II and Yu, 1991). Here we use the definition from (Čulik II and Yu, 1991).

An *ωω-finite automaton* (ωω-FA) A is a quintuple (K, S, δ, K_L, K_R), where

- K is the finite set of states;
- S is the input alphabet;
- $\delta : K \times S \cup \{\varepsilon\} \to 2^K$ is the transition function;
- $K_L \subseteq K$ is the set of left (accepting) states; and
- $K_R \subseteq K$ is the set of right (accepting) states.

An ωω-FA A can be represented by a diagram in the usual way with the left states indicated by L and the right states by R, see Fig. 1.

A bi-infinite word v is said to be recognized by A if there is a mapping $\mathbb{Z} \to K$, i.e. a bi-infinite sequence of states

$$\ldots q_{-2}, q_{-1}, q_0, q_1, q_2, \ldots$$

and a configuration c in v such that, for all $j \in \mathbb{Z}$,

(1) $\delta(q_j, c_j) = q_{j+1}$; and
(2) there exist $m, n \in \mathbb{Z}$, $m \leq j \leq n$, such that $q_m \in K_L$ and $q_n \in K_R$.

In other words, v is said to be recognized by A if there is a bi-infinite computation of A on a configuration c in v such that there is a left state appearing arbitrarily early, and there is a right state appearing arbitrarily late in the computation. Such a computation will be called an *accepting computation.*

The set of bi-infinite words recognized by A is denoted $L(A)$. We call $L(A)$ an *ωω-regular set.* Clearly, every ωω-regular set over S corresponds to a shift-invariant subset of $S^{\mathbb{Z}}$.

Let $w \in \Sigma^\star$. By w^ω we denote the one-way infinite word (ω-word) obtained by the infinite repetition of w. By ${}^\omega w$ we denote the reverse of $(w^R)^\omega$, i.e. the infinite repetition of w to the left. For example, the bi-infinite word (ωω-word) of infinitely many a's followed by infinitely many b's is written as ${}^\omega ab^\omega$. For ${}^\omega aa^\omega$ we also write ${}^\omega a^\omega$ or $a^{\omega\omega}$.

Example.

Let $A = (K, S, \delta, K_L, K_R)$ be an ωω-FA, where $K = \{0, 1\}$, $S = \{a, b\}$, $K_L = \{0\}$, $K_R = \{1\}$, and δ is given in Figure 1. The set of bi-infinite words recognized by A consists of the simple bi-infinite word ${}^\omega a\, b^\omega$.

The set of finite (one-way), infinite and bi-infinite words over S is denoted by $S^\star$, S^ω and $S^{\omega\omega}$, respectively. Finite or one-way infinite words can be considered as special cases of bi-infinite words in the following sense: A special quiescent symbol, usually 0, is specified such that a one-way infinite word (ω-word) is a bi-infinite word with infinitely many consecutive quiescent symbols on the left end, and a finite word is a bi-infinite word with a finite consecutive nonquiescent subword.

In an $\omega\omega$-FA, a left (right) state that is not in a cycle can be changed into a non-left (non-right) state without affecting the set of bi-infinite words recognized by the $\omega\omega$-FA. A state which cannot be reached from any left state or from which no right state can be reached is useless – it does not contribute to the recognition of any bi-infinite word. We say that an $\omega\omega$-FA is *reduced* if it satisfies the following conditions:

(i) every left state is in a cycle;
(ii) every right state is in a cycle;
(iii) every state can be reached from some left state;
(iv) from every state some right state can be reached.

Obviously, for any given $\omega\omega$-FA we can construct a reduced one that recognizes the same set of bi-infinite words.

An *$\omega\omega$-sequential machine* ($\omega\omega$-SM) is a 5-tuple $M = (K, S, \gamma, K_L, K_R)$, where

- K is the set of states,
- S is the input-output alphabet,
- $\gamma \subseteq K \times S \times S \times K$ is the transition relation,
- $K_L \subseteq K$ is the set of left states, and
- $K_R \subseteq K$ is the set of right states.

An $\omega\omega$-sequential machine M can be represented by a labeled directed graph with nodes K, an edge from node q to node p labeled a, b for each transition (q, a, b, p) in γ, the nodes in K_L indicated by L, and the nodes in K_R by R.

Machine M computes a relation $\rho(M)$, called $\omega\omega$-SM relation, between bi-infinite sequences of configurations $S^{\mathbb{Z}}$. Configurations x and y are in relation $\rho(M)$ if and only if there is a bi-infinite sequence q of states of M, such that, for every $i \in \mathbb{Z}$, there is a transition from q_{i-1} to q_i labeled by x_i,y_i, and there exist $m, n \in \mathbb{Z}$ such that $m \leq i \leq n$, $q_m \in K_L$ and $q_n \in K_R$.

We give the closure and decidability results for $\omega\omega$-regular set which will be useful later. From (Čulik II and Yu, 1991, corollary 2.6) and its proof immediately follow the following

Theorem 2

The family of $\omega\omega$-regular sets is effectively closed under boolean operations.

By modifying the proof for the closure of $\omega\omega$-regular sets under intersection, we get

Theorem 3

If ρ is an $\omega\omega$-SM relation and R is an $\omega\omega$-regular set, then the restriction $\rho_M = \{(u, v) \mid u \in R, (u, v) \in \rho\}$ of ρ to R is effectively an $\omega\omega$-SM relation.

Since our $\omega\omega$-SM is a special case of an $\omega\omega$-transducer of (Čulik II and Yu, 1991) we have the following special case of Theorem 2.2 of (Čulik II and Yu, 1991).

Theorem 4

The family of $\omega\omega$-regular sets is effectively closed under $\omega\omega$-SM relations.

Theorem 5

Given $\omega\omega$-FA A, B it is decidable whether

(a) $L(A) = \emptyset$,
(b) $L(A) = S^{\mathbb{Z}}$,
(c) $L(A) = L(B)$,
(d) $L(A) \subseteq L(B)$.

Proof Assume A is reduced. Clearly, $L(A) \neq \emptyset$ iff there is a path from a state in K_L to a state in K_R, which is easy to test; (b), (c) and (d) follow from (a) and Theorem 2. □

A relation $R \subseteq S^{\mathbb{Z}} \times S^{\mathbb{Z}}$ is called *shift-invariant* if $(c, d) \in R$ iff $(\sigma(c), \sigma(d)) \in R$, R is called *strongly shift-invariant* if $(c, d) \in R$ iff $(\sigma^i(c), \sigma^j(d)) \in R$ for all $i, j \in \mathbb{Z}$. Clearly, every $\omega\omega$-SM defines a shift-invariant relation on $S^{\mathbb{Z}}$. Note, however, that a relation on bi-infinite words over S corresponds to a strongly shift-invariant relation on $S^{\mathbb{Z}}$. Hence, two $\omega\omega$-SM might be equivalent on bi-infinite words, that is on $S^{\omega\omega}$, without being equivalent on $S^{\mathbb{Z}}$.

The proof of Lemma 2.4 of (Čulik II and Yu, 1991) is constructive. Thus we have the following representation lemma for $\omega\omega$-regular sets.

Lemma 1

A set of bi-infinite words is $\omega\omega$-regular if and only if it can be presented by $D_1^R F_1 \cup D_2^R F_2 \cup \ldots \cup D_n^R F_n$ where $D_1, \ldots, D_n, F_1, \ldots, F_n$ are ω-regular sets and D^R denotes the reversal of D. Given an $\omega\omega$-FA A such a canonical expression for $L(A)$ can be constructed.

3. Global cellular automata

Now we are ready to introduce our main definition.

Definition 1

A Global Cellular Automaton (GCA) *is an $\omega\omega$-SM $M = (K, S, \gamma, K_L, K_R)$ which is*

(a) *complete, i.e.* $\mathrm{dom}(\rho(M)) = S^{\mathbb{Z}}$;
(b) *single-valued, i.e. $\rho(M)$ is a function;*

Note that for every $\omega\omega$-SM M, $\rho(M)$ is shift-invariant.

The (global) function defined by GCA M is denoted by G_M. That is, $G_M : S^{\mathbb{Z}} \to S^{\mathbb{Z}}$, $G_M(c) = d$ iff $(c, d) \in \rho(M)$. It follows from Theorem 1 that any GCA function which is not a global CA-function cannot be continuous.

Now, we show that the definition of GCA is effective, that is given an $\omega\omega$-SM we can test whether M is a GCA.

Lemma 2

Given an $\omega\omega$-SM $M = (K, S, \gamma, K_L, K_R)$ *it is decidable whether* M *is complete, i.e. whether the domain of* M *is* $S^{\mathbb{Z}}$.

Proof The domain of every $\omega\omega$-SM is clearly a shift-invariant subset of $S^{\mathbb{Z}}$, corresponding to an $\omega\omega$-regular set R. By omitting the outputs of M we can easily construct an $\omega\omega$-FA A such that $L(A) = R$. By theorem 5 we can test whether $L(A) = S^{\mathbb{Z}}$. □

Using the terminology of L-system theory we call *coding* on $S^\star$ a letter-to-letter morphism on $S^\star$, i.e. $c : S^\star \to S^\star$ such that $c(a) \in S$ for each $a \in S$.

We recall that $c^{\omega\omega}$ denotes the bi-infinite word over alphabet S corresponding to $c \in S^{\mathbb{Z}}$ (and to all its shifts). In (Čulik II and Yu, 1991) it has been stated that the Nivat theorem for finite transducers (Berstel, 1979) can be restated for $\omega\omega$-finite transducers. For $\omega\omega$-SM we have the following simple Nivat-like representation:

Theorem 6

Let $\pi \subseteq S^{\omega\omega} \times S^{\omega\omega}$. *There exists an* $\omega\omega$-SM M *such that* $R(M) = \pi$ *iff there effectively exists an* $\omega\omega$*-regular set* R *and codings* $g, h : S^{\omega\omega} \to S^{\omega\omega}$ *such that* $\pi = \{(g(w), h(w)) \mid w \in R\}$.

As an auxiliary tool we will use one-way ω-SM. An ω-SM M has initial states rather than left states. The definition of the relation $\rho(M)$ on S^ω is an obvious modification of the definition of the relation defined by an $\omega\omega$-SM.

Lemma 3

Given a (one-way) ω-SM M, *it is decidable whether* M *is single-valued (on* S^ω*).*

Proof The family of ω-SM relations is clearly closed under composition and inversion. Hence, we can construct ω-SM N such that $\rho(N) = (\rho(M))^{-1} \circ \rho(M)$. Clearly, $\rho(N)$ is a restriction of identity iff M is single-valued. By lemma 1 there effectively exist ω-FA A and codings g, h such that $\rho(N) = \{(g(w), h(w)) \mid w \in L(A)\}$. Clearly, $\rho(N)$ is a restriction of identity iff $g(w) = h(w)$ for all $w \in L(A)$. The latter condition is easy to test. □

Let M be a $\omega\omega$-SM. Clearly, the single-valuedness of $R(M)$ is a necessary condition for the single-valuedness of $\rho(M)$. However, it is not sufficient;

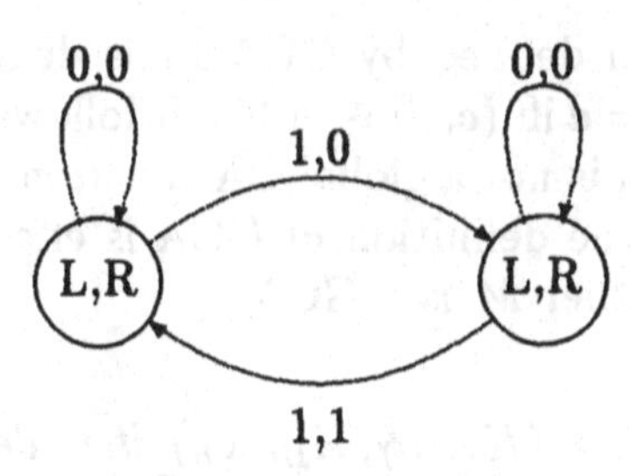

Figure 2. An $\omega\omega$-SM M_1

consider, for example, M that defines the union of the identity and shift σ on $S^{\mathbf{Z}}$. We will use the ω-regular sets (sets of one-way infinite strings) to test the single-valuedness of $\omega\omega$-SM.

Lemma 4

Given an $\omega\omega$-SM M *it is decidable whether* $\rho(M)$ *is single-valued.*

Proof We will construct an (one-way) ω-SM $\hat{M}$ so that $\hat{M}$ is single-valued on $(S\times S)^\omega$ iff M is single-valued on $S^{\omega\omega}$. Let $M = (K, S, \gamma, K_L, K_R)$. We first construct ω-SM $M' = K\times K, S\times S, \gamma', I, F)$ where $I = \{(q,q) \mid q \in K\}$ is the initial set of states, $F = K_R \times K$ is the set of final (right) states, and $((p,q),(a,b),(a',b'),(p',q')) \in \gamma'$ iff $(p,a,a',p') \in \gamma$ and $(q',b,b',q) \in \gamma$. Clearly, M' simulates a computation of M on an ω-word with "two tracks" obtained by folding a bi-infinite word over S. However, M' only tests the condition for the right states (K_R). In order to test the condition for the left states (K_L) we restrict the relation $\rho(M')$ to the ω-regular set R defined by ω-FA $A = (K \times K, S \times S, \gamma_A, I, F_A)$ where I is as above, right states are defined by $F_A = K \times K_L$ and the transition relation γ_A as follows: $((p,q),(a,b),(p',q')) \in \gamma_A$ iff for some $a', b' \in S$ $(p,a,a',p') \in \gamma$ and $(q',b,b',q) \in \gamma$. Similarly, like for bi-infinite word (theorem 4), ω-regular sets are clearly closed under ω-SM relations, so there effectively exists an ω-SM $\hat{M}$ with the properties described above. By lemma 1 we can test whether $\hat{M}$ is single-valued. □

Corollary 1

Given an $\omega\omega$-SM M *it is decidable whether* M *is a* GCA.

Proof By lemma 2 and lemma 4. □

Example.

$\omega\omega$-SM M_1 shown in Fig. 2 is not single-valued, starting from any position $k \in \mathbf{Z}$ we can map either every odd 1 to 0 and every even 1 to 1 or vice versa, so M_1 is two-valued.

Lemma 5

Every 1-dimensional CA *(with arbitrary neighborhood* r*) can be implemented by a complete, single-valued (simple)* $\omega\omega$-SM.

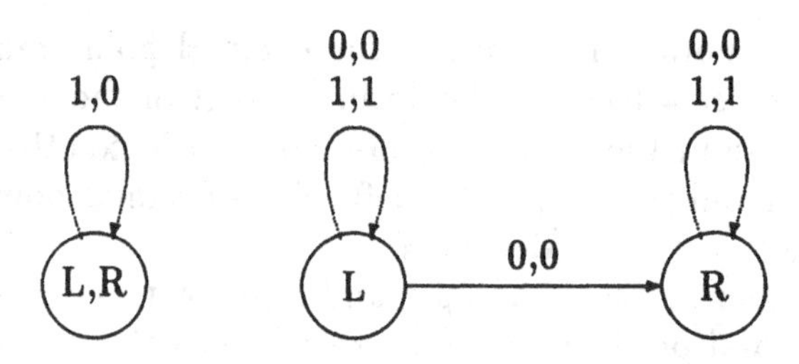

Figure 3. An $\omega\omega$-SM M_2

Proof CA $A = (s, r, f)$ can be simulated by $\omega\omega$-SM M with $2r+1$ states. For simplicity we show the construction for $r = 1$. We construct $M = (S \times S \times S, S, \gamma, S, S)$ where $((p,q,r), q, f(p,q,r), (q,r,s)) \in \gamma$ for each $p, q, r, s \in S$. M is nondeterministic but clearly single-valued, and complete. It is easy to verify that $G_M(c) = G_f(c)$ for all $c \in S^{\mathbb{Z}}$.

Note that M is a simple $\omega\omega$-SM in the terminology introduced in the last section. □

Theorem 7

The family of the global functions of 1-dimensional CA *is properly included in the family of the functions defined by* GCA.

Proof The inclusion follows by Lemma 5. Consider the $\omega\omega$-SM M_2 shown in Fig. 3. It maps the string $1^{\omega\omega}$ to $0^{\omega\omega}$, otherwise it is an identity. Clearly, it is complete and single-valued and thus a GCA. Clearly, G_{M_2} is not the global function of any CA. □

Definition 2

Let $M_i = (K^i, S, K^i_L, K^i_R, \gamma^i)$, *for* $i = 1,2$ *be* $\omega\omega$-SM. *The union of* M_1 *and* M_2 *is denoted by* $M_1 + M_2$. *Assuming* $K^1 \cap K^2 = \emptyset$, *we define* $M_1 + M_2 = (K^1 \cup K^2, S, K^1_L \cup K^2_L, K^1_R \cup K^2_R, \gamma^1 \cup \gamma^2)$. *Clearly,* $\rho(M_1 + M_2) = \rho(M_1) \cup \rho(M_2)$.

Lemma 6

Let $M_1, M_2, \ldots, M_n$ *be single-valued* $\omega\omega$-SM *such that* $\bigcup_{i=1}^{n} \mathrm{dom}(M_i) = S^{\mathbb{Z}}$, *and* $\mathrm{dom}(M_i) \cap \mathrm{dom}(M_j) = \emptyset$, *for* $i \neq j$, *i.e. the domains of* $M_1, \ldots, M_n$ *give a partition of* $S^{\mathbb{Z}}$. *Then the* $\omega\omega$-SM $M_1 + M_2 + \ldots + M_n$ *is a* GCA.

Proof The first property of the partition assures that the union machine is complete, the second one that it is single-valued. □

Example.

$\omega\omega$-SM M_3 shown in Fig. 4 has the domain ${}^{\omega}0(0+1)^{\star}.(0+1)^{\omega}$. Using the binary representation of rational numbers with the decimal point dividing the integer and the fractional part and with an infinite number of leading zeroes, M_3 multiplies (syntactically correct) input by 3 and produces the output in the same notation (in $\rho(M_1)$ the decimal point remains in the same position). We use $\bullet$ to show the decimal point in the diagrams.

$\omega\omega$-SM M_4 shown in Fig. 4 accepts any input with exactly one decimal point and infinite number of 1's to the left of the decimal point. It always produces ${}^{\omega}1.0^{\omega}$ as a representation of ∞.

$\omega\omega$-SM M_5 shown in Fig. 4 accepts any input with less or more than one decimal point and produces ${}^{\omega}.^{\omega}$ (an error message).

Clearly, $M_3 + M_4 + M_5$ is a GCA.

Corollary 2

Let $R_1, \ldots, R_n \subseteq S^{\omega\omega}$ be pair-wise disjoint $\omega\omega$-regular sets, and let $M_1, \ldots, M_{n+1}$ be single-valued $\omega\omega$-SM such that $R_i \subseteq \mathrm{dom}^{\omega\omega}(M_i)$ for $i = 1, \ldots, n$ and $R_{n+1} = R_1 \cup \ldots \cup R_n \subseteq \mathrm{dom}^{\omega\omega}(M_{n+1})$. Let $\hat{M}_i$ be the restriction of M_i to R_i for $i = 1, \ldots, n+1$. Then $\hat{M}_1 + \hat{M}_2 + \ldots + \hat{M}_{n+1}$ is a GCA.

Proof By theorem 2, R_{n+1} is an $\omega\omega$-regular set. Hence the result follows by lemma 6. □

Since by lemma 5 every CA can be implemented by a single-valued $\omega\omega$-SM we can always combine several CA working on different domains into one GCA.

There are other ways how to combine two or more CA's into one GCA. Consider, for example, two CA A_1 and A_2 over alphabet (states) $\{a, b\}$. Then we can easily implement a GCA M over alphabet $\{0, 1, a, b\}$ that preserves 0 and 1, simulates CA A_1 on every subconfiguration in $\{a, b\}^*$ between two 0's (with the neighborhood extended e.g. cyclically) and simulates CA A_2 on the other subconfigurations in $\{a, b\}^*$, i.e. between 0 and 1, 1 and 0, or 1 and 1.

4. Decision Problems

We have considered some decision problems about $\omega\omega$-SM in the previous section. Now, we will study decision problems about GCA. Clearly, every problem that is undecidable for 1-dimensional CA is also undecidable for GCA. So we will consider only those problems that are decidable for CA.

The (1 step) equivalence problem for global CA functions is trivially decidable, they are equivalent iff they are identical. This is not the case for GCA, however, the problem is still decidable.

Theorem 8

The equivalence problem for GCA *is decidable.*

Proof Given two GCA, i.e. two complete and single-valued $\omega\omega$-SM M and N we want to test whether $G_M(c) = G_N(c)$ for all $c \in S^{\mathbf{Z}}$. Note that the equivalence of M and N on $S^{\omega\omega}$ does not imply the equivalence of G_M

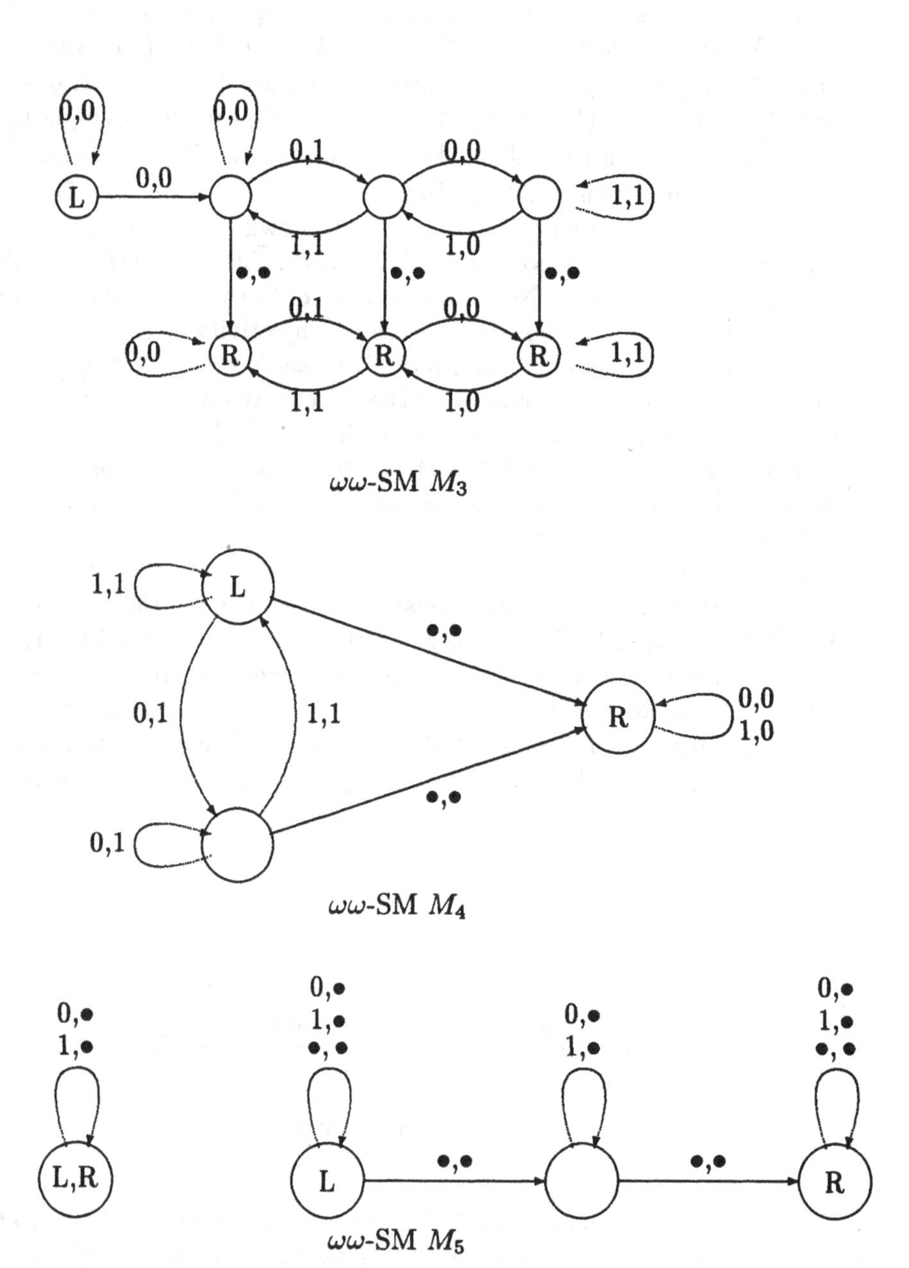

Figure 4. GCA $M_3 + M_4 + M_5$

and G_N on $S^{\mathbf{Z}}$, e.g. the identity and shift are equivalent on $S^{\omega\omega}$ but not on $S^{\mathbf{Z}}$. To prevent shifting we will replace M and N by (one-way) ω-SM $\hat{M}$ and $\hat{N}$, respectively, as in the proof of lemma 4. Clearly, $G_M \equiv G_N$ on $S^{\mathbf{Z}}$ iff $\hat{M} \equiv \hat{N}$ on $(S \times S)^\omega$. The equivalence problem of single-valued ω-transducer is shown to be decidable in (Čulik II and Pachl, 1981), $\hat{M}$ and $\hat{N}$ are (length-preserving) ω-transducers. □

The injectivity and surjectivity are well known decidable problems for 1-dimensional CA, see (Amoroso and Patt, 1972; Čulik II, 1987); we will extend these results to GCA. Note that because of the shift-invariance, the injectivity of GCA M on $S^{\mathbf{Z}}$ clearly implies the injectivity of G_M on $S^{\omega\omega}$. The converse is less obvious but holds, too. The case $k \geq 1, c \neq \sigma^k(c), G_M(c) = G_M(\sigma^k(c)) = d$ seems to violate only the injectivity on $S^{\mathbf{Z}}$ but not on $S^{\omega\omega}$. However, since G_M is shift-invariant we have $(\sigma^k(c), \sigma^k(d)) \in \rho(M)$ and the single-valuedness of M implies that $\sigma^k(d) = d$. Thus G_M on $S^{\omega\omega}$ maps a string which does not have a period of length k to one that has and therefore cannot be injective.

For GCA neither injectivity implies surjectivity nor surjectivity implies injectivity. To show the former consider G defined on the $\omega\omega$-words in ${}^\omega 01^*0^\omega$ by the $\omega\omega$-SM M_f from Fig. 5 otherwise as identity. Clearly, G is injective and not surjective $0^{\omega\omega}$ is not in the range of G. To show the latter consider G defined on strings from ${}^\omega 01^+0^\omega$ by the inverse (interchanged inputs and outputs) M_f^{-1} of $\omega\omega$-SM M_f from Fig. 5, and again as identity elsewhere. Clearly, M_f^{-1} is surjective but not injective since $G_{M_f}^{-1}({}^\omega 010^\omega) = G_{M_f}^{-1}({}^\omega 0^\omega) = {}^\omega 0^\omega$.

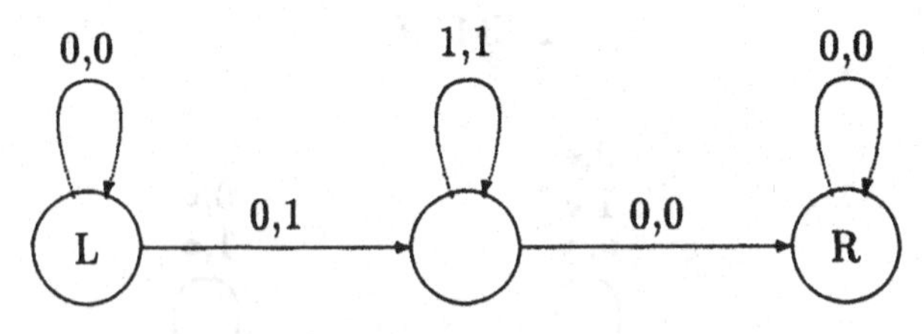

Figure 5. An $\omega\omega$-SM M_f

Following the terminology for 1-dimensional CA, see e.g. (Čulik II *et al.*, 1990) we call a GCA M *reversible* if there exists an "inverse" GCA N such that $G_M(c) = d$ iff $G_N(d) = c$ for all $c, d \in S^{\mathbf{Z}}$.

The following extends a well known result for 1-dimensional CA, see e.g. (Čulik II *et al.*, 1990, Theorem 33).

Theorem 9

A GCA M is reversible iff G_M is bijective.

Proof Assume that G_M is bijective. Let M^{-1} be obtained by interchanging input and output symbols at every transition. Since G_M is bijective M^{-1} is single-valued and complete so it is a GCA. Clearly, M_{-1} defines the inverse of G_M.

The converse is obvious. □

Theorem 10

Given a GCA M *it is decidable whether* G_M *is injective (on* $S^{\mathbb{Z}}$*).*

Proof Clearly, a GCA M is injective iff its inverse M^{-1} is single-valued. The latter is decidable by lemma 4. □

Theorem 11

Given a GCA M *it is decidable whether* G_M *is surjective (on* $S^{\mathbb{Z}}$*).*

Proof By theorem 4 we can construct $\omega\omega$-FA that accepts $G_M(S^{\mathbb{Z}})$. Clearly, G_M is surjective iff $G_M(S^{\mathbb{Z}}) = S^{\mathbb{Z}}$, which is decidable by theorem 5. □

In the theory of CA computation on (pseudo) finite configurations are frequently considered. Now we will study GCA working on (pseudo) finite configurations. Let us reserve 0 for so called quiescent symbol. A configuration c in $S^{\mathbb{Z}}$ is called *pseudo-finite* if there are m, n such that $c = 0$ for all $i \leq m$ and all $i \geq n$. $\omega\omega$-regular set ${}^{\omega}0S^{\star}0^{\omega}$ is the set of bi-infinite words corresponding to (pseudo) finite configurations. We say that $\omega\omega$-SM $M = (K, S, K_L, K_R, \gamma)$ is a 0-GCA if $0 \in S$, M is single-valued, G_M is defined on all (pseudo) finite configurations, i.e. $\mathrm{dom}(\rho^{\omega\omega}(M)) \subseteq^{\omega} 0S^{\star}0^{\omega}$ and G_M preserves (pseudo) finite configurations, i.e. $G_M^{\omega\omega}({}^{\omega}0S^{\star}0^{\omega}) \subseteq^{\omega} 0S^{\star}0^{\omega}$.

Lemma 7

It is decidable whether a given $\omega\omega$-SM *is a* 0-GCA.

Proof Follows from theorem 2 and 4 and lemma 4. □

Using a simple modification of the proofs of theorems 10 and 11 we have the following

Theorem 12

Given a 0-GCA M *it is decidable whether* M *is (a) injective, (b) surjective, on (pseudo) finite configurations.*

References

Amoroso S. and Patt Y. N. Decision procedures for surjectivity and injectivity of parallel maps for tessellation structures. *J. of Comp. and System Sciences* **Vol. no. 6**: 448-464, 1972.

Beauquier D. Ensembles reconnaissables de mots bi-infinis. Limite et déterminisme. in: M. Nivat and D. Perrin, eds. Automata on Infinite Words *Lecture Notes in Computer Science*, Springer-Verlag **Vol. no. 192**: 28-46, 1984.

Berstel J. *Transductions and context-free languages.*, Teubner, Stuttgart, 1979.

Čulik II K. On invertible cellular automata. *Complex Systems* **Vol. no. 1**: 1035-1044, 1987.

Čulik II K. An aperiodic set of 13 Wang tiles. *Discrete Applied Mathematics*, to appear.

Čulik II K. , Hurd L. and Yu S. Computation Theoretic Aspect of Cellular Automata. *Physica D* **Vol. no. 45**: 357-378, 1990.

Čulik II K. and Pachl J. Equivalence problems for mappings of infinite strings. *Inf. and Control* **Vol. no. 49**: 52-63, 1981.

Čulik II K. and Yu S. Cellular automata, $\omega\omega$-regular sets, and sofic systems. *Discrete Applied Mathematics* **Vol. no. 32**: 85-101, 1991.

Čulik II K. , Pachl J. and Yu S. On the limit sets of cellular automata. *SIAM J. Comput.* **Vol. no. 18**: 831-842, 1989.

Hedlund G. A. Transformations commuting with the shift. In: *Topological Dynamics*, eds. Auslander J. and Gottschalk W. G. (Benjamin, New York, 1968), 259; Endomorphisms and automorphisms of the shift dynamical system. *Math. Syst. Theor.* **Vol. 3 no. 320**, 1969.

Hopcroft J. E. and Ullman J. D. *Introduction to automata theory, languages, and computation.* Addison-Wesley, 1979.

Kelley J. L. *General Topology.*, American Book Company, 1955.

Nivat M. and Perrin D. Ensembles reconnaissables de mots biinfinis. *Proceedings of the 14th Annual ACM Symposium on Theory of Computing* 47-59, 1982; also: *Canad. J. Math.* **Vol. no. 38**: 513-537 (revised version), 1986.

AN INTRODUCTION TO AUTOMATA ON GRAPHS

E. RÉMILA
Grima, IUT Roanne,
20 avenue de Paris,
42334 Roanne cedex, France.
LIP, ENS Lyon, CNRS Ura 1398,
46 Allée d'Italie,
69364 Lyon Cedex 07, France

Abstract. The formalism of graph automata is presented. The notion of neighborhood vector is especially described. Afterwards, graph automata and classical cellular automata are compared.

1. Introduction

Cellular graph automata (C. G. A.'s for short) were first introduced by P. Rosenstiehl under the name of "intelligent graphs" (Rosensthiel, 1966), (Rosensthiel *et al.*, 1973), surely because a finite automata network is able to know some properties about its underlying structure. P. Rosenstiehl gives algorithms for finding Eulerian paths, spanning trees and Hamiltonian cycles of "intelligent graphs", under the condition that each node has a fixed degree.

In order to relax that condition, A. Wu and A. Rosenfeld (Wu and Rosenfeld, 1979a) (Wu and Rosenfeld, 1979b) introduce a new notion: the nodes with lower degree than the maximum degree of the graph are connected to special nodes (the ♯ nodes), always remaining in a special state. Their paper gives a fairly comprehensive treatment of C. G. A.'s and shows how we can find various properties about the structure of the underlying graph : radius, center, area, cut nodes, blocs ...

In this paper we give a tutorial presentation of the formalism of graph automata introduced by A. Wu and A. Rosenfeld. If those authors present the power of graph automata, our purpose here is to show the motivations and justifications of their definitions. In particular, we point out the notion of neighborhood vector. From basic examples, we see the interest of this notion.

Afterwards, we compare graph automata and classical cellular automata. This comparizon yields to the notion of Cayley Graph automata.

2. Definition of graph automata

Informally, a computation step of a cellular graph automaton can be described as follows: a finite automaton is placed on each vertex of an undirected (finite or infinite) graph G and these automata are connected according to the edges of G. Through the edges, they exchange messages and the state of each automaton changes according to rules of transition. Formally, we have the following definitions:

2.1. REGURALIZATION OF A GRAPH : D-GRAPH

Let $G_0 = (V_0, E_0)$ be any graph whose each vertex has at most d neighbors. A *d-graph* which regularizes G_0 is a pair (G, g) where :

- i - $G = (V, E)$ is a connected undirected graph, which have two types of vertices:
  - the vertices of degree d,
  - the vertices of degree 1 (called the $\sharp$-vertices)

  The subgraph of G obtained by deleting the $\sharp$-vertices is G_0.
- ii - g is a mapping from the set of ordered pairs (v, v') of vertices of V which are the extremities of a same edge of G into $\mathbf{Z}_d = \{1, 2, \ldots, d\}$, such that, for each vertex v of degree d, the partial function g_v, defined by $g_v(v') = g(v, v')$, is a one-to-one mapping.

Informally, a regularization is a technical tool which permits to consider only two kinds of vertices (the vertices of the second kind are virtual) and to have the edges numbered.

In the following, we only study d-graphs. This is justified by the method of regularization. We now give some complementary definitions.

Let v and v' be the extremities of an edge e of G and let state $i = g(v, v')$. We say that v' is the i-th neighbor of v.

We define a mapping H from the set of vertices of degree d of G into $(\mathbf{Z}_d)^d$ by : $H(v) = (t_1, t_2, \ldots, t_d)$, where, for each integer i such that $1 \le i \le d$, if $g(v, v') = i$ then $g(v', v) = t_i$.

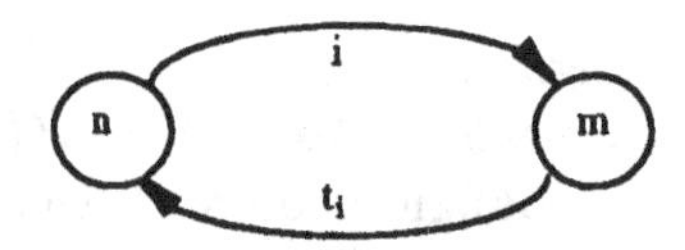

Figure 1. Numbering of directed edges

$H(v)$ is called the *neighborhood vector* of v. It gives relations between numbers of directed edges whose origin is v and numbers of directed edges whose extremity is v (see figure 1).

The subgraph of G obtained by deleting $\sharp$-vertices is called the *underlying graph* of G. It is denoted by $U(G)$. A vertex of degree k in $U(G)$ is called an *k-vertex*.

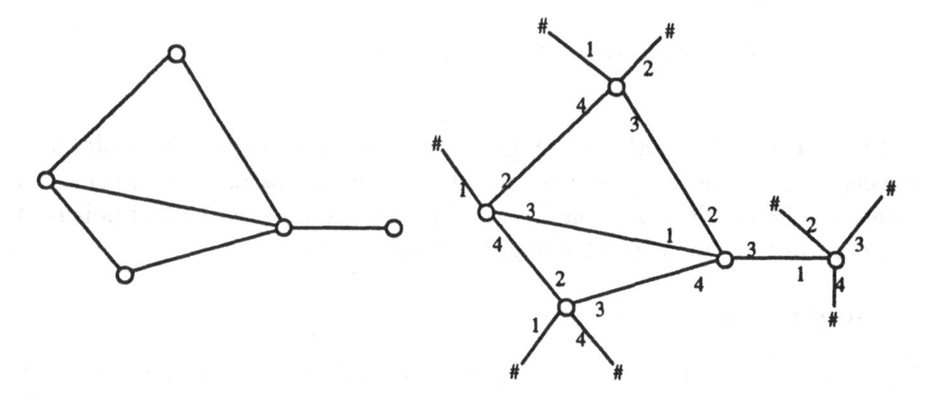

Figure 2. From a graph to a d-graph

2.2. D-GRAPH AUTOMATON

Informally, a finite automaton is placed at each vertex of $U(G)$. At each time step, each vertex reads its neighbor vector, senses the states of its neighbors, and changes its own state according to these data and its preceding state (in the next section, we will see why the neighbor vector is essential).

Formally, a *deterministic d-graph automaton* M is a 4-uple (G, g, Q, δ) where :

- a - (G, g) is a d-graph,
- b - Q is a finite set of *states*, such that the symbol $\sharp$ is an element of Q,
- c - δ is a transition function from $Q \times \mathbf{Z}_d^d \times Q^d$ into Q.

A *configuration* c of G is a mapping from the vertices of G into Q, such that $c(v) = \sharp$, for each $\sharp$-vertex. We say that we have a *step of computation* from a configuration c to a configuration c' (which is denoted by $c \longrightarrow c'$) if, for each vertex v of degree d of G, $c'(v)$ equals $\delta(c(v), H(v), c(v_1), \ldots, c(v_d))$,

where v_i denotes the i-th neighbor of v.
A state q_0 is *quiescent* if, for each vertex v, $\delta(q_0, H(v), q_0, \ldots, q_0) = q_0$.
An *initial* configuration is a configuration such that each vertex of $U(G)$, except a distinguished vertex v_{dist} (which is in a special state q_{start}), is in a same quiescent state q_0.
A configuration c of G, is *realizable* by the automaton **A**, if there exists a sequence of configurations $c_0, c_1, \ldots, c_p$ such that :
- a - c_0 is initial,
- b - $c_p = c$,
- c - $c_i \longrightarrow c_{i+1}$ for each integer i, such that $0 \leq i < p$.

In this case, we say that **A** realizes c, from c_0, in p time units (or at $t = p$ for short).

3. Interest of the neighborhood vector

The use of the neigborhood vector is the most surprizing difference between the classical formalism of cellular automata and the formalism presented above. The example below shows the power of the formalism of the graph automata, with the neighborhood vector.

3.1. ORIENTATION OF A CYCLE

Take a 2-graph formed by three neighboring nodes, say a, b and c, in such a way that the underlying graph is a cycle of length 3 (see figure 3), a is the distinguished node, $H(a) = (1, 1)$, $H(b) = (1, 2)$ and $H(c) = (2, 2)$.

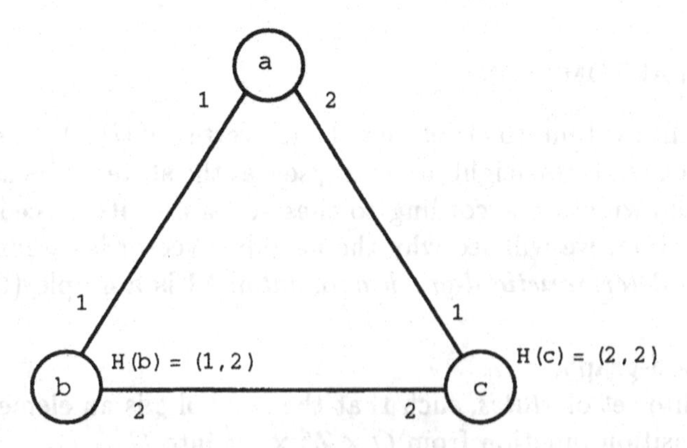

Figure 3. Construction of an orientation. If neighborhood vectors are not used, then vertices b and c always are in a same state.

We want to direct this cycle (i. e. to obtain a configuration such that the state of each vertex indicates (by a correct encoding) the number of the successors of each vertex, for a fixed orientation of the cycle). We claim that this is impossible if the transition function does not depend on the the neighborhood vectors. Assume that the transition function δ does not depend on the neigborhood vector. At $t = 0$, vertex a is in state q_{start} and vertices b and c both are in a same quiescent state q_0. Thus at $t = 1$, vertices b and c both are in a state q_1 such that $q_1 = \delta(q_0, q_{start}, q_0)$. Repeating this argument, one proves that vertices b and c both are in the same state at $t = 2, t = 3, \ldots$ and so on. If number 1 is encoded in the common state for the number of the successor, then vertex a is the common successor of the two other cells which is a contradiction, and if number 2 is encoded, then vertex a has no predecessor, which is another contradiction.

At the opposite, if the transition function δ depends on the neigborhood vector then at $t = 1$, vertex b is in a state q_1 such that $q_1 = \delta(q_0, (1,2)q_{start}, q_0)$ and vertex c is in a state q'_1 such that
$q'_1 = \delta(q_0, (2,2)q_{start}, q_0)$
and consequently δ can be chosen in such a way that $q_1 \neq q'_1$. Thus an orientation of a cycle can be constructed at $t = 1$.

Remark : if the neighborhood vector of each vertex is $(1,2)$, then the difficulty presented here desappears. But if the structure of the underlying graph is not analyzed, a same vector cannot a priori be attributed to each vertex.

3.2. EMISSION AND RECEPTION OF MESSAGES

Generally, we limit ourselves to the case where each state is a d+1-uple $(m_0, m_1, \ldots, m_d)$ (the $\sharp$-state is seen as the d+1-uple $(\sharp, \ldots, \sharp)$) ; m_0 is called the *memory component* of q and m_i (for $1 \leq i \leq d$) is called the i-th*message component* of q (informally, m_0 contains the information stored in the cell and m_i contains the information which is sent to the i-th neighbor).

We also limit ourselves to the case where the state
$q' = \delta(q, H(v), q_1, \ldots, q_d)$ of vertex v after one step only depends on the d+1-uple $(m_0, m'_{t_1}, m'_{t_2}, \ldots, m'_{t_d})$, where m'_{t_i} denotes the t_i-th message component of q_i. In this case, we say that vertex v in state q, receives m'_{t_i} from its i-th neighbor and the i-th neighbor of v sends m'_{t_i} to v (m'_{t_i} is the t_i-th message component of the state of the vertex whose t_i-th neighbor is v).

Remark that vector $H(v)$ is absolutely indispensable to decide what component of q_i equals m'_{t_i}. This vector $H(v)$ plays role of a decoding vector when a set of states is received from neighboring.

If c is a configuration of G such that $c(v) = (m_0, \ldots, m_d)$, we also say that v has m_0 in its memory, when G is in c.
Thus, using the neighborhood vector, we have given a precise sense to the

expressions: "vertex v' sends a message to vertex v", "vertex v' receives a message from vertex v" and "vertex v has m_0 in its memory".

4. Comparizons of cellular automata and graph automata

Why is there no neighborhood vector in the formalism of cellular automata? Because the neigborhood vector is implicit and uniform in a cellular antomaton: take, for example, a bidimensional cellular automaton. If we say that the northern (respectively western, southern, eastern) neighbor is the first (respectively second, third, fourth) neighbor, then the neighborhood vector is always (3, 4, 1, 2) (uniformity of the neighborhood vector), and consequently, the transition function can be defined without reference to this vector (implicitly of the neighborhood vector).

Notice that the implicitly and uniformity of the neighborhood vector is a consequence of the regularity of the networks used for cellular automata.

The example below clearly points out the differences and the analogies between cellular automata and graph automata.

4.1. LEADER ELECTION IN A LINE

We consider (G, g_1) and (G, g_2) which are 2-graphs whose underlying graph is a line of four vertices, say a, b, c and d, in an order given by the line.

In (G, g_2) the neighborhood vector is uniformly $(2, 1)$ (which means that each graph automaton constructed from (G, g_2) is (equivalent to) a cellular automaton : "1" can be interpreted as "left" and "2" as "right"). In (G, g_1), we have $H(a) = H(d) = (2, 1)$ and $H(b) = H(c) = (2, 2)$ (see figure 4).

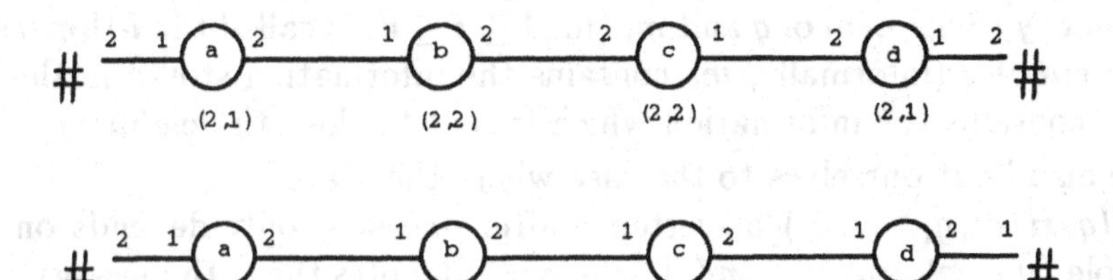

Figure 4. The 2-graphs (G, g_1) and (G, g_2).

A leader can easily be elected on (G, g_2). It suffices to use a graph automaton such that $\delta(q, (2, 1), q', q'')$ is equal to a state "leader" if and only if $q' = \sharp$. With only one step of computation, vertex a becomes the unique leader and always remains leader.

At the opposite, if the four cells of (G, g_1) are in the same state at $t = 0$, then an obvious induction proves that, at each time step, cells a and d are in a same state and cells b and c are in a same state. This yields that a leader will never be elected.

5. Automata on Cayley graphs

In the example above, we have seen the interest of a uniform neighborhood vector. The question which naturally arises now is : what are the graph structures which admit a uniform neighborhood vector ? A partial answer has been given by Z. Róka (Róka, 1994) introducing cellular automata on Cayley graphs.

Cayley graphs are representations of groups : let $S = \{g_1, g_1, \ldots, g_n\}$ be a set and $(r_1, r_2, \ldots, r_p)$ and $(r'_1, r'_2, \ldots, r'_p)$ be two lists of strings of $\{g_1, g_2, \ldots, g_n, g_1^{-1}, g_2^{-1}, \ldots, g_n^{-1}\}^*$. The group

$$G = \left\langle g_1, g_2, \ldots, g_n \mid r_1 = r'_1, r_2 = r'_2, \ldots, r_p = r'_p \right\rangle$$

is the group generated by S such that the equalities which are true in G are exactly those which can be deduced from equalities of the set $\{r_1 = r'_1, r_2 = r'_2, \ldots, r_p = r'_p\}$. This group is unique up to isomorphism. Well-known examples are $\mathbb{Z} =< a| >$, $\mathbb{Z}/n\mathbb{Z} =< a|a^n = e >$ and $\mathbb{Z}\mathbb{Z}^2 =< a, b|ab = ba >$.

A group G, given as above, can be described as a directed graph (called Cayley graph) whose vertices are the elements of G.

An edge is placed from an element g to an element g' if there exists a generator g_i such that $gg_i = g'$. Moreover, this edge is labeled by g_i. Hence, each vertex of a Cayley graph defined by a group with n generators has n incoming edges and n outcoming edges, one of each label.

For example, the Cayley graph deduced from $< a, b, c|abc = e, acb = e >$ is constructed on the triangular lattice of the plane, as shown in Figure 5. Consequently, if a piece P of a triangular lattice is taken as underlying

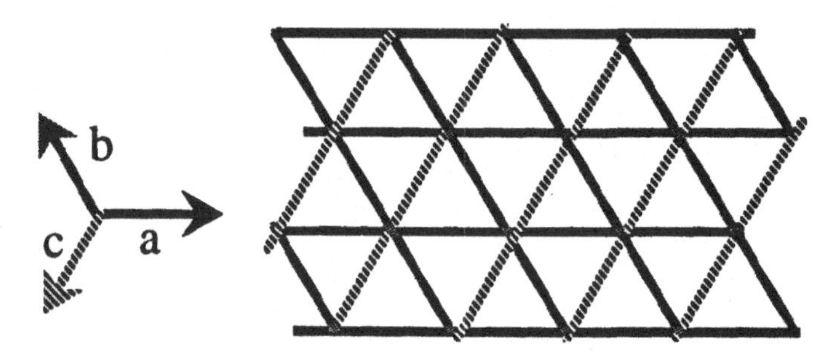

Figure 5. Cayley graph $G_{triangle}$.

graph, a graph automaton with uniform neighborhood vector can easily be constructed from P: if the outcoming edge labeled by a (respectively b, c) takes number 1 (respectively 2, 3) and the incoming edge labeled by a (respectively b, c) takes number 4 (respectively 5, 6), then the uniform neighborhood vector is (4, 5, 6,1, 2, 3).

This kind of construction can be done from any piece of any Cayley graph,which proves that one can construct a graph automaton with uniform neighborhood vector on each graph which is a part of a Cayley graph.

References

Cole S. Real-time computation by n-dimensional iterative arrays of finite-state machine *IEEE Trans. Comput.* Vol. no. C-18: 349–365, 1969.

Róka Zs. Automates cellulaires sur les graphes de Cayley Ph.D Thesis, Université Lyon I et Ecole Normale Supérieure de Lyon, 1994.

Rosenstiehl P. Existence d'automates finis capables de s'accorder bien qu'arbitrairement connectés et nombreux. *Internat. Comp. Centre* Vol. no. 5: 245–261, 1966.

Rosenstiehl P., Fiksel J.R. and Holliger A. Intelligent graphs: Networks of finite automata capable of solving graph problems. Reed R. C. , Ed, *Graph Theory and computing*, Academic Press, New York 210–265, 1973.

Wu A. and Rosenfeld A. Cellular graph automata I. *Information and Control* Vol. no. 42: 305–328, 1979

Wu A. and Rosenfeld A. Cellular graph automata II. *Information and Control* Vol. no. 42: 330–353, 1979.

References

Achasova S., Bandman O., Markova V. and Piskunov S. *Parallel substitution algorithm* World Scientific, 1994.

Albert J. and Čulik K. A simple universal cellular automaton and its one-way and totalistic version. *Complex Systems.* **Vol. no.1**: 1–16, 1987.

Allouche J.-P. and Reder C. Oscillations spatio-temporelles engendrées par un automate cellulaire. *Discr. Appl. Math.* **Vol. no. 8**: 215–254, 1984.

Allouche J.-P. Automates finis en théorie des nombres. *Exposition. Math.* **Vol. no. 5**: 239–266, 1987.

Allouche J.-P. Transcendence of the Carlitz-Goss Gamma function at rational arguments. *Number Theory.* **Vol. no. 60**: 318–328, 1996.

Allouche J.-P., von Haeseler F., Peitgen H.-O., Petersen A. and Skordev G. Automaticity of double sequences generated by one-dimensional linear cellular automata. To appear in *Theoret. Comput. Sci.*

Allouche J.-P., von Haeseler F., Peitgen H.-O. and Skordev G. Linear cellular automata, finite automata and Pascal's triangle. *Discrete Appl. Math.* **Vol. no. 66**: 1–22, 1996.

Allouche J.-P. and Skordev G. Schur congruences, Carlitz sequences of polynomials and automaticity. preprint, *Inst. Dyn. Syst. Univ. Bremen.* **Vol. no. 399**, 1997.

Amoroso S. and Patt Y.N. Decision procedures for surjectivity and injectivity of parallel maps for tesselation structures. *Journal of Computer and Systems Sciences.* **Vol. no. 6**: 448–464, 1972.

Aso H. and Honda N. Dynamical Characteristics of Linear Cellular Automata. *Journal of Computer and System Sciences.* **Vol. no. 30**: 291–317: 1985.

Atrubin A. An iterative one-dimensional real-time multiplier. *Term paper for App. Math. 298*, Stanford University, 1962.

Auger P., Bardou A., Coulombe A. and Degonde J. Computer simulation of different mechanisms of ventricular fibrillation. *Proceedings of the 9th IEEE/EMBS International Conference*: 171–174, 1987.

Auger P., Bardou A., Coulombe A. and Degonde J. Computer simulation of different electrophysiological mechanisms of ventricular fibrillation. *Comput. in cardiol..* **Vol. no. 87**: 527–530, 1987.

Auger P., Coulombe A., Govaere M.-C., Chesnais J.-M., Von Euw D. and Bardou A. Computer simulation of mechanisms of ventricular fibrillation and defibrillation. *Innov. Technol. Biol. Med..* **Vol. no. 10**: 299–312, 1989.

Auger P., Bardou A., Coulombe A., Govaere M.-C., Rochette L., Schreiber M.-C., Von Euw D. and Chesnais J.-M. Computer simulation of ventricular fibrillation and of defibrillating electric shocks. Effects of antiarrhythmic drugs. *Mathl. Comput. Modelling.* **Vol. no. 14**: 576–581, 1990.

Auger P., Coulombe A., Dumée P., Govaere M.-C., Chesnais J.-M. and Bardou A. Computer simulation of cardiac arrhythmias and of defibrillating electric shocks. Effects of antiarrhythmic drugs. in *Non linear coherent structures in Physics and Biology.* Lecture Notes in Physics **Vol. no. 393**, Remoissenet, M. and Peyrard, M. Eds. Heidelberg: Springer Verlag, 133–140, 1990.

Auger P., Cardinal R., Bril A., Rochette L. and Bardou A. Interpretation of cardiac mapping by means of computer simulations: Applications to calcium, lidocaïne and to brl 34915. *Acta Biotheoretica.* **Vol. no. 40**: 161–168, 1992.

Balzer R. Studies concerning minimal time solutions to the firing squad synchronization problem Ph.D. thesis, Carnegie Institute of Technology, 1966.

Balzer R. An 8-states minimal time solution to the firing squad synchronization problem. *Information and Control.* **Vol. no.10**: 22–42, 1967.

Banks E. Universality in Cellular Automata. *I.E.E.E. Ann. Symp. Switching and Automata Theory.* Santa Monica, **Vol. no. 11**: 194–215, 1970.

Banks E. Information processing and transmission in cellular automata. *PhD thesis*,

Mass. Inst. of Tech., Cambridge, Mass, 1971.

Banks J., Brooks J., Cairns G., Davis G. and Stacey P. On Devaney's definition of chaos. *American Mathematical Monthly.* **Vol. no. 99**: 332–334, 1992.

Bardos C. and Bessis D., editors, *Bifurcation Phenomena in Mathematical Physics and Related Topics.* D. Reidel, Dordrecht, 1980.

Bardou A., Auger P., Cardinal R., Dumée P., Birkui P., Von Euw D. and Govaere M.-C. Theoretical study by means of computer simulation of conditions in which extrasystoles can trigger ventricular fibrillation. validation by epicardial mappings. *Journal of Biological Systems.* **Vol. no. 1**: 147–158, 1993.

Bardou A., Auger P., Achour S., Dumée P., Birkui P. and Govaere M.-C. Effect of myocardial infarction and myocardial ischemia on induction of cardiac reentries and ventricular fibrillation. *Acta Biotheoretica.* **Vol. no. 43**: 363–372, 1995.

Bardou A., Auger P., Chassé J.-L. and Achour S. Effects of local ischemia and transient conduction blocks on the induction of cardiac reentries. *International Journal of Bifurcation and Chaos.* **Vol. no. 6**: 1657–1664, 1996.

Bartlett R. and Garzon M. Monomial Cellular Automata. *Complex Systems.* **Vol. no. 7**: 367–388, 1993.

Bartlett R. and Garzon M. Bilinear Cellular Automata. *Complex Systems.* **Vol. no. 9**: 455–476, 1995.

Bass F.M. A new product growth for model consumer durables. *Management Sciences.* **Vol. no.15**:215, 1969.

Beauquier D. Minimal automaton for a factorial, transitive, rational language. *Theoretical Computer Science.* **Vol. no. 67**: 65–73, 1989.

Beauquier D. Ensembles reconnaissables de mots bi-infinis. Limite et déterminisme. in: M. Nivat and D. Perrin, eds. Automata on Infinite Words *Lecture Notes in Computer Science*, Springer-Verlag **Vol. no. 192**: 28-46, 1984.

Beeler G. and Reuter R. Reconstruction of the action potential of ventricular myocardial fibres. *J. Physiol.*. **Vol. no. 268**:177–210, 1977.

Ben-Naim E. and Krapivsky P.L. Cluster approximation for the contact process. *J. Phys. A: Math. Gen.* **Vol. no.27**:L481, 1994.

Berlekamp E., Conway V., Elwyn R. and Guy R. *Winning way for your mathematical plays, volume 2*, Academic Press, 1982.

Berstel J. *Transductions and context-free languages.*, Teubner, Stuttgart, 1979.

Berthé V. Automates et valeurs de transcendance du logarithme de Carlitz. *Acta Arith.*. **Vol. no. 66**: 369–390, 1994.

Bigger T. and Sahar D. Clinical types of proarrhythmic response to antiarrhythmic drugs. *Amer. J. Cardiol.* **Vol. no. 59**: 2E–9E, 1987.

Blanchard F., Kůrka P. and Maas A. Topological and measure-theoretic properties of one-dimensional cellular automata. *Physica D.* **Vol. no. 103**: 86–99, 1997.

Boccara N. and Cheong K. Automata network sir models for the spread of infectious diseases in populations of moving individuals. *J. Phys. A: Math. Gen.*. **Vol. no.25**:2447, 1992.

Boccara N. and Cheong K. Critical behaviour of a probabilistic automata network sis model for the spread of an infectious disease in a population of moving individuals. *J.Phys. A: Math. Gen.*. **Vol. no.26**: 3707, 1993.

Boccara N., Cheong K. and Oram M. Probabilistic automata network epidemic model with births and deaths exhibiting cyclic behaviour. *J. Phys. A: Math. Gen.*. **Vol. no. 27**:1585, 1994.

Boccara N. Fukś H. and Geurten S. A new class of automata networks. *Physica D.* **Vol. no.103**, 1997. To appear.

Braga G., Cattaneo G., Flocchini P. and Vogliotti C. Pattern growth in elementary cellular automata. *Theoretical Computer Science.* **Vol. no. 145**: 1–26, 1995.

Bucher B. and Čulik II K. On real time and linear time cellular automata. *R.A.I.R.O. Informatique théorique/Theoretical Informatics.* **Vol.18 no.4**:307–325, 1984.

Burks E. *Theory of Self-reproduction*, University of Illinois Press, Chicago, 1966.

Burks E. *Essays on Cellular Automata* University of Illinois Press, 1972.

Byl J., Self-reproduction in small cellular automata. *Physica D.* **Vol. no. 34**: 295–299, 1989.

Caianiello E.R. and de Luca A. Decision equation for binary systems: application to neuronal behavior. *Kybernetic.* **Vol. no. 3**: 33-40, 1966.

Cardinal R., Vermeulen M., Shenasa F., Roberge P., Page P., Hélie F. and Savard P. Anisotropic conduction and functional dissociation of ischemic tissue during reentrant ventricular tachycardia in canine myocardial infarction. *Circulation.* **Vol. no. 77**: 1162–1176, 1988.

Cattaneo G., Finelli M. and Margara L. Investigating topological chaos by elementary cellular automata dynamics. Submitted to *Theoretical Computer Science.* 1998.

Cattaneo G. and Margara L. Generalized sub-shifts in elementary cellular automata: the strange case of chaotic rule 180. Accepted in *Theoretical Computer Science.* 1998.

Cayley A., Theory of groups. *American Journal of Mathematics.* **Vol. no. 1**: 50–52, 1878.

Chandra C., Kozen D. and Stockmeyer L. Alternation. *J. ACM.* **Vol.28 no.1**:114–133, 1981.

Chang C., Ibarra O. and Palis M. Efficient simulations of simple models of parallel computation by space-bounded TMs and time-bounded alternating TMs. *Theoretical Computer Science.* **Vol. no.68**:19–36, 1989.

Chang C., Ibarra O. and Vergis A. On the power of one-way communication. *J. ACM.* **Vol. no.35**: 697–726, 1988.

Chassé J.-L., Auger P., Bardou A. and Achour S. Some mechanisms of initiation of ventricular fibrillation: Transient blocks and ischemic zones. *IEEE/Engin. Med. Biol., Soc. 17th Conf. Proceed.* : 1755–1758, 1995.

Chiaso D., Michaels D. and Jalife J. Supernormal excitability as a mechanism of chaotic dynamics of activation in cardiac purkinje fiber. *Circ. Res.* **Vol. no. 90**: 525–545, 1990.

Choffrut C. and Čulik K. On real time cellular automata and trellis automata. *Acta Informatica.* **Vol. no. 21**: 393–407, 1984.

Chomsky N. and Schützenberger M. The algebraic theory of context-free languages. in: *Computer programming and formal languages*, North Holland, Amsterdam, 118–161, 1963 .

Christol G. Ensembles presque périodiques k-reconnaissables. *Theoret. Comput. Sci.* **Vol. no. 9**: 141–145, 1979.

Christol G., Kamae T., Mendès France M. and Rauzy G. Suites algébriques, automates et substitutions. *Bull. Soc. Math. France* **Vol. no. 108**: 401–419, 1980.

Cobham A. On the base-dependence of sets of numbers recognizable by finite automata. *Math. Systems Theory.* **Vol. no. 3**: 186–192, 1969.

Cobham A. Uniform tag sequences. *Math. Systems Theory.* **Vol. no. 6**: 164–192, 1972.

Codd E.F. *Cellular Automata*, Academic Press, New York, 1968.

Codenotti B. and Margara L. Transitive cellular automata are sensitive. *American Mathematical Monthly.* **Vol. no. 103**: 58–62, 1996.

Cole S. Real-time computation by n-dimensional iterative arrays of finite-state machine. *IEEE Trans. Comput.***Vol. no. C-18**: 349–365, 1969.

Collet P. and Eckmann J.P. *Iterated Maps of the Interval as Dynamical Systems.* Birkhauser, Basel, 1980.

Conley C. Isolated invariant sets and the Morse index. *American Mathematical Society, CBMS Regional Conference*, **Ser. 38**, 1978.

Cook S. A taxonomy of problems with fast parallel algorithms. *Information and Control.* **Vol. no.64**: 2–22, 1985.

Cosnard M., Goles E. and Moumida D. Bifurcation structure of a discrete neuronal equation. *Discrete Applied Mathematics.* **Vol. no. 21**: 21-34, 1988.

Cosnard M., Goles E., Moumida D. and de St. Pierre T. Dynamical behavior of a neural automaton with memory. *Complex Systems.* **Vol. no. 2**: 161-176, 1988.

Cosnard M., Tchuente M. and Tindo G. Sequences Generated by Neuronal Automata with Memory. To appear, 1998.

Coven E.M. and Paul M.E. Sofic systems. *Israel J. of Mathematics.* **Vol. 20 no. 2**: 165–177, 1975.

Crannell A. The role of transitivity in Devaney's definition of chaos. *American Mathematical Monthly.* **Vol. no. 102**: 768–793, 1995.

Čulik II K. On invertible cellular automata. *Complex Systems.* **Vol. no. 1**: 1035-1044, 1987.

Čulik II K. Variation of the firing squad synchronization problem. *Information Processing Letter.* **Vol. no. 30**: 152-157, 1989.

Čulik II K. An aperiodic set of 13 Wang tiles. To appear in *Discrete Applied Mathematics.* 1998.

Čulik II K., Dube S. Fractal and Recurrent Behavior of Cellular Automata. *Complex Systems.* **Vol. no. 3**: 253–267, 1989.

Čulik K., Gruska J. and Salomaa A. Systolic trellis automata (for VLSI). *Res. Rep. CS-81-34* Dept. of Comput. Sci., University of Waterloo, **Vol. no. 34**, 1981.

Čulik K. and Hurd L. Computation theoretic aspects of cellular automata. *Physica D.* **Vol. no.32**: 357–378, 1990.

Čulik II K. , Hurd L. and Yu S. Computation Theoretic Aspect of Cellular Automata. *Physica D.* **Vol. no. 45**: 357-378, 1990.

Čulik II K., Ibarra O. and Yu S. Iterative Tree Arrays with Logarithmic Depth. *Intern. J. Computer Math.* **Vol. no.20**: 187–204, 1986.

Čulik II K. and Pachl J. Equivalence problems for mappings of infinite strings. *Inf. and Control.* **Vol. no. 49**: 52-63, 1981.

Čulik II K., Pachl J. and Yu S. On the limit sets of cellular automata. *SIAM Journal on Computing.* **Vol. 18 no. 4**: 831–842, 1989.

Čulik K. and Yu S. Undecidability of CA classification schemes. *Complex Systems.* **Vol. no. 2**: 177–190, 1988.

Čulik II K. and Yu S. Cellular automata, $\omega\omega$-regular sets, and sofic systems. *Discrete Applied Mathematics.* **Vol. no. 32**: 85-101, 1991.

Dekking F. , Mendès France M. and van der Poorten A. Folds!. *Math. Intelligencer.* **Vol. no. 4**: 130–138, 173–181, 190–195, 1982.

Deligne P. Intégration sur un cycle évanescent. *Invent. Math.* **Vol. no. 76**: 129–143, 1984.

Delorme M. and Mazoyer J. Reconnaissance de Langages sur Automates Cellulaires. *Research report LIP-IMAG,* **Vol. no. 94-46**, 1994.

Delorme M. and Mazoyer J. An overview on language recognition on one dimensional cellular automata. in *Semigroups, Automata and Langages.* World Scientific, J.Almeida Ed: 85-100, 1996.

Delorme M., Mazoyer J. and Tougne L. Discrete parabolas and circles on 2D cellular automata. To appear in *Theoretical Computer Science.*

Devaney R. *Introduction to chaotic dynamical systems.* Addison-Wesley, second ed., 1989.

Dickman R. and Jense, I. Time-dependent perturbation theory for nonequilibrium lattice models. *Phys. Rev. Lett.* **Vol. no. 67**:2391, 1991.

Dubacq J.-C.. How to simulate Turing machines by invertible one-diensional cellular automata. *International Journal of Foundations of Computer Science.* **Vol.6 no. 4**: 395–402, 1995.

Durand B. Global properties of 2D-cellular automata. in *Cellular Automata and Complex Systems*, GolèsE. and Martinez S. Eds, Kluwer, 1998.

Dyer C. One-way bounded cellular automata. *Information and Control.* **Vol. no.44**: 54–69, 1980.

Eckmann J-P. and Ruelle D. Ergodic theory of chaos and strange attractors. *Rev. Modern Physics.* **Vol. no. 57**: 617–656, 1985.

Fischer P.C. Generation of primes by a one dimensional real time iterative array. *Journal of A.C.M..* **Vol. no. 12**: 388–394, 1965.

Fischer R. Sofic systems and graphs. *Monatshefte für Mathematik*. **Vol. no. 80**: 179–186, 1975.

Fokas A., Papadopoulou E. and Saridakis V. Particles in soliton cellular automata. *Complex Systems*. **Vol. no. 3**: 615–633, 1989.

Fukś H. and Boccara N.: Cellular automata models for diffusion of innovations. in *Instabilities and Nonequlibrium Structures VI*, editor E. Tirapegui, Kluwer, Dordrecht, to appear.

Gajardo A. Universality in a 2-dimensional cellular space with a neighborhood of cardinality 3. *Preprint*, 1998.

Garzon M. Cyclic Automata. *Theoretical Computer Science*. **Vol. no. 53**: 307–317, 1987.

Garzon M.Cayley Automata. *Theoretical Computer Science*. **Vol. no. 103**: 83–102, 1993.

Garzon M. *Models of Massive Parallelism*. Springer, 1995.

Gleason A. Semigroups of shift register counting matrices. *Math. Systems Theory*. **Vol. no. 25**: 253–267, 1992, Revised by Kochman F., Neuwirth L.

Godsil C. and Imrich W. Embedding graphs in Cayley graphs. *Graphs and Combinatorics*. **Vol. no. 3**: 39–43, 1987.

Goles E. and Martinez S. *Dynamical Behaviour and Applications*. Kluwer Academic Publishers, 1990.

Goles E., Maass A. and Martinez S. On the limit set of some universal cellular automata. *Theoretical Computer Science*. **Vol. no. 110**: 53–78, 1993.

Golze U. Differences between 1- and 2-dimensional cell spaces. in Lindenmayer A. and Rozenberg G., editors, *Automata, Languages and Development* North-Holland, 369–384, 1976.

Goto E. A minimal time solution to the firing squad synchonization problem. Course notes for applied mathematics Harvard University, 1962.

Greenberg J., Greene C. and Hastings S. A combinatorial problem arising in the study of reaction-diffusion equations. *SIAM J. Algebraic Discrete Methods*. **Vol. no. 1**: 34–42, 1980.

Greenberg J., Hassard B. and Hastings S. Pattern formation and periodic structures in systems modeled by reaction-diffusion equations. *Bull. Am. Math. Soc.* **Vol. no. 84**: 1296–1327, 1978.

Greenberg J. . and Hastings S. Spatial patterns for discrete models of diffusion in excitable media. *SIAM J. Appl. Math.* **Vol. no. 34**: 515–523, 1978.

Guckenheimer J. and Holmes P. *Nonlinear oscillations, dynamical systems, and biforcations of vector fields*. Springer Verlag, Berlin, 1983.

Gutowitz H.A. Local Structure Theory for Cellular Automata. Ph.D. thesis, Rockefeller University, New York, New York, 1987.

Gutowitz H. A hierarchical classification of cellular automata. *Physica D.* Special issue, **Vol. no. 45**: 130–156, 1990.

Gutowitz H.A., Victor J.D. and Knight B.W. Local structure theory for cellular automata. *Physica D.* **Vol. no. 28**:18, 1987.

Hägerstrand T. The Propagation of Innovation Waves. Lund Studies in Geography. Gleerup, Lund, Sweden, 1952.

Hägerstrand T. On Monte Carlo simulation of diffusion. *Arch. Europ. Sociol.* **Vol. no. 6**:43, 1965.

Harris T.E. *Ann. Prob.* **Vol. no. 2**:969, 1974.

Hartmanis J. and Stearns R. On the computational complexity of algorithms. *Trans. Amer. Math. Soc.* **Vol. 117 no. 5**: 285–306, 1965.

Haynes K.E. Mahajan V. and White G.M. Innovation diffusion: A deterministic model of space-time integration with physical analog. *Socio-Econ. Plan. Sci.* **Vol. no. 11**:23, 1977.

Head T. Linear CA: injectivity from ambiguity. *Complex Systems*. **Vol. 3 no. 4**: 343–348, 1989.

Hedlund G. Transformations commuting with the shift. In: *Topological Dynamics*, eds. Auslander J. and Gottschalk W. G. (Benjamin, New York, 1968), 259; Endomorphisms and automorphisms of the shift dynamical system. *Math. Syst. Theor.* **Vol. 3 no. 320**, 1969.

Hedlund G. Endomorphisms and automorphisms of the shift dynamical system. in Z. Kohavi, editor, *The theory of machines and computation* Academic Press, New York, 320–375, 1971.

Heen O. Economie de ressources sur Automates Cellulaires. *Diplôme de Doctorat*, Université Paris 7 (in french), 1996.

Hélie F., Cossette F., Vermeulen M. and Cardinal R. Differential effects of lignocaine and hypercalcemia on anisotropic conduction and reentry in the ischemically damaged canine ventricle. *Cardiovascular research.* **Vol. no. 29**: 359–372, 1995.

Hemmerling A. Real time recognition of some langages by trellis and cellular automata and full scan Turing machines. *EATCS.* **Vol. no. 29**: 35-39, 1986.

Hennie F. *Iterative arrays of logical circuits* MIT Press, Cambridge, Mass., 1961.

Hodgkin A. and Huxley A. A quantitative description of membrane current and its application to conduction and excitation in nerve. *J. Physiol.* **Vol. no. 117**: 500–544, 1952.

Hodgkin A. and Huxley A. A quantitative description of membrane current and its application to conduction and excitation in nerve. *Bull. Math. Biol.* **Vol. no. 52**: 25–71, 1990.

Holmes P. Poincaré, celestial mechanics, dynamical system theory and chaos. *Physics Report (Review Section of Physics Letters).* **Vol. no. 193**: 137–163, 1990.

Hopcroft J.E. An n log n algorithm for minimizing states in a finite automaton. *Math. Systems Theory.* **Vol. no. 3**: 189–196, 1971.

Hopcroft J. and Ullman J. *Introduction to automata theory, languages, and computation.* Addison-Wesley, 1979.

Horowitz L., Zipes D., Campbell R., Morganroth J., Podrid B., Rosen M. and Woosley R. Proarrhythmia, arrhythmogenesis or aggravation of arrhythmia-a status report. *Amer. J. Cardiol.* **Vol. no. 59**: 54E–56E, 1987.

Hurd L. Formal language characterizations of cellular automata limit sets. *Complex Systems.* **Vol. 1 no. 1**: 69–80, 1987.

Hurley M. Attractors, persistence and density of their basin. *Trans. Amer. Math. Soc.* **Vol. no. 269**: 247–27, 1982.

Ibarra O. and Jiang T. *On one-way cellular arrays, SIAM J. on Computing.* **Vol. no. 16**: 1135–1154, 1987.

Ibarra O. and Jiang T. Relating the power of cellular arrays to their closure properties. *Theoretical Computer Science.* **Vol. no. 57**: 225–238, 1988.

Ibarra O., Jiang T. and Vergis A. On the power of one-way communication. *J.A.C.M.* **Vol.35 no. 3**: 697-726, 1988.

Ibarra O., Kim S. and Moran S. Sequential machine characterizations of trellis and cellular automata and applications. *SIAM J. Computing.* **Vol. no.14**: 426–447, 1985.

Ibarra O. and Palis M. Two-dimensional systolic arrays: characterizations and applications. *Theoretical Computer Science.* **Vol. no. 57**: 47–86, 1988.

Ibarra O., Palis M. and Kim S. Some results concerning linear iterative (systolic) arrays. *J. of Parallel and Distributed Computing.* **Vol. no. 2**: 182–218, 1985.

Ibarra O., Kim S. and Palis M. Designing Systolic Algorithms using sequential machines. *IEEE Trans. on Computers.* **Vol.C35 no. 6**: 31–42, 1986.

Inoue K. and Nakamura A. Some properties of two-dimensional on-line tessalation acceptors. *Information Sciences.* **Vol. no. 13**: 95-121, 1977.

Ishii S. Measure theoretic approach to the classification of cellular automata. *Discrete Applied Mathematics.* **Vol. no. 39**: 125–136, 1992.

Jen E. Linear cellular automata and recurrence systems in finite fields. *Comm. Math. Physics.* **Vol. no. 119**: 13–28, 1988.

Jensen I. and Dickman R. Nonequilibrium phase transitions in systems with infinitely many absorbing states. *J. Stat. Phys.* **Vol. no. 48**:1710, 1993.

Kari J. Reversibility and surjectivity problems of cellular automata. *Journal of Computer*

and System Sciences. Vol. no. **48**: 149–182, 1994.
Kelley J. L. *General Topology.* American Book Company, 1955.
Kitagawa T. Dynamical Systems and Operators Associated with a Single Neuronic Equation. *Mathematical Biosciences.* Vol. no. **18**, 1973.
Knudsen C. Chaos without nonperiodicity. *American Mathematical Monthly.* Vol. no. **101**: 563–565, 1994.
Korec I. Pascal triangles modulo n and modular trellises. *Computers and Artificial Intelligence.* Vol. no. **9**: 105–113, 1990.
Kosaraju S. On some open problems in the theory of cellular automata. *IEEE Trans. on Computers.* Vol. C no. **23**: 561–565, 1974.
Kůrka P. Languages, equicontinuity and attractors in cellular automata. *Ergod. Th. & Dynam. Sys.* Vol. no. **17**: 417–433, 1997.
Langton C. Self-Reproduction in Cellular Automata. *Physica.* Vol. no. **100**: 135–144, 1984.
LaSalle J.-P. Stability Theory for Difference Equations. MAA Studies in Math., *American Mathematical Society.* 1976.
Levin L. Theory of computation: How to start. *SIGACT News.* Vol. **22** no.**1**: 47–56, 1991.
Liggett T.M. Interacting particle systems. Springer-Verlag, New York, 1985.
Lind D. and Marcus B. *Introduction to Symbolic Dynamics and Coding.* Cambridge University Press, 1995.
Lindgren K. and Nordahl M. Universal computation in simple one dimensional cellular automata. *Complex Systems.* Vol. no. **4**: 299–318, 1990.
Litow B. and Dumas P. Additive cellular automata and algebraic series. *Theoret. Comput. Sci.* Vol. no. **119**: 345–354, 1993.
Luo C. and Rudy Y. A model of the ventricular action potential. depolarization, repolarization and their interaction. *Circ. Res.* Vol. no. **68**: 1501–1526, 1991.
MacEachern S. and Berliner L. Aperiodic chaotic orbits. *American Mathematical Monthly.* Vol. no. **100**: 237–241, 1993.
Machí A. and Mignosi F. Garden of Eden configurations for cellular automata on Cayley graphs of groups. *SIAM Journal on Discrete Mathematics.* Vol. no. **6**: 44–56, 1993.
Machtley M. and Young P. *An introduction to the general theory of algorithms*, Elsevier, 1978.
Mahajan V., Muller E. and Bass F.M. New product diffusion models in marketing: A review and directions for research. *Journal of Marketing.* Vol. no. **54**:1, 1990.
Mahajan V. and Peterson R.A. Models for Innovation Diffusion. Number 07-048 in Quantitative Applications in the Social Sciences. Sage Publications, Beverly Hills, 1985.
Margara L. Cellular Automata and Chaos. PhD dissertation, Università di Pisa, Dipartimento di Informatica, 1995.
Martin O., Odlyzko A. and Wolfram S. Algebraic properties of cellular automata. *Comm. Math. Phys.* Vol. no. **93**: 219–258, 1984.
Martin B. A universal automaton in quasi-linear time with its s-n-m form. *Theoretical Computer Science.* Vol. no. **124**: 199–237, 1994.
MazoyerJ. A six states minimal time solution to the firing squad synchronization problem. *Theoretical Computer Science.* Vol. no. **50**: 183–238, 1987.
Mazoyer J. Entrées et sorties sur lignes d'automates *Algorithmique Parallèle*, Masson, Cosnard C., Nivat M. and Robert Y., Edts, 47–61, 1992.
Mazoyer J. Cellular automata, a computational device. in *CIMPA School on Parallel Computation.* Temuco (Chile), 1994.
Mazoyer J. Computations on one-dimensional cellular automaton. *Annals of Mathematics and Artificial Intelligence.* Vol. no. **16**: 285–309, 1996.
Mazoyer J. Parallel language recognition on a plane of automata. Unpublished paper, 1998.
Mazoyer J. and RapaportI. Inducing an order on cellular automata by a grouping oper-

ation. in *STACS'98*, LNCS **Vol. 1373**: 116–127, 1998.

Mazoyer J. and Terrier V. Signals in one dimensional finite automata, Tech. Report 1994-51, LIP-ENS Lyon, 1994, accepted in *Theoretical Computer Science.*

McCullough W. and Pitts W. A logical calculus of the ideas immanent in nervous activity. *Bull. Math. Biophys.* **Vol. no. 5**: 115–133, 1943.

Mehlhorn K. *Data Structures and Algorithms, vol. II: Graph Algorithms and NP-Completeness.* Springer-Verlag, 1984.

Mendès France M. and Yao J.-Y. Transcendence and the Carlitz-Goss gamma function. *J. Number Theory.* **Vol. no. 63**: 396–402, 1997.

Minsky M. *Finite and infinite machines* Prentice Hall, 1967.

Moe G., Rheinboldt W. and Abildskov J. R. A computer model of atrial fibrillation. *Am. Heart J.* **Vol. no. 67**: 200–220, 1964.

Moore E. F. Machine models of self-reproduction. *Proc. Symp. Appl. Math.* **Vol. no. 14**: 17–33, 1962.

Morita K. A Simple Construction Method of a Reversible Finite Automaton out of Fredkin Gates, and its Related Model. *Transactions of the IEICE.* **Vol. no. E**: 978–984, 1990.

Morita K. and Imai K. A Simple Self-Reproducing Cellular Automaton with Shape-Encoding Mechanism. *Artificial Life V.* 1996.

Mosconi J. La constitution de la théorie des automates. *Thèse d'Etat*, Université Paris 1, 1989.

Myhill J. The converse of moore's garden-of-eden theorem. *Proc. AMS.* **Vol. no. 14**: 685–686, 1963.

Nagami H., Kitahashi and Tanaka Characterization of Dynamical Behaviour Associated with a Single Neuronic Equation. *Mathematical Biosciences.* **Vol. no. 32**, 1976.

Nasu M. Homomorphisms of graphs and their global maps. in Saito N. and Nishiziki T., editors, *Graph theory and algorithms*, Springer Verlag, **LNCS 108**:159–170, 1981.

Nasu M. Textile Systems for Endomorphisms and automorphisms of the shift. Memoires, *American Mathematical Society.* 1995.

Nivat M. and Perrin D. Ensembles reconnaissables de mots biinfinis. *Proceedings of the 14th Annual ACM Symposium on Theory of Computing* 47-59, 1982; also: *Canad. J. Math.* **Vol. no. 38**: 513-537 (revised version), 1986.

Noble D. *The initiation of the heart beat.* Oxford University Press, 1975.

Nourai F. and Kashef S. A universal four states cellular computer. *IEEE Transaction on Computers.* **Vol.C24 no. 8**: 766–776, 1975.

Packard N. Complexity of growing patterns in cellular automata. in *Dynamical Systems and Cellular Automata*, Demongeot M. T. J., Goles E., ed., Academic Press, 1985.

Papadimitriou C.H. *Computational Complexity.* Addison-Wesley, 1994 .

Paterson N. Tape bounds for time-bounded Turing machines. *Journal of Computer and System Sciences.* **Vol. no. 6**: 116–124, 1972.

Peitgen H.-O., Rodenhausen A. and Skordev G. Self-similar Functions Generated by Cellular Automata. *Research Report 426*, Bremen University, 1998.

Pertsov A., Davidenko J., Salomonsz R., Baxter W. and Jalife J. Spiral waves of excitation underlie reentrant activity in isolated cardiac muscle. *Circ. Res.* **Vol. no. 72**: 631–650, 1993.

Reimen N., Superposable Trellis Automata. *LNCS*,**Vol. no. 629**: 472–482, 1992.

Richardson D.Tesselations with local transformations. *Journal of Computer and System Sciences.* **Vol. no. 5**: 373–388, 1972.

Rogers E.M. *Diffusion of Innovations.* The Free Press, New York, 1995.

Rogers H. *Theory of Recursive Functions and Effective Computability*, MIT Press, 1967.

Róka Zs. Automates cellulaires sur les graphes de Cayley Ph.D Thesis, Université Lyon I et Ecole Normale Supérieure de Lyon, 1994.

Róka Zs.One-way cellular automata on Cayley graphs. *Theoretical Computer Science.* **Vol. no. 132**: 259–290, 1994.

Róka Zs. Simulations between Cellular Automata on Cayley Graphs. to appear in *The-*

oretical Computer Science. 1998.

Rosenberg A. Real-time definable languages. *J. ACM*. **Vol.14 no. 4**: 645–662, 1967.

Rosenstiehl P. Existence d'automates finis capables de s'accorder bien qu'arbitrairement connectés et nombreux. *Internat. Comp. Centre*. **Vol. no. 5**: 245–261, 1966.

Rosenstiehl P., Fiksel J.R. and Holliger A. Intelligent graphs: Networks of finite automata capable of solving graph problems. Reed R. C. , Ed, *Graph Theory and computing*, Academic Press, New York 210–265, 1973.

Ruelle D. Strange attractors. *Math. Intelligencer*. **Vol. no. 2**: 126–137, 1980.

Ruelle D. Small random perturbations of dynamical systems and the definition of attractors. *Commun. Math. Phys*. **Vol. no. 82**: 137–15, 1981.

Ruzzo W. On uniform circuit complexity. *Journal of Computer and System Sciences*. **Vol. no. 22**: 365–383, 1981.

Salon O. Suites automatiques à multi-indices. *Séminaire de Théorie des Nombres de Bordeaux*. **Exposé 4**: 4-01–4-27; followed by an Appendix by J. Shallit, 4-29A–4-36A, 1986-1987.

Salon O. Suites automatiques à multi-indices et algébricité. *C. R. Acad. Sci. Paris, Série I*. **Vol. no. 305**: 501–504, 1987.

Savitch W. Relationships between nondeterministic and deterministic complexities. *Journal of Computer and System Sciences*. **Vol. no. 4**: 177–192, 1970.

Saxberg B. and Cohen R. Cellular automata models of cardiac conduction. in *Theory of the heart.*, Glass L., Hunter P. and McCulloch A., Eds, Springer Verlag, New-York: 437–476, 1991.

Serizawa T. Three state neumann neighbor cellular capable of constructing self reproducing machines. *System and Computers in Japan*. **Vol. 18 no. 4**: 33–40, 1987.

Signorini J. Programmation par configurations des ordinateurs cellulaires à très grande échelle. Ph.D. thesis, Université Paris 8, 1992.

Smith III A. Cellular automata theory. *Technical Report 2*. Stanford University, 1960.

Smith III A. Cellular Automata and formal languages. *Proc. 11th IEEE Ann. Symp. Switching Automata Theory*. **vol. 14 no. 4**: 216–224, 1970.

Smith III A. Cellular automata complexity trade-offs. *Information and Control*. **Vol. no. 18**: 466–482, 1971.

Smith III A. Simple computation-universal spaces. *Journal of ACM*. **Vol. no. 18**: 339-353, 1971.

Smith III A. Two-dimensional formal languages and pattern recognition by cellular automata. *Proceedings of the 12-eme Annual IEEE Symposium on Switching and Automata Theory.*: 144–152, East Lansing, Michigan, 1971.

Smith III A. Real-time language recognition by one-dimensional cellular automata. *Journal of Computer and System Sciences*. **Vol. no. 6**: 233–253, 1972.

Smith III A. Simple Non-trivial Self-Reproducing Machines. *Proceedings of the Second Artificial Life Workshop*, Santa Fe: 709–725, 1991.

Smith J., Ritzenberg R. and Cohen R. Percolation theory and cardiac conduction. *Computers in Cardiology*. Washington, DC: IEEE Computer Society Press: 201–204, 1984.

Solomon J. and Selvester R. Simulation of measured activation sequence in the human heart. *Am. Heart J*. **Vol. no. 85**: 518–523, 1973.

Sutner K. Classifying circular cellular automata. *Physica D*. **Vol. no. 45**: 386–395, 1990.

Sutner K. De Bruijn graphs and linear cellular automata. *Complex Systems*. **Vol.5 no. 1**: 19–30, 1991.

Sutner K. Linear cellular automata and Fischer automata. To appear in *Parallel Computing*.

Sutner K. The size of power automata. Submitted, 1997.

Szathmary V. and Osvald R. An intercative computer model of propagated activation with analytically defined geometry of ventricles. *Computers and Biomed. Res*.**Vol. no. 27**: 27–38, 1994.

Terrier V. Temps réel sur automates cellulaires. Ph.D. thesis, Université Lyon 1, 1991.

Terrier V. Language recognizable in real time by cellular automata. *Complex systems*.

Vol. no. 8: 325–336, 1994.

Terrier V. On real time one-way cellular automata. *Theoretical Computer Science*. **Vol. no. 141**: 331–335, 1995.

Terrier V. Language not recognizable in real time by one-way cellular automata. *Theoretical Computer Science*. **Vol. no. 156**: 281–287, 1996.

Terrier V. Two dimensional cellular automata recognizers. To appear in *Theoretical Computer Science*.

Thatcher J. Universality in the von Neumann cellular model. Tech. Report 03105-30-T, University of Michigan, 1964.

Turing A. On computable numbers, with an application to the Entscheidungsproblem. *P. London Math. Soc.* **Vol. no. 42**: 230–265, 1936.

Umeo H., Morita K. and Sugata K. Deterministic one-way simulation of two-way real-time cellular automata and its related problems. *Inform. Process. Lett.* **Vol. no. 14**: 159–161, 1982.

Uspensky V. and Semenov A. *Algorithms : main ideas and applications*. Kluwer, 1993.

Van Capelle F. and Durrer D. Computer simulation of arrhythmias in a network of coupled excitable elements. *Circ. Res.* **Vol. no. 47**: 454–466, 1980.

Velbit V., Podrid P., Lown B., Cohen B. and Graboys T. Aggravation and provocation of ventricular arrhythmias by antiarrhythmic drugs. *Circulation*. **Vol. no. 65**: 886–894, 1982.

von Neumann J. The General and Logical Theory of Automata. in *Collected Works, Vol. 5*, Taub. A., Eds, New York: Pergamon Press: 288–328, 1963.

von Neumann J. *The Computer and the Brain*, Yale University Press, New Haven, 1958.

von Neumann J. *Theory of Self-reproducing automata*, University of Illinois Press, Chicago, 1966.

von Neumann J. *Probabilistic logic and the synthesis of reliable organisms from unreliable components*, Shannon and McCarthy eds, Princeton University Press, Princeton, 1956.

Wagner K. and Wechsung G. *Computational Complexity*. Reidel, 1986.

Waksman A. An optimum solution to the firing squad synchronization problem. *Information and Control*. **Vol. no. 9**: 66–78, 1966.

Weiss B. Subshifts of finite type and sofic systems. *Monatshefte für Mathematik*. **Vol. no. 77**: 462–474, 1973.

Wiener N. *Cybernetics, or control and communication in the animal and the machine*, M.I.T. Press, New Haven, 1961.

Wiggins S. *Global Bifurcations and Chaos*. Springer, Berlin, 1988.

Willson S. Cellular automata can generate fractals. *Discrete Applied Mathematics*. **Vol. no. 8**: 91–99, 1984.

Wolfram S. Twenty problems in the theory of cellular automata. *Physica Scripta*. **Vol. no. 79**: 170–183, 1985.

Wolfram S. Universality and complexity in cellular automata. *Physica D*. **Vol. no. 10**: 91–125, 1984.

Wolfram S. *Theory and Applications of Cellular Automata*, World Scientific, 1986.

Wu A. and Rosenfeld A. Cellular graph automata I. *Information and Control*. **Vol. no. 42**: 305–328, 1979.

Wu A. and Rosenfeld A. Cellular graph automata II. *Information and Control*. **Vol. no. 42**: 330–353, 1979.

Yu S. Regular languages. in Rozenberg G. and Salomaa A., editors, *Handbook of Formal Languages*, Springer Verlag, **volume 1, chapter 2**, 1997.

Yamaguti M. and Hata M. On the Mathematical Neuron Model. *Lecture Notes in Biomathematics*. **Vol. no. 45**: 171-177, 1982.

Yoshizawa S. Periodic Pulse Generated by an Analog Neuron Model. *Lecture Notes in Biomathematics*. **Vol. no. 45**: 155-170, 1982.

Yunes J.B. Seven states solutions to the Firing Squad Synchronization Problem. *Theoretical Computer Science*. **Vol. 127 no. 2**: 313–332, 1993.

Zhou X., Daubert J., Lown B., Wolf P., Smith W. and Ideker R. Size of the critical mass for defibrillation. *Circulation.* **Vol. no. 80**: II–531, 1989.

LIST OF AUTHORS

INDEX